Edited by
Javier Garcia-Martinez and
Elena Serrano-Torregrosa

The Chemical Element

Related Titles

Apotheker, Jan / Simon Sarkadi, Livia (eds.)

European Women in Chemistry

2011.

ISBN: 978-3-527-32956-4

Rehder, Dieter

Chemistry in Space

From Interstellar Matter to the Origin of Life
2010
ISBN: 978-3-527-32689-1

Armaroli, Nicola / Balzani, Vincenzo

Energy for a Sustainable World

From the Oil Age to a Sun-Powered Future
2010
ISBN: 978-3-527-32540-5

Anastas, P. T., Horvath, I. T.

Green Chemistry for a Sustainable Future

ISBN: 978-0-470-50351-5

Desai, P.

One Planet Communities

A real-life guide to sustainable living

ISBN: 978-0-470-71546-8

Olah, G. A., Goeppert, A., Prakash, G. K. S.

Beyond Oil and Gas: The Methanol Economy

2010
ISBN: 978-3-527-32422-4

Garcia-Martinez, Javier (ed.)

Nanotechnology for the Energy Challenge

2010
ISBN: 978-3-527-32401-9

Cocks, F. H.

Energy Demand and Climate Change

Issues and Resolutions

2009
ISBN: 978-3-527-32446-0

Rojey, A.

Energy and Climate

How to achieve a successful energy transition

ISBN: 978-0-470-74427-7

Coley, D.

Energy and Climate Change

Creating a Sustainable Future

ISBN: 978-0-470-85313-9

Edited by
Javier Garcia-Martinez and Elena Serrano-Torregrosa

The Chemical Element

Chemistry's Contribution to Our Global Future

WILEY-VCH Verlag GmbH & Co. KGaA

The Editors

Prof. Javier Garcia-Martinez
University of Alicante
Inorganic Chem. Deptarment
Carretera San Vicente s/n.
03690 Alicante
Spanien

Dr. Elena Serrano-Torregrosa
Dept. Inorganic Chemistry
University of Alicante
Campus de San Vicente
03690 Alicante
Spanien

Library of Congress Card No.:
applied for

British Library Cataloguing-in-Publication Data
A catalogue record for this book is available from
the British Library.

**Bibliographic information published by the
Deutsche Nationalbibliothek**
The Deutsche Nationalbibliothek lists this
publication in the Deutsche Nationalbibliografie;
detailed bibliographic data are available on the
Internet at <http://dnb.d-nb.de>.

© 2011 Wiley-VCH Verlag & Co. KGaA,
Boschstr. 12, 69469 Weinheim, Germany

Typesetting   Toppan Best-set Premedia Limited
Printing and Binding   betz-druck GmbH,
Darmstadt
Cover Design   Formgeber, Eppelheim

Printed in the Federal Republic of Germany
Printed on acid-free paper

ISBN Print: 978-3-527-32880-2

ISBN ePDF: 978-3-527-63566-5
ISBN ePub: 978-3-527-63565-8
ISBN Mobi: 978-3-527-63567-2
ISBN oBook: 978-3-527-63564-1

The Chemical Element: Chemistry's Contribution to Our Global Future

Every year several books are published dealing with chemistry, but this book is different and takes the reader far from the expected esoteric and academic chemistry to a chemistry that embraces our continuing existence on planet Earth. By placing chemistry at the centre of challenges and solutions for our planet, it provides a much-needed perspective on the role and importance of science for development and demonstrates the critical linkage between research in chemistry, policy, industry, education and concrete actions for sustainable development. The book is inspired by the United Nations declaration of 2011 as the International Year of Chemistry (IYC), and clearly spells out the role and importance of chemistry for meeting the United Nations Millennium Development Goals.

The International Union of Pure and Applied Chemistry (IUPAC) and the United Nations Educational, Scientific and Cultural Organisation (UNESCO) were designated by the United Nations General Assembly as lead agencies for promoting and coordinating the IYC. The objectives of the Year are to:

- increase the public appreciation and understanding of chemistry in meeting world needs,

- encourage the interest of young people in chemistry,

- generate enthusiasm for the creative future of chemistry,

- celebrate the role of women in chemistry or major historical events in chemistry, including the centenaries of Mme. Curie's Nobel Prize and the founding of the International Association of Chemical Societies.

Through the Year, the world is celebrating the art and science of chemistry, and its essential contributions to knowledge, environmental protection, improvement of health and economic development. The critical over-arching need in this context is for the responsible and ethical use of chemical research, and its applications and innovations, for equitable sustainable development.

In January 2011, the official launch of the IYC took place at UNESCO Headquarters in Paris. This meeting set the themes for the Year by associating "chemistry" with the words "progress of civilization, solutions for global challenges, climate change, creating a sustainable future, nutrition, food production, water, health and disease, global health, energy solutions for the future, materials of tomorrow, economic and social aspects . . .". The chapters of this book mirror these themes and present the reader with a comprehensive view of what "chemistry" means for our lives and our futures.

This book is therefore to be highly recommended to a wide readership including individuals concerned for sustainable development, politicians, young people, scientists, teachers, and global strategists. It is a must for every chemist who can use it as a tool in teaching students or in informing non-scientists about the possibilities of this fundamental science. Most of all, we hope that this book will be used to show young people that "chemistry" is exciting and meaningful, and that many will be enticed and inspired to take up careers in this field of scientific endeavour.

We congratulate the editors and authors of this marvelous book, published specially as part of the celebration of the IYC.

Nicole Moreau   Julia Hasler
President, IUPAC   UNESCO Focal Point for IYC

Contents

Introduction

"*The future of humanity is uncertain, even in the most prosperous countries, and the quality of life deteriorates; and yet I believe that what is being discovered about the infinitely large and infinitely small is sufficient to absolve this end of the century and millennium*", wrote Primo Levi in his essay "News from the Sky". The challenge now is to apply all that knowledge to secure the future of humanity, improve our quality of life and tackle the challenges we have been facing for millennia.

With this aim, 192 heads of state and government joined in 2000 to agree on eight very specific and achievable goals, known as the Millennium Development Goals (MDG). "*Time is short. We must seize this historic moment to act responsibly and decisively for the common good*", reminded United Nations Secretary-General Ban Ki-moon. Only four years before the deadline we gave ourselves to achieve these goals, the United Nations has declared 2011 as the International Year of Chemistry (IYC), which aims to "*overcome the challenges facing today's world, for example in helping to address the United Nations Millennium Goals.*"

This book is a celebration of the many contributions of chemistry to our wellbeing, coinciding with the IYC, and also a roadmap of the tools we have at our finger tips to make a significant contribution to the lives of those who are not benefiting from the technological advances of our time. We try to provide at the same time a comprehensive review of the current status of some critical issues and a description of the technological possibilities we have today to overcome some of our most urgent needs. Our generation is the first one that has the financial resources and technological tools to significantly mitigate the suffering that many are sentenced to, from hunger to curable diseases, from unsafe water, and polluted air to poverty.

This book is divided into nine chapters that represent the biggest and most urgent challenges of our time in which chemistry can provide a significant contribution. Because of the scope and aim of this book, the authors are leaders in their fields and a broad representation of what chemistry is today. In general, each chapter covers one MDG by recognizing the present and future contributions of chemistry to this MDG. The chapters are excellent reviews of the current state of the subject, from the point of view of the world leaders in each field, but above all, a glimpse into the future.

Chapter 1, written by scientists of the International Organization for Chemical Sciences in Development (IOCD), summarizes the scope of the book by highlighting the possible state-of-the-art contributions of chemistry to human advancement through the classification of the MDG. Chemistry's contributions to human advancement include benefits in the health, agricultural and industrial sectors of developing countries, thereby improving the quality of life for the vast majority of people on the planet: food supply, medicines, construction materials, new jobs and clean water.

Chapters 2 and 3 are devoted to hunger and poverty, respectively. As mentioned in Chapter 1, the World Bank defines poverty in very crude terms: "*Poverty is hunger. Poverty is lack of shelter. Poverty is being sick and not being able to see a doctor. Poverty is not having access to school and not knowing how to read. Poverty is not having a job, is fear for the future, living one day at a time. Poverty is losing a child to illness brought about by unclean water. Poverty is powerlessness, lack of representation and freedom.*" These are major problems and chemistry can provide real solutions to every one of them, such as food even in poor soils using better fertilizers, shelters using more sustainable materials, new medicines for pandemic illnesses, and jobs and opportunities for many, as described in detail in these chapters.

Chemistry education's contribution to our global future, directly related to the second, seventh and eighth MDGs, is analyzed in Chapter 4. The central question of the chapter is focused on how scientists and citizens can do better in the decades following the IYC to answer the question: Has education about the nature and role of chemistry succeeded in creating the public climate needed to support the fundamental and applied research required to tackle these IYC global challenges?

The contribution of chemical science to health (fourth to sixth MDGs) is illustrated in Chapter 5. More specifically, the authors concentrate primarily on various aspects involved in drug discovery and development, as well as their research activities concerning the first commercial human synthetic vaccine against bacterial infections causing the death of more than half a million infants each year.

Chapter 6 is focused on green chemistry as a tool to integrate the principles of sustainable development into country policies and programmes and reverse the loss of environmental resources and reduce the biodiversity loss caused by the industries (seventh MDG).

Chapter 7, entitled Water: Foundation for a Sustainable Future, resumes the chemical contribution to water, as one of the principles of sustainable development ranging from poverty and health (Goals 1, 4–6) to environmental sustainability (Goal 7). Many of the MDGs are related to health and thus indirectly related to water and sanitation.

To quote Kofi Annan: "*For future scientific research to unleash the potential of life-changing technologies, the greatest challenge will be to provide clean and affordable energy to the poor*". Chapter 8 provides a comprehensive and updated view of the many research activities for achieving energy security and sustainability and ending energy poverty. A significant burden on the shoulders of many nations is lack of enough energy to unleash their economic potential.

Chapter 9 deals with some of the most dramatic consequences of the bad applications of technologies that lead to ozone layer depletion and climate change. Whereas the former has been significantly mitigated by the use of alternative more benign solutions, climate change is one of the most serious threats to our well-being, safety and economic growth. Some of the solutions that are being investigated today to deal with CO_2 emissions, from reducing its production to its storage and reuse, are described by some of the leading experts in the field.

This book is intended to serve a very large audience interested in the roles of science and technology in global issues. For helping with new concepts, the book includes boxes with simple and concise explanations of key ideas and multiple examples, tables and figures.

What we managed to achieve so far is truly amazing, for example, turning air into bread by reacting nitrogen with hydrogen to produce ammonia and then fertilizers, which are responsible for the survival of 40% of our planet's human population. It is astonishing that approximately half of the nitrogen atoms in each human body have come at some point through the Haber–Bosch process. But there is much more waiting for us to be discovered. Only time will tell how human creativity and ingenuity will solve the problems we are facing. No doubt, this is an amazing endeavour worth taking.

Elena Serrano Torregrosa and Javier Garcia Martinez
Alicante (Spain), February 2011

List of Contributors

Berhanu M. Abegaz
Department of Chemistry
University of Botswana
Private Bag 0022
Gaborone
Botswana

Fabio Aricò
Cà Foscari University
Department of Environmental
Sciences
Dorsoduro 2137
30123 Venice
Italy

Arlin Briley
University of South Florida
Department of Civil and
Environmental Engineering
4202 East Fowler Ave, ENB 118
Tampa, FL 33620
USA

Glenn Carver
University of Cambridge
Centre for Atmospheric Science
Lensfield Road
Cambridge CB2 1EW
UK

Gabriele Centi
University of Messina and
INSTM/CASPE
Dip. di Chimica Industriale ed
Ingegneria dei Materiali
V.le F. D'Alcontres 31
98166 Messina
Italy

Omatoyo K. Dalrymple
University of South Florida
Department of Civil and
Environmental Engineering
4202 East Fowler Ave, ENB 118
Tampa, FL 33620
USA

Jessica Fanzo
Columbia University
The Earth Institute
61 Route 9 W
Palisades, NY 10964
USA

Mari-Carmen Gomez-Cabrera
Department of Physiology
University of Valencia
Avenida Vicente Andrés Est.
46100 Burjassot (Val.)
Spain

and

Alimentos Mundi
University of Valencia
Avenida Vicente Andrés Est.
46100 Burjassot (Val.)
Spain

Julia Hasler
Programme Specialist
Division of Basic and Engineering
Sciences Natural Sciences Sector
UNESCO
1 rue Miollis
75732 Paris cedex 15
France

Joniqua A. Howard
University of South Florida
Department of Civil and
Environmental Engineering
4202 East Fowler Ave, ENB 118
Tampa, FL 33620
USA

Peter Mahaffy
King's University College
Department of Chemistry
9125 50th Street
Edmonton, AB
Canada T6B 2H3

Cecilia Martínez-Costa
Department of Pediatrics
University of Valencia
Avenida Vicente Andrés Est.
46100 Burjassot (Val.)
Spain

and

Alimentos Mundi
University of Valencia
Avenida Vicente Andrés Est.
46100 Burjassot (Val.)
Spain

Stephen A. Matlin
International Organization for
Chemical Sciences in Development
IOCD
Flat 4, 50 Netherhall Gardens
London NW3 5RG
UK

Con Robert McElroy
Cà Foscari University
Department of Environmental
Sciences
Dorsoduro 2137
30123 Venice
Italy

James R. Mihelcic
University of South Florida
Department of Civil and
Environmental Engineering
4202 East Fowler Ave, ENB 118
Tampa, FL 33620
USA

Nicole J. Moreau
30 avenue Jean Jaures
Charenton F-94220
France

Siglinda Perathoner
University of Messina and INSTM/
CASPE
Dip. di Chimica Industriale ed
Ingegneria dei Materiali
V.le F. D'Alcontres 31
98166 Messina
Italy

Roseline Remans
Columbia University
The Earth Institute
61 Route 9 W
Palisades, NY 10964
USA

René Roy
Université du Québec à Montréal
PharmaQAM
Department of Chemistry
2101
rue Jeanne-Mance
Montreal, QC
Canada H2X 2J6

Jeffrey Sachs
The Earth Institute, Columbia
University
405 Low Library, MC 4335
535 West 116th Street
New York, NY 10027

Pedro Sanchez
Columbia University
The Earth Institute
61 Route 9 W
Palisades, NY 10964
USA

Juan Sastre
Department of Physiology
University of Valencia
Avenida Vicente Andrés Est.
46100 Burjassot (Val.)
Spain

and

Alimentos Mundi
University of Valencia
Avenida Vicente Andrés Est.
46100 Burjassot (Val.)
Spain

Ken D. Thomas
University of South Florida
Department of Civil and
Environmental Engineering
4202 East Fowler Ave, ENB 118
Tampa, FL 33620
USA

Maya A. Trotz
University of South Florida
Department of Civil and
Environmental Engineering
4202 East Fowler Ave, ENB 118
Tampa, FL 33620
USA

Pietro Tundo
Cà Foscari University
Department of Environmental
Sciences
Dorsoduro 2137
30123 Venice
Italy

1
Chemistry for Development

Stephen A. Matlin and Berhanu M. Abegaz

1.1
Chemistry, Innovation and Impact

The foundations of modern chemistry were laid in the 18th and 19th centuries and further extended in the 20th century. They encompassed the development of a theoretical framework for understanding and explaining the physical and chemical properties of atoms and molecules, together with the invention of increasingly sophisticated techniques for interacting with these entities in order to study and influence their structures and behaviors. These developments have given humanity a degree of mastery over its physical environment that surpasses the sum of achievements over the entire previous period of human history.

Chemistry's contributions to human advancement need to be seen in terms of its own core role as a physical science, but also as a "platform science" in the context of its relationships within the group of "natural sciences" that includes physics and biology. Chemistry provides the basis for understanding the atomic and molecular aspects of these disciplines and, through its interfaces with a range of pure and applied sciences, underpins the dramatic advances seen in recent decades in such diverse fields as medicine, genetics, biotechnology, materials and energy. Hence, this discussion of the role of chemistry in the process of development is framed in the broader context of the roles of science, technology and innovation more generally.

Innovation, which may operate in both technological and social fields [1], encompasses not only the birth of an idea or a discovery, but its application in practice – taking the outputs of research and invention and using them to put new goods, services or processes into use. While innovation is sometimes represented as a straightforward linear system (Figure 1.1), in reality this is an over-simplified model and innovation needs to be treated as a complex, highly nonlinear ecosystem, full of interdependences and feedback loops.

The Chemical Element: Chemistry's Contribution to Our Global Future, First Edition.
Edited by Javier Garcia-Martinez, Elena Serrano-Torregrosa.
© 2011 Wiley-VCH Verlag GmbH & Co. KGaA. Published 2011 by Wiley-VCH Verlag GmbH & Co. KGaA.

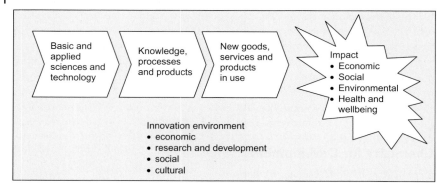

Figure 1.1 The chain of scientific innovation – from ideas to impact.

Chemistry may be involved not only in the initial stages of research (e.g., in areas such as agrochemicals and pharmaceuticals: chemical synthesis of new molecules for testing), but also in intermediate stages (e.g., product development, quality control) and in the evaluation of impact (e.g., health status assessment, environmental monitoring), thus contributing in key ways at every stage of the technological innovation chain.

Throughout the modern period of its development, chemistry has contributed enormously both to broad improvements in human wellbeing (including enhancements of health and quality of life) and to wealth creation for individuals and nations. Some landmark examples are summarized in Table 1.1. Early developments in electrochemistry and synergies with physics and engineering led to methods for producing electrical energy, which has impacted on virtually every aspect of human activity. Electrochemistry also provided the basis for the industrial transformation of many materials and, in particular, for the production of metals such as aluminum and important feedstocks such as caustic soda and chlorine. Industrial organic chemistry built on mid-19th century processes for manufacturing dyestuffs, but by the 20th century had expanded to include the synthesis of pharmaceuticals. In parallel with advances in public health (measures for reducing the spread of infectious diseases through improved water, sanitation and vaccination; and for improving health through ensuring optimal nutrition – in all of which areas chemistry has played a major role), pharmaceutical chemistry has contributed enormously to improving life expectancy and the quality of life through the treatment of infectious diseases and metabolic disorders and the control of pain. Chemistry has contributed to many of the advances in agriculture (e.g., fertilizers, plant growth regulators, pesticides) which have been characterized as a "green revolution" and which have helped to feed the world's population while it grew from about 1 billion to 6 billion during the 20th century. Moreover, chemistry has given the world a wide array of new materials, including polymers, plastics, semiconductors and superconductors, with applications from fabrics and structural materials to information and communications technologies and medical imaging.

Table 1.1 Landmark examples of chemistry breakthroughs contributing to health and wealth.

Date	Scientist	Breakthrough	Impact	Refs
		Industrial chemistry, electrochemistry, power and light		
1800	Alessandro Volta	Discovered that a continuous flow of electricity was generated when using certain fluids as conductors to promote a chemical reaction between the metals or electrodes.	Mass production of portable power sources enabled a vast range of applications– from automobiles to radios. The nickel–metal hydride battery, commercialized in 1990, provided a high energy density and absence of toxic metals. It has found numerous applications include mobile phones and laptop computers	[2]
1802	William Cruickshank	Designed the first electric battery capable of mass production.		
1839	William Grove	Invented the H_2/O_2 fuel cell.		
1859	Gaston Platé	Invented the first rechargeable battery, based on lead-acid chemistry.		
1806	Humphry Davy	Connected a very powerful electric battery to charcoal electrodes and produced "the most brilliant ascending arch of light ever seen".	Invention of the electric light bulb paved the way for replacement of polluting combustion processes for lighting and enabled mass lighting in homes, workplaces and public spaces.	[2]
1879	Thomas Edison	Discovered that a carbon filament in an oxygen-free bulb glowed but did not burn up.		
1820	André-Marie Ampère	Observed that wires carrying an electric current attracted or repelled one another.	Invention of the electric generator transformed life in industrialized countries, impacting on transport, work and leisure.	[2]
1831	Michael Faraday	Demonstrated that a copper disc was able to provide a constant flow of electricity when revolved in a strong magnetic field.		
1800	William Cruickshank	First description of electrolysis of brine	Electrolysis became an extremely important method of transforming materials and especially for the production of inorganic chemicals and compounds, either for use in their own right (e.g., see entry on aluminum below) as a source of feedstocks for the manufacture of other compounds, including organics. For example, the chlorine by-product from the electrolysis of brine was the starting point for the manufacture of organic compounds including solvents, pesticides and plastics	[2–4]
1833	Michael Faraday	Formulation of the laws that govern the electrolysis of aqueous solutions		
1861	Ernest Solvay (1838–1922)	Patented Solvay Process for manufacture of industrial soda using carbon dioxide, brine and ammonia. After the first commercial plant for the electrolysis of brine was built in 1891, caustic soda was increasingly produced directly by this method.		
1897	Herbert Dow	Dow persuaded a group of Cleveland investors to back him in building a chloralkali business in Midland, to be known as The Dow Chemical Co.		

Table 1.1 *Continued*

Date	Scientist	Breakthrough	Impact	Refs
1825	Hans Christian Oersted	Metallic aluminum first made by heating potassium amalgam with aluminum chloride.	Due to its low density, high thermal and electrical conductivity, non-magnetic character, high ductility and the capacity of the metal and its alloys to be cast, rolled, extruded, forged, drawn, and machined, aluminum became one of the most important and ubiquitous metals in the 20th century.	[5] [6]
1886	Charles Hall Paul Heroult	Electrolytic processes for the production of aluminum.		
1840s	August Hoffman	Discovery of aniline and process for its synthesis from benzene	Aniline dyes became the basis for the development of the dyestuffs industry in the 19th century, leading to major growth in chemical industries in the UK, France and Germany.	[7]
1856	William Perkin	Invention of mauve dye (aniline purple).		

Medicinal chemistry and medicine

Date	Scientist	Breakthrough	Impact	Refs
1853	Charles Gerhardt	First synthesis of acetylsalicylic acid	Studies of microorganisms and the physiological effects of chemicals and work on the structural modification of natural products and synthetic chemicals in the 19th century laid the foundations for the pharmaceutical industry in the 20th century. Major classes of therapeutic agents soon emerged, including analgesics, anaesthetics, anti-infectives and anti-tumor agents.	[8–18]
1897	Felix Hoffmann	Investigated acetylsalicylic acid as a less-irritating replacement for salicylate medicines, e.g., for treating rheumatism.		
1860–1864	Louis Pasteur	Demonstrated that fermentation is caused by specific microorganisms and formulated the germ theory of disease – providing the basis for biotechnology and anti-microbial chemotherapy.	Growing understanding of the chemistry of metabolic processes and of the structures and functions of proteins and nucleic acids all contributed to the evolution of pharmacology and molecular biology as distinct sciences and to drug targeting and rational drug design.	
1909	Paul Ehrlich	Synthesis of anti-syphilis organo-arsenical drug, Salvarsan.		
1928 1940	Alexander Fleming Howard Florey Ernst Chain	Discovery of penicillin, the first of a family of β-lactam antibiotics, and development of large-scale process for its production		
1932	Gerhard Domagk	Began testing Prontosil, leading to development of the sulfonamide antibiotics	Understanding of the metabolic roles of vitamins and hormones paved the way for a range of drug therapies for metabolic disorders and for development of hormonal contraceptives.	

Table 1.1 *Continued*

Date	Scientist	Breakthrough	Impact	Refs
1865	Heinrich Lissauer	Used potassium arsenite to treat chronic myelogenous leukemia – the first instance of effective chemotherapy for malignant disease. The modern era of cancer chemotherapy began with the study of the cytotoxic effects of nitrogen mustards on lymphoid tissues by Alfred Gilman, Louis Goodman and coworkers in the 1940s.	The pharmaceutical industry now employs well over half a million people and generates global sales in excess of US$ 700 billion per year. In the UK alone, the industry provided employment for 67 000 workers in 2007.	
1912	Casimir Funk	Published the "vitamine theory", based on observations of the effects of depriving animals of small amounts of essential dietary chemicals. Paved the way for the modern understanding of vitamins and their roles as key biochemical catalysts.		
1901 1915 1921– 22 1951	Jokichi Takamine Edward Kendall Frederick Banting John McLeod Charles Best Carl Djerassi	First isolation of epinephrine. First isolation in crystalline form of thyroxine from thyroid gland. First isolation of insulin and demonstration of its capacity to treat diabetes. Synthesized norethindrone, the first effective oral contraceptive.		
1949	Linus Pauling, Harvey Itano, S. J. Singer, Ibert Wells	Publication of "Sickle Cell Anemia, a Molecular Disease" – the first proof of a human disease caused by an abnormal protein and the dawn of molecular genetics.		
1953	James Watson Francis Crick	Discovery of the double helix structure of DNA – the foundation of molecular biology.		
1955 1958	Frederick Sanger Max Perutz John Kendrew	First determination of the complete amino acid sequence of a protein – insulin. First three-dimensional structures of proteins solved by X-ray crystallography – hemoglobin and myoglobin.		

Table 1.1 *Continued*

Date	Scientist	Breakthrough	Impact	Refs
		Dentistry		
1826	Auguste Taveau	First to use amalgam as a dental restorative material	The development of safe and effective materials for dental restoration and anesthesia transformed dentistry, which had hitherto been a crude and extremely painful procedure.	[19]
1844–1846	Horace Wells William Morton	First uses of nitrous oxide and ether as general anesthetics for dental extractions		
1901	Frederick McKay	Began investigating the cause of widespread brown staining of teeth in Colorado Springs, which he discovered was associated with a dramatic absence of dental caries. Work by chemists in the 1930s eventually traced the cause to fluoride in drinking water.	Water fluoridation and the development of fluoride-containing toothpastes have further contributed to a dramatic improvement in oral health in many countries.	
1905	Alfred Einhorn	First synthesis of procaine (novocaine), a synthetic analog of cocaine without its addictive properties and the first safe local anesthetic for dentistry.		
		Agrochemistry		
1909	Fritz Haber	Invented the Haber process for nitrogen fixation, later scaled up by Carl Bosch. Fixation of nitrogen as ammonia, which can then be oxidized to make nitrates and nitrites, made possible the industrial production of many classes of compounds including nitrate fertilizers and explosives.	The Haber process now produces 100 million tons of nitrogen fertilizer per year, consuming 3–5% of world natural gas production (ca. 1–2% of the world's annual energy supply) and generating fertilizer which is responsible for sustaining one-third of the Earth's population.	[20] [21] [22] [23] [24] [25]
1874 1939	Othmar Zeidler Paul Müller	First synthesis of DDT. Insecticidal properties discovered. DDT became the first commercial organochlorine insecticide. Prior to its banning on environmental grounds, DDT was a major weapon in the fight to eliminate malaria.	The development of plant growth promoters, crop protection agents and agents promoting animal health contributed to an agrochemicals industry with global annual sales of over of US$ 100 billion and, together with advances in plant breeding and agronomy methods, produced the green revolution of the 1960s–1980s which has helped feed the burgeoning population of the planet.	
1951	Geigy Chemical Co.	Introduction of carbamate insecticides		
1960s	Michael Elliott	Development of synthetic pyrethroid insecticides.		

Table 1.1 *Continued*

Date	Scientist	Breakthrough	Impact	Refs
		Analytical chemistry		
1901	Michael Tswett	First use of an adsorption column for the separation of plant pigments marked the birth of chromatography – later to develop into a family of 2- and 3-dimensional techniques involving combinations of gas, liquid and solid phases, for analytical and preparative scale separation of compounds.	The pioneering studies by a range of scientists, including botanists, physicists and physical chemists, led to the development of extremely powerful sets of techniques for separating chemical species, identifying them and measuring their concentrations. The evolving and often intertwined fields of analytical and separation sciences have been of fundamental importance, not only to the advance of chemistry itself but also to a wide range of areas including clinical and environmental sciences.	[26] [27] [28] [29] [30] [31]
1800	William Herschel	Discovered infrared radiation.		
1854	August Beer	Extending earlier work by by Pierre Bouguer and Johann Lambert, published what became known as the Beer-Lambert Law, defining the relationship between the extent of absorption of light and the properties of the material through which it is traveling.		
1859	Robert Bunsen Gustav Kirchoff	Developed the first spectroscope.		
1895	Wilhelm Röntgen	First systematic studies of X-rays, which later became the basis of X-ray medical diagnosis and X-ray crystallography.		
1913	Joseph Thomson	Invented mass spectrometry.		
1938	Isidor Rabi	First described and measured nuclear magnetic resonance in molecular beams. NMR became the basis of techniques for molecular structure elucidation and also medical imaging.		

Table 1.1 *Continued*

Date	Scientist	Breakthrough	Impact	Refs
		Polymers and plastics		
1839	Charles Goodyear	Discovered the process of vulcanization of natural rubber by heating with sulfur	Materials based on synthetic plastics and polymers became ubiquitous in the 20th century, finding applications in clothing, products from containers and appliance casings to non-stick pans, thermal and electrical insulators, components of transport machinery, medical and surgical devices, in space exploration and in much else.	[32] [33] [34] [35] [36]
1855	Alexander Parkes	Created the first plastic by treating cellulose treated with nitric acid.		
1909	Leo Hendrik Baekeland	Bakelite, the first synthetic polymer plastic, made from phenol and formaldehyde.		
1934	Wallace Carothers	First synthesis of a synthetic fiber, nylon, by co-polymerization of hexamethylene diamine and adipic acid		
		Solid state chemistry		
1833	Michael Faraday	Described first semiconductor effect, noting that electrical conductivity in silver sulfide increased with increasing temperature	Semiconductors based on silicon became the basis of solid state electronic devices including computers and provided the foundation for the modern digital age.	[37]

The value added by these products of chemistry and related sciences has contributed to the rapid growth in world GDP [38], especially in the industrially advanced countries during the second half of the 20th century (Figure 1.2). Knowledge-intensive and technology-intensive industries are estimated [39] to have accounted for 30% of global economic output, or some US$15.7 trillion, in 2007.

1.2
Poverty and Disparities in Life Expectancy

The benefits from advances in chemistry and other sciences have not been evenly distributed globally. The least industrially/technologically advanced countries have remained the poorest and people in the low- and middle-income countries (LMICs) have often fared worse than those in high-income countries (HICs), as illustrated by the dramatic relationship between poverty and life expectancy: the poor die young. Life expectancies around the world have increased very markedly over the course of the last century, but as they have done so the disparities between populations have grown larger [40]. However, the relationship between life expectancy

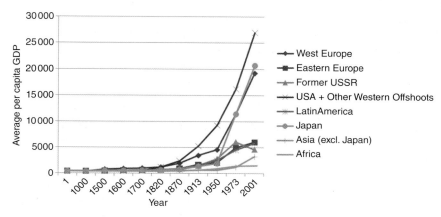

Figure 1.2 Per capita GDP: Regional and world averages, 1–2001 AD (millions 1990 international Geary-Khamis dollars). Data from [38], Table 8c.

and the average per capita income of the country is not a straightforward one and income is not the only factor involved. The economist Easterlin [41] concluded that much of the decline in mortality in the 20th century had its origin in technical progress – and in this context, "technical progress" refers to a combination of technological advances and their diffusion and uptake in different countries and the capacities of the countries themselves to conduct and apply research. Much of the variation in life expectancies seen between countries is explained by differences in the rate of this technical progress – for example, it explains two thirds of the variation in the decline in infant mortality over a 25 year period, whereas change in income explains only 9% [42, 43].

1.3
The Millennium Development Goals

In response to the unacceptable levels of poverty (Box 1.1) and growing disparities in health and wellbeing between people in different countries, the world's

Box 1.1   Poverty

What is poverty?

Poverty is hunger. Poverty is lack of shelter. Poverty is being sick and not being able to see a doctor. Poverty is not having access to school and not knowing how to read. Poverty is not having a job, is fear for the future, living one day at a time. Poverty is losing a child to illness brought about by unclean water. Poverty is powerlessness, lack of representation and freedom.

The World Bank [46]

governments met at the Millennium Summit [44] in New York on 6–8 September 2000, issuing the Millennium Declaration which led to agreement on a series of Millennium Development Goals (MDGs) [45] that were set for 2015 (Table 1.2). The targets were acknowledged to be extremely ambitious – but it was recognized that, for the first time in history, mankind had the capacity to substantially reduce or eliminate many sources of human suffering and to offer every person on the planet a basic level of existence that would be free from hunger, disease and discrimination in access to opportunities for development.

As stated in the report of the Task Force on Science Technology and Innovation of the Millennium Project [47]:

> "Since their adoption at the United Nations Millennium Summit in 2000, the Millennium Development Goals have become the international standard of reference for measuring and tracking improvements in the human condition in developing countries. The Goals are backed by a political mandate agreed by the leaders of all UN member states. They offer a comprehensive and multidimensional development framework and set clear quantifiable targets to be achieved by 2015."

The latest assessment shows that uneven progress has been made towards meeting the targets. Unmet commitments, inadequate resources, lack of focus and accountability and insufficient dedication to sustainable development have created shortfalls in many areas and without a major push forward many of the MDG targets are likely to be missed in most regions [48].

To achieve the goals will require a collective global effort harnessing political will, available resources and innovation in all areas, including the application of science and technology.

1.3.1
Goal 1: Reducing Poverty and Hunger

Economic growth, especially in the world's most populace country, China, resulted in hundreds of millions of people being lifted out of poverty during the last quarter of the 20th century [46]. Nevertheless, at the end of the century, out of a global population of 6 billion there were more than one billion people living on less than $1 a day, more than three billion living on less than $2 a day and nearly a billion suffering from hunger or severe malnutrition.

While many economically advanced countries produce an excess of food, some of which goes to waste, halving the proportions of those suffering poverty or hunger by 2015 is not merely a matter of redistributing available food. To overcome the net food shortage, allow for the expanding world population (already approaching 7 billion by 2010), ensure food security and independence from aid handouts, and respond to the agricultural impacts of climate change, it is necessary to expand agriculture throughout the world. Better applications of existing technologies and development of innovative new ones are essential

Table 1.2 Millennium development goals.

Goals	Targets
Goal 1 Eradicate extreme poverty and hunger	Halve, between 1990 and 2015, the proportion of people whose income is less than $1 a day Halve, between 1990 and 2015, the proportion of people who suffer from hunger
Goal 2 Achieve universal primary education	Ensure that, by 2015, children everywhere, boys and girls alike, will be able to complete a full course of primary schooling
Goal 3 Promote gender equality and empower women	Eliminate gender disparity in primary and secondary education preferably by 2005 and in all levels of education no later than 2015
Goal 4 Reduce child mortality	Reduce by two-thirds, between 1990 and 2015, the under-five mortality rate
Goal 5 Improve maternal health	Reduce by three-quarters, between 1990 and 2015, the maternal mortality ratio
Goal 6 Combat HIV/AIDS, malaria and other diseases	Have halted by 2015 and begun to reverse the spread of HIV/AIDS Have halted by 2015 and begun to reverse the incidence of malaria and other major diseases
Goal 7 Ensure environmental sustainability	Integrate the principles of sustainable development into country policies and programmes and reverse the loss of environmental resources Halve, by 2015, the proportion of people without sustainable access to safe drinking water and basic sanitation Have achieved, by 2020, a significant improvement in the lives of at least 100 million slum dwellers
Goal 8 Develop a global partnership for development	Develop further an open, rule-based, predictable, non-discriminatory trading and financial system (includes a commitment to good governance, development, and poverty reduction – both nationally and internationally) Address the special needs of the least developed countries (includes tariff-and quota-free access for exports, enhanced program of debt relief for HIPC and cancellation of official bilateral debt, and more generous ODA for countries committed to poverty reduction) Address the special needs of landlocked countries and small island developing states (through the Program of Action for the Sustainable Development of Small Island Developing States and 22nd General Assembly provisions) Deal comprehensively with the debt problems of developing countries through national and international measures in order to make debt sustainable in the long term In cooperation with developing countries, develop and implement strategies for decent and productive work for youth In cooperation with pharmaceutical companies, provide access to affordable, essential drugs in developing countries In cooperation with the private sector, make available the benefits of new technologies, especially information and communications

[49]–amounting to a second "green revolution" in which chemistry must play multiple important roles. Critical areas include improving plant varieties and methods for the efficient production, processing and preservation of foods that are healthy and nutritious.

The poverty goal is often referred to as an overarching goal, as it is intimately associated with the problems that are tackled in the other goals, including lack of gender equality, poor education and illiteracy, unacceptably high rates of maternal, neonatal and child mortality and of deaths from infectious diseases, lack of access to improved water and sanitation, and poor environment. However, it would be wrong to focus excessively on achieving this goal in the hope that the others will be met as a consequence. Other areas, such as education and health, have also been stressed as fundamental enablers of progress and the barriers to reaching each goal are varied and complex in nature, requiring individual attention. The reality is that effort is necessary across the whole range of issues highlighted in the MDGs. It must also be stressed that the MDGs are by no means comprehensive or complete, if a permanent shift in the trajectory of human development is to be achieved–for example, the MDGs make no direct reference to overcoming the challenges of unmet needs for reproductive health or the burgeoning rates of non-communicable diseases in LMICs.

1.3.2
Goal 2: Achieving Universal Primary Education

Education is a fundamental enabler of many other aspects of human development. Access to literacy, numeracy and knowledge transforms the lives of individuals, leading to better health and enhancing economic and social advancement, as well as contributing to national economic development. Yet, in the year 2000, there were more than 110 million children of primary school age out of school [50], a very low standard of education available for many children officially attending schools in some countries, and many hundreds of millions of adults who were illiterate. Moreover, education has exhibited a very high degree of gender discrimination, with a majority of those lacking access being girls and women. The high importance of education warrants the goal of ensuring universal primary schooling as a first step towards enabling access to secondary and further education.

1.3.3
Goal 3: Promoting Gender Equality and Empowering Women

Discrimination in access to education and health services and economic, social and political opportunities is experienced by girls and women in every part of the world. The fundamental right of females to equality of status, opportunity and treatment in all areas of human endeavor was established by a series of UN conventions and intergovernmental declarations during the 20th century [51–54], but at the close of the millennium the reality of women's and girls' experiences fell very far short of these standards in many parts of the world, and especially in many low- and middle-income countries. The MDGs set specific targets for moving towards gender equity, to help drive the process forward. Among the areas highlighted for urgent action was access to education – and it is notable that, even for girls enrolled in education, they often experience barriers in access to science and technology education in many countries [55].

1.3.4
Goals 4 and 5: Reducing Maternal and Under-Five Child Mortality

The chances of a woman dying during pregnancy or childbirth or in the immediate post-partum period, or of a child dying in the first few years of life, can be a hundred-fold greater in some of the world's poorest countries than in some of the richest. The causes, which link to poverty, poor nutrition, lack of education and inadequate availability of and access to effective health services including emergency obstetric care, may be varied. However, they are well understood and it is unacceptable in the 21st century that women should continue to die in large numbers simply because they become pregnant or that infants should die because they are not provided with the means of survival. The latest assessments [56] show some progress, but maternal and child mortality levels in many LMICs remain unacceptably high (Figure 1.3) and many countries are still off track to meet the MDG targets.

1.3.5
Goal 6: Combating HIV/AIDS, Malaria and Other Diseases

The development of antibiotics and vaccines has contributed to a massive reduction in mortality and morbidity due to infectious diseases in high-income countries during the last hundred years. However, many LMICs continued to experience major problems with communicable diseases – especially those caused by tropical parasitic infections such as malaria, leishmaniasis, trypanosomiasis, schistosomiasis and Guinea Worm. The advent of the HIV/AIDS epidemic, which began spreading rapidly in countries in Africa and elsewhere in the 1980s and 1990s, transformed the situation into one of crisis, compounded by the concomitant resurgence of tuberculosis in increasingly drug-resistant forms. Meeting the targets for halting and rolling back the spread of these diseases requires not just better access to existing technologies but also, in many cases, innovations in the form of new diagnostics, drugs, vaccines and delivery systems – all areas where chemical sciences must make a core contribution.

1.3.6
Goal 7: Ensuring Environmental Sustainability

The broad concept of "sustainable development" – recognizing the finite nature of the world's physical and biological resources and the importance of protecting and preserving them while engaging in human activity on the planet – emerged during

(a)

(b)

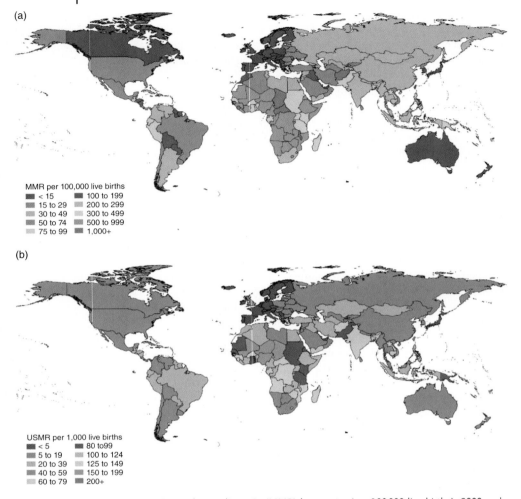

MMR per 100,000 live births
- < 15
- 15 to 29
- 30 to 49
- 50 to 74
- 75 to 99
- 100 to 199
- 200 to 299
- 300 to 499
- 500 to 999
- 1,000+

USMR per 1,000 live births
- < 5
- 5 to 19
- 20 to 39
- 40 to 59
- 60 to 79
- 80 to 99
- 100 to 124
- 125 to 149
- 150 to 199
- 200+

Figure 1.3 (a) Maternal mortality ratio (MMR) by country (per 100 000 live births), 2008 and (b) under-5 mortality rate (U5MR) by country (per 1000 live births), 2008 (from [56]). (Please find a color version of this figure in the color plates.)

the second half of the last century and was the focus of attention in world summits in Rio de Janeiro in 1992 [57] and in Johannesburg in 2002 [58]. The MDG targets represent an attempt to make some headway with these extremely challenging problems.

While there was still dispute at the end of the 20th century about the degree of climate change that the world would experience, there is now conclusive evidence that global warming is a real phenomenon and that climate change is already having, and will continue to have, increasingly severe impacts on many aspects of the human condition, including health, agriculture, the availability of fresh water, human habitation and, especially for some low-lying countries, even their very existence [59].

A further important aspect of the changing human environment has been a major shift during the last century from rural to urban dwelling. In 2007, for the first time, the proportion of human beings living in urban dwellings reached 50% and the transition continues [60]. Since most of the increase in the world population expected to occur during the next half century (from 6 billion in 2000 to 9 billion in 2050) will take place in LMICs, and since cities in many of these countries already have a high proportion of their inhabitants living in slum conditions without adequate water or sanitation, the challenges for city planners and technologists are enormous.

1.3.7
Goal 8: Developing a Global Partnership for Development

Globalization (the increasingly rapid and less restricted movement of people, goods, services and information around the world) brings with it a growing global interdependence of people and economies. This has resulted in a pressing need for global systems governing a wide range of human activities that impact on health, trade, the environment and much else. The eighth MDG calls for effective global partnerships among all the relevant actors to address a range of concerns that were seen to be important at the opening of the new millennium, including the rules governing access to health technologies and to information and communications technologies.

One aspect of health technology that has attracted much attention has been the issue of how the rules governing intellectual property rights should be applied justly and humanely in the field of medicine, considering the high costs of anti-retroviral drugs for the treatment of people living with HIV/AIDS and other life-threatening diseases [61]. The eighth MDG looks to governments and pharmaceutical companies to cooperate in providing access to affordable essential drugs in LMICs.

1.4
Science, Technology and Development

Advances in science and technology (S&T) enabled countries in Europe and North America to industrialize rapidly in the 19th and 20th centuries. For example, industrialization in Belgium drew on the Solvay process for manufacture of soda, which helped to establish Belgium as one of the world's leading countries in the chemical industry sector (Box 1.2).

While this process in Europe and North America was under way, from as early as the end of the 19th century a number of less developed countries were beginning to recognize the importance of S&T, either for economic growth or to address serious health challenges such as epidemics. Some notable examples include:

Box 1.2 Chemistry and the industrialization of Belgium

The industrial revolution began in Belgium with the development of British-style machine shops at Liège (*ca.* 1807), and Belgium became the first country in continental Europe to be transformed economically. Like its English counterpart, the Belgian industrial revolution initially centered on iron, coal and textiles. During the 19th century the chemical industry added momentum to Belgium's industrial and economic development.

Ernest Solvay (1838–1922) developed a process for manufacturing industrial soda (sodium carbonate), in which carbon dioxide (from limestone) is mixed with sodium chloride solution and ammonia. The Solvay process, which was cheaper and more efficient than the old Leblanc process, was patented in 1861 and Solvay opened his first factory at Couillet in Belgium in 1863. Industrial soda is used in manufacturing glass, steel and detergents and demand was huge. By 1900, the Solvay process was used to manufacture 95% of the world's industrial soda, and Solvay had an extensive business empire, with factories in Europe and the USA. Today, around 70 plants using the Solvay process are in operation around the world.

As demand increased, in 1898 Solvay started producing caustic soda directly by the electrolysis of brine, a process which also yields chlorine and hydrogen and gradually replaced the older method. The chlorine was used to produce hydrochloric acid, sodium hypochlorite and, later, organic derivatives including vinyl chloride (the monomeric precursor of PVC plastics); while the hydrogen was oxidized to hydrogen peroxide. Solvay is now the world's leading producer of peroxygen-based products, with a global network of peroxygen production plants.

In 2007, on a per capita basis Belgium was the number one producer of chemicals in the world and the share of the chemical industry in Belgium's economy was even bigger than that in Germany [62, 63].

- A number of research institutions established branches or offshoots in less industrialized countries. In particular, France's Institut Pasteur, created in 1887 for the prevention and treatment of infectious diseases through research, teaching and public health initiatives, established an international network which currently counts 30 members spread over the five continents [64]. Early members of the network included Pasteur institutes in Vietnam (1891), Tunisia (1893), Algeria (1894), Madagascar (1898), Morocco (1911), Iran (1920), Senegal (1923) and French Guiana (1940).

- Conceived in 1896 by Jamsetji Nusserwanji Tata and finally born in 1909, the Indian Institute of Science was an early example of a research institute established in the British colonial period [65]. The South African Chemical Institute [66] was founded in 1912.

- Brazil's Federal Seropathy Institute, established in 1900 to produce serums and vaccines against the plague, in 1908 became the Oswaldo Cruz Institute and later the Oswaldo Cruz Foundation [67]. It has made huge contributions to combating disease, including through the production of medicines by its Farmanguinhos branch [68].

- Rubber Research Institutes were established to support the expanding demand for rubber products. For example, the origin of research on rubber in Sri Lanka dates from 1909, when a group of planters in the Kalutara District met and agreed to engage a chemist to study the coagulation of rubber [69]. Similarly, Rubber Research Institutes emerged in other rubber-growing countries, including Nigeria [70] (1900), Malaya [71] (1925) and India [72] (1946).

Since the mid-20th century, the importance of science and technology for development has increasingly been recognized by international agencies [73–75], development assistance partners [76] and the governments of LMICs [77–79].

Within the UN family, UNESCO is the UN specialized agency mandated to build institutional and human capacity in the basic and engineering sciences, which are seen as a prerequisite for social and economic development. UNESCO's activities focus principally on third-level, but also second-level, education and on research in mathematics, physics, chemistry, biology, biotechnology and basic medical sciences [80]. The UNESCO Science Prize is awarded biennially to "a person or group of persons for an outstanding contribution they have made to the technological development of a developing member state or region through the application of scientific and technological research (particularly in the fields of education, engineering and industrial development)." The first prize was awarded to Robert Simpson Silver in 1968 for his discovery of a process for the demineralization of sea water; several subsequent prizes have also been chemistry-related [81]. A partnership between UNESCO and L'Oréal, *Awards For Women in Science*, forms a core element of UNESCO national and international activities to foster gender equality and equity in science [80].

The work of UNESCO is reinforced by the recognition by the UN Development Programme (UNDP) of the importance of technology for the progress of the least developed countries [82].

A number of nongovernmental organizations (NGOs) have been established to promote the roles in development of S&T in general. Some notable examples include:

- **Academy of Sciences for the Developing World (**TWAS: **originally known as the** Third World Academy of Sciences**):** TWAS is an international NGO founded in 1983 in Trieste, Italy by a distinguished group of scientists from the South under the leadership of the late Nobel laureate Abdus Salam of Pakistan. Its principal aim is to promote scientific excellence and capacity in the South for science-based sustainable development [83].

- Third World Organization for Women in Science **(TWOWS):** Established as an NGO in Trieste in 1989, TWOWS is the first international forum to unite eminent women scientists from the South with the objective of strengthening their role in the development process and promoting their representation in scientific and technological leadership [84].

- International Association of Science and Technology for Development **(IASTED):** A non-profit organization devoted to promoting economic and cultural advancement, IASTED was established in 1977. It organizes multidisciplinary conferences for academics and professionals, in both industrialized and developing countries, mainly in the fields of engineering, science, and education [85].

Perspectives on the nature of the development process itself have changed markedly during the last half-century. In the period of the 1950s–1970s, on the HIC side much of the development process was driven by and centered around post-colonial relationships and geopolitical cold-war maneuvrings, while LMICs themselves were beginning to seek ways to develop their own resources and capacities. There has been a movement away from HIC-driven approaches towards national self-determination and South–South cooperation and mutual reinforcement. Gradually there was a shift from a utilitarian perspective, which primarily focused on economic advancement as the main goal and saw human resources development, including S&T capacities, as a means of achieving this, to a human rights perspective which saw human development, equity and well-being as the primary objectives, with economic development being an important mechanism for enabling all people to achieve certain standards of health and freedom from want of basic needs as an inalienable right.

A series of world conferences in the 1990s, covering education, health, population, and sustainable development, culminated in the Millennium Declaration [86] in 2000. Reflecting the shifts in approach, the work of the Commission on Macroeconomics and Health, which reported to the World Health Organization in 2001, emphasized that health is an essential prerequisite for development, rather than the converse [87].

The role of science and innovation as drivers of development was examined in detail by the Task Force on Science, Technology and Innovation of the UN Millennium Project [47]. The Task Force, led by Calestous Juma and Lee Yee-Cheong, identified the important roles that science and technology can play in achieving the MDGs. It stressed the importance of S&T policies tailored to the specific needs and circumstances of each country and the need to create international partnerships that allow mutual learning.

The report of the Task Force outlined key areas for policy action, including:

- focusing on platform (generic) technologies
- improving infrastructure services as a foundation for technology
- improving higher education in science and engineering and redefining the role of universities
- promoting business activities in science, technology, and innovation
- improving the policy environment
- focusing on areas of underfunded research for development.

In a further study [88], Juma proposed that international development policy should be directed at building technical competence in developing countries rather than conventional relief activities. He argued that institutions of higher learning, especially universities, should have a direct role in helping to solve development challenges [89].

1.5
Chemistry and Development

Within the broader domain of S&T, chemistry has emerged as a key discipline able to contribute to development [90]. A number of NGOs, as well as programs within existing bodies, have been established to promote this contribution internationally. Two, in particular, are notable as examples of efforts at the global level to address major development needs and to build capacities for relevant chemistry in LMICs.

1.5.1
Chemical Research Applied to World Needs

At its meeting in Munich in 1973, the International Union of Pure and Applied Chemistry (IUPAC) considered ways in which it could foster opportunities for international cooperation. The result was the establishment of Chemical Research Applied to World Needs (CHEMRAWN) as a mechanism through which member nations of IUPAC could aid in identifying and solving important chemistry problems that have a direct impact on world needs [91]. The initial purposes proposed for CHEMRAWN were:

1) To identify human needs amenable to solution through chemistry with particular attention to those areas of global or multinational interest.

2) To serve as an international body and forum for the gathering, discussion, advancement and dissemination of chemical knowledge deemed useful for the improvement of humankind and our environment.

3) To serve as an international, nongovernmental source of advice for the benefit of governments and international agencies with respect to chemistry and its application to human needs.

The major activity of the CHEMRAWN Committee has been to organize a series of conferences, designed to identify and focus attention on world needs and to make recommendations for action to the global scientific community [92]. The highly ambitious nature of these conferences envisaged (Box 1.3) at the outset of CHEMRAWN illustrates the complexity and the importance of the potential roles of chemistry in development. In particular, four key elements remain the bedrock of achieving chemistry's potential in development almost four decades after the vision was first enunciated:

• A systems approach is essential to understanding and responding to human needs.

• Many interlocking systems are involved, requiring approaches that cross boundaries between S&T disciplines and social, economic, environmental and political sectors as well as needing engagement between governments, industries and academia.

Box 1.3   Chemical Research Applied to World Needs (CHEMRAWN)

"It was envisioned that CHEMRAWN activities would provide the basis for treating chemical-based human needs as systems. Thus, CHEMRAWN conferences by their very nature would be highly interdisciplinary and would take into account the social, economic, environmental and political factors, as well as the technical components involved. It was planned that these international conferences would attract world leaders from governments, industries and academia, and that the goal and focal point of the conference activities would be an attempt by recognized and influential world leaders to take an initial step toward developing a sense of future direction that would be of value to the world chemical community. Such direction would be provided in recommendations set forth in conference proceedings and made available to participants and policymakers and governments, industries, and academic institutions worldwide. Further, it was determined that CHEMRAWN conferences would provide continuity in areas where there is a persistent need."

Bryant Rossiter, first Chair of the CHEMRAWN Committee, quoted in [92].

- Engagement with politicians and the creation of a supportive policy environment are essential for advancing and sustaining the development agenda.

- All countries, including the less economically advanced, can contribute to the development process.

1.5.2
International Organization for Chemical Sciences in Development

The International Organization for Chemical Sciences in Development (IOCD) was the first international NGO specifically devoted to enhancing the role of the chemical sciences in the development process and involving chemists working in LMICs [93–96]. Its origins lay in a program established by the Special Programme of Research, Development and Research Training in Human Reproduction (HRP) at the World Health Organization (WHO) in the 1970s. Since many contraceptives appropriate for use in LMICs were not of major interest to the pharmaceutical companies whose markets were mainly in HICs, HRP-WHO sponsored a program to develop novel contraceptives outside the traditional pharmaceutical industry channels. In a project coordinated by the Belgian chemist Pierre Crabbé, the skills of groups of chemists in LMICs were engaged to synthesize compounds for biological evaluation. Over a number of years, several hundred novel steroids were synthesized, formulated and tested [97–99]. The success of this program [96] led to the idea that it might serve as a model for developing other drugs or even pesticides, while simultaneously stimulating capacity building in LMICs and enabling chemists in these countries to contribute to key S&T areas for development [100].

Building on this idea, Crabbé invited distinguished scientists from more than a dozen countries to meet at UNESCO, Paris in 1981, to consider how to give sustained support to the research work of chemists in developing countries. They recognized that many barriers hinder the efforts by scientists in LMICs to carry out research, including inadequate laboratory equipment, lack of up-to-date books and journals, long periods of isolation from mainstream scientific activities, and so on. The vision of how these barriers might be lowered was to engage scientists from LMICs in collaborative research with scientists from HICs. IOCD was established to take forward the model [101]. Initially housed at UNESCO in Paris, IOCD soon moved to Mexico City, where it was given support by the Ministry of Health. The first group of elected officers were Glenn Seaborg (Nobel Laureate chemist, Berkely University, USA) as President; C.N.R. Rao, (Head of the Indian Institute of Science, Bangalore, India) and Sune Bergström (Nobel Laureate chemist, Karolinska Institute, Sweden) as Vice Presidents; and Elkan Blout, (Dean of the Harvard School of Public Health, USA) also as a Vice President and Treasurer. The involvement of these eminent scientists and of a range of other high-profile scientists from LMICs and HICs in the IOCD Advisory Council (including the father of the "green revolution", Nobel Laureate Norman Borlaug), was important in the early years in securing funding from a range of international organizations and foundations and in attracting prominent scientists to serve as leaders of IOCD's scientific Working Groups (Figure 1.4).

The first two IOCD Working Groups were aimed at the development of compounds for male fertility regulation and to treat tropical diseases. Modest grants were provided to facilitate the purchase of laboratory supplies and support research students in the collaborating LMIC laboratories, with collaborators in HIC laboratories assisting to overcome barriers to supply and providing back-up such as advanced spectroscopic and analytical services. While the work inevitably proceeded slowly in the first few years, a key spin-off was the establishment of networks of chemists collaborating across countries and many of the contacts and collaborations survived long after the projects themselves came to an end. IOCD was able to sponsor a number of site visits and training exchanges and a key event was a meeting of all the scientists involved in the IOCD programs in Oaxtepec, Mexico in 1986.

Figure 1.4 IOCD scientists meeting at Berkeley, California in 1986. From left to right: Carlos Rius, IOCD's first secretary; Pierre Crabbé, founder; Elkan Blout, first treasurer and one of three founding vice presidents; Carl Djerassi, one of the inspirations behind IOCD; Sune Bergström, a founding vice president; Sydney Archer, leader of the Tropical Diseases Working Group; (unknown); Glenn Seaborg, IOCD's first president and associate director of the Lawrence Berkeley National Laboratory; C.N.R. Rao, a founding IOCD vice president; and Joseph Fried, leader of the Male Fertility Regulation Working Group. (Please find a color version of this figure in the color plates.)

Pierre Crabbé was tragically killed in a car accident in 1987, but under its new Executive Secretary, Robert Maybury, IOCD continued to work and to grow, adding additional working groups on plant chemistry and on environmental analytical chemistry, and later a group on bioprospecting. The emphasis has gradually shifted away from active project funding for chemistry research programs to capacity building activities through organizing training workshops, supporting attendance at scientific meetings and supporting the networking efforts of scientists in Africa. Most recently, IOCD has adopted two long-term projects to help reinforce scientific capacities in LMICs: one on Books for International Development, which organizes the collection and transfer of books and journals to developing countries, with each shipment containing tens of thousands of items; and one on micro-scale chemistry, which helps support an international program that provides low-cost, small-scale equipment to enable students to gain hands-on practical skills in experimental chemistry even in very resource-poor settings [101].

1.6
Science and Technology for National Development

1.6.1
Investments in Research and Development

From 2002 to 2007, world R&D expenditure increased by 44%, from an estimated 788.5 billion PPP$ (purchasing power parity dollars) to 1137.9 billion PPP$. In relative terms, 1.7% of the world's Gross Domestic Product (GDP) was devoted to R&D in 2007 [55]. A number of Asian economies, including those of the Republic of Korea, Taiwan, Hong Kong and Singapore, became highly successful during the last half century and were able to maintain growth rates of 8–10% over a number of years. While many factors have been considered to contribute to the success of these "Asian tigers", important common threads have been an emphasis on higher education and on balanced investments across a range of business and technology sectors. This has enabled the economies to grow rapidly and to shift away from dependence on the export of raw materials and primary products, towards the production of high value-added products of S&T.

Given the importance of S&T for economic advancement and competitiveness in all countries [102], it is not surprising that, in recent years, there has been an increased focus on targeting specific levels of national investment in research and development (R&D) as a key driver of innovation. In 2002, the European Union (EU) set a target of reaching a level of 3% gross expenditure on R&D (GERD: also known as "research intensity") as a percentage of GDP by 2010. Of this 3%, it was projected that one third would come from public sector investment and two thirds from the private sector. By 2007, only Finland and Sweden had passed the 3% target, Austria, Denmark and Germany had reached 2.5% and France had reached 2%, while ten of the 27 EU member states were still investing below 1% [103, 104] (Figure 1.5a).

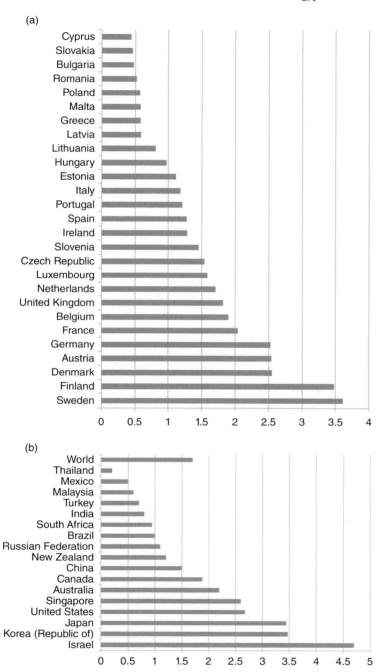

Figure 1.5 Gross domestic expenditure on R&D as a percentage of GDP for 2007.
(a) European Union countries (data from [104]), (b) other countries (data from [104]).

Outside the European Union, gross expenditure on R&D among economically advanced countries and emerging economies also varies widely (Figure 1.5b):

- Recognizing the weakness of its performance in science, technology and innovation, the African Union (AU) has initiated efforts to increase its investments in R&D, spearheaded by the New Partnership for Africa's Development (NEPAD). At the first NEPAD Ministerial Conference on Science and Technology in November 2003, Ministers of Science and Technology of 20 AU countries reaffirmed their commitment to increasing public spending on R&D to at least 1% of GDP within five years and the AU commitment to this 1% target has been reiterated on a number of occasions and member countries are still working towards it [103, 105]. To date, only South Africa regularly measures and reports data on its research intensity, which had risen to 0.95% by 2007 – of which 56% came from the business enterprise sector [106]. Data on R&D investments in other African countries appears only sporadically and, for the few countries where information is available, suggests a range from <0.1% (Algeria) to 1% (Tunisia) in North Africa and 0.5% (Mozambique) or less in the rest of Africa. *Africa's Science and Technology Consolidated Plan of Action 2006-2010* was first elaborated in 2005 by the African Union/NEPAD and is being implemented with assistance from UNESCO, which has adopted three flagship projects: (i) capacity building in S&T and innovation policy; (ii) enhancing science and technology education; and (iii) the African Virtual Campus. NEPAD has instituted the African Science, Technology & Innovation Indicators Initiative (ASTII) and the establishment of the African Observatory for Science, Technology and Innovation (AOSTI). ASTII aims at the development and adoption of African common science, technology and innovation indicators, while AOSTI will ensure that the STI indicators and information gathering as well as collation, compilation and validation are standardized [107].

- Among the emerging economies, China has demonstrated a dramatic rate of increase in GERD, which almost tripled to 1.5% between 1996 and 2007 [55] (Figure 1.6).

- Israel, Japan and Korea all invest more than 3% of GDP in R&D. The USA has long recognized the strategic economic importance of investing in R&D but its research intensity has not kept pace with this leading group. In April 2009, President Barack Obama announced that the USA will devote more than 3% of its GDP to R&D, with policies that invest in basic and applied research, create new incentives for private innovation, promote breakthroughs in energy and medicine, and improve education in math and science. This represents the largest commitment to scientific research and innovation in American history [108].

The term "Innovative Developing Countries" (IDCs) has begun to be used to describe a number of countries which have been making strong advances in strengthening their S&T to support their own development. These include Argentina, Brazil, China, India, Indonesia, Malaysia, South Africa and Thailand. At a

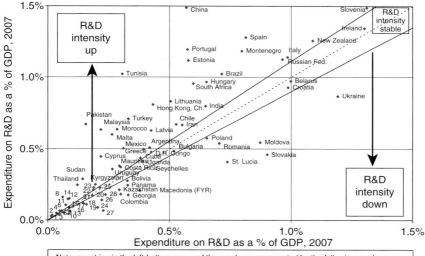

Figure 1.6 Changes in R&D intensity, 1996–2007.

meeting in 2005 to mark the 60th anniversary of South Africa's Council for Scientific and Industrial Research, leaders of science institutions from a number of these IDCs reached a consensus [109] on finding ways for S&T to play a part in sustainable development (Box 1.4). Some of the IDCs are now becoming significant development assistance partners, especially in Africa, including providing support for building higher education and research capacity [110].

1.6.2
Outputs from Investments in Research and Development

Not surprisingly, the higher levels of investments in R&D seen in many of the HICs and emerging economies correlate, at least to a degree, with science, technology and innovation outputs.

In particular, there is a strong correlation between levels of R&D investments and the densities of research workers [55] (Figure 1.7). For example, Asia represented 41.4% of world researchers in 2007 compared to 35.7% in 2002. This rise was mainly due to the increasing share of researchers in China from 14.0% to 20.1% between 2002 and 2007, reflecting the major increase in China's investments in R&D.

More funding and more researchers generally lead to higher outputs of scientific papers (Figure 1.8). The countries of the European Union, the United States and Japan collectively represent almost 70% of global R&D expenditure and these

Box 1.4   The Tshwane Consensus on Science and Development

The Emergence of Innovative Developing Countries (IDCs)

Many challenges remain. The world needs new and sustainable energy sources, protection from emerging diseases, and lower cost infrastructure. Moreover, the S&T environment has changed significantly in the last ten years. Knowledge production has been internationalized, access to money and skills has become increasingly competitive, and global technology and markets are changing with breathtaking speed.

A new set of actors has emerged in the quest to meet these challenges. Following sustained investment in education, research infrastructure and manufacturing in a number of developing countries, the IDCs have achieved high levels of economic progress and overall improvements in human wellbeing. How can these successes be generalized, and what role do the IDCs have in contributing to sustainable development?

The S&T leaders concluded that the IDCs can play a crucial role in developing innovative and appropriate solutions to global challenges, and at the same time strengthening their own S&T expertise. These leaders urged IDCs to coordinate their efforts, in order to increase investment in S&T aimed at the problems of developing countries. In particular, the leaders stressed the need for:

- developing nations, especially the poorest, to devote a proportion of their resources to S&T

- S&T leadership in developing countries to be strengthened and to define a clear set of priorities; this leadership needs to make a persuasive statement to the public that the scientific effort is essential and useful

- the political leadership of developing countries to press for a greater role in decision making on global development programs, including bilateral and multilateral aid; and to

insist that a proportion of these resources be devoted to research and nurturing local scientific and technical capacity

- the benefits of S&T need to be extended to all; S&T efforts need to be increasingly directed to the creation of affordable and accessible products and services for poor people

- the strengthening of mechanisms, such as academies of S&T, for advising high levels of government on issues of S&T

- access to careers in S&T to be widened, and at the same time systems that reward and offer S&T careers to the most talented to be developed

- the broadening of the science education base within schools, technical colleges, universities, science councils, academies of sciences, government departments and industry; these institutions are fundamental to development and wealth creation. It is clear that an environment of excellent research is necessary to attract and retain young talent in scientific careers.

Although it is highly desirable for all countries, and especially developing countries to have functioning S&T systems, the symposium noted that this is not presently the situation, and in the interim several steps needs to be taken, including:

- the establishment of regional networks between national systems to overcome the lack of a critical mass, which is presently limiting the success of S&T in many developing countries

- the implementation of appropriate performance measures at all levels and for different types of S&T institutions in order to get the most out of the available resources

- the introduction of appropriate tax incentives and grants to encourage private sector participation in R&D; additional private sector resources for S&T could be accessed by addressing sources of market failure, including the preconditions for the entry of Technology Risk Capital.

- the close networking of universities, research councils and industry in order to promote innovation, entrepreneurship and wealth generation; mission-oriented clusters of institutions focused on identified priority issues must be established to aim at discovery, development and delivery.

In conclusion, the leaders noted that it is time for a number of important initiatives. It is time for developing and post-colonial societies to "name the ghosts" of science, technology, and higher education. While benefiting many people, S&T has also systematically excluded many groups. Governments and industries often use technologies in a way that harms both workers and the natural environment. Openness about these spectres will help to assure more equitable and constructive practices in the future.

It is also time for the IDCs to act collectively and think globally. An effective response to a number of shared global challenges, such as global climate change, infectious diseases and the loss of biodiversity, can only be achieved with the involvement of all countries, and especially the developing countries. The S&T systems of the innovative developing countries can play a crucial role in building such national capacity, and in shaping their own futures.

Extract from Tshwane Consensus [109]

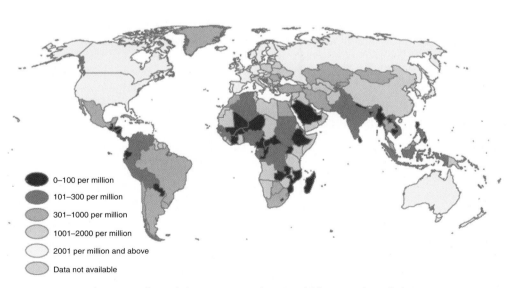

0–100 per million

101–300 per million

301–1000 per million

1001–2000 per million

2001 per million and above

Data not available

Figure 1.7 Researchers per million inhabitants, 2007 or latest available year. (Please find a color version of this figure in the color plates.)

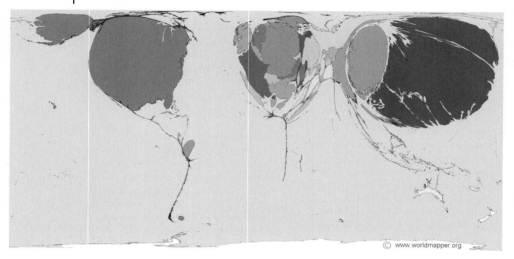

© www.worldmapper.org

Figure 1.8 Scientific publications by countries, 2001. Territory size shows the proportion of all scientific papers published in 2001 written by authors living there. Scientific papers cover physics, biology, chemistry, mathematics, clinical medicine, biomedical research, engineering, technology, and earth and space sciences [111]. (Please find a color version of this figure in the color plates.)

countries publish roughly three times more scientific papers per person living there than in any other region [55, 111]. Bibliometric analyses of country outputs can be a useful tool for uncovering strengths and weaknesses in particular areas of science [112, 113]. For example, a 2003 study [114] in Malaysia noted that more papers were produced by Malaysian scientists in physical chemistry (10.16% of the total) than in any other area of science between 1955 and 2002, followed by agriculture (5.14% of the total).

The largest outputs of patents are predominantly made by HICs [111] (Figure 1.9). In 2002, 312 000 patents were granted around the world. More than a third of these were granted in Japan and just under a third were granted in the United States.

Migration of highly skilled workers ("brain drain") has a powerful effect on these outputs. The loss of such workers from the LMICs substantially lowers their capacities to innovate. One study [115] of migration identified 15 countries, especially in the Latin America/Caribbean and African regions, for which the percentage of highly skilled workers among the migrants was in the range 33–83% (Figure 1.10).

Migrants can bring a range of economic benefits to the receiving countries, including higher rates of innovation. Productivity gains in a number of destination places have been traced to the contributions of foreign students and scientists to the knowledge base. Data from the USA show that between 1950 and 2000, skilled migrants boosted innovation: a 1.3% increase in the share of migrant university graduates increased the number of patents issued per capita by a massive 15%, with marked contributions from science and engineering graduates and without any adverse effects on the innovative activity of local people [116, 117].

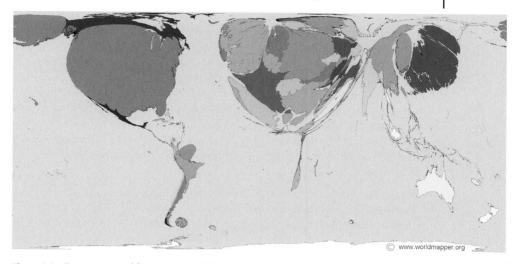

Figure 1.9 Patents granted by countries, 2002. Territory size shows the proportion of all patents worldwide that were granted there [111]. (Please find a color version of this figure in the color plates.)

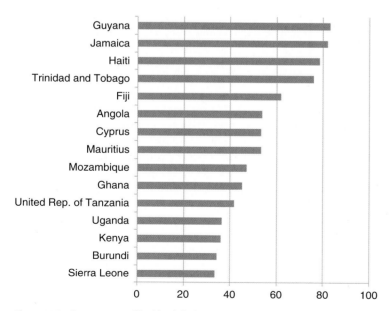

Figure 1.10 Percentages of highly skilled expatriates among total expatriates from selected non-OECD countries.

1.6.3
Connecting Science, Technology and Innovation

Scientific output in the form of publications and patents does not necessarily translate directly into innovation. The innovation environment in which scientists, inventors and entrepreneurs work plays a substantial role in determining how successful a country becomes in translating novel ideas into practical processes and products that contribute to national development [118].

A Rand study [119] which examined the scientific capability of 29 representative countries to adopt 16 technology applications divided the countries into four capability categories:

- **Advanced** – Australia, Canada, Germany, Israel, Japan, Korea, USA
- **Proficient** – China, India, Poland, Russia
- **Developing** – Brazil, Chile, Colombia, Indonesia, Mexico, South Africa, Turkey,
- **Lagging** – Cameroon, Chad, Dominican Republic, Egypt, Fiji, Georgia, Iran, Jordan, Kenya, Nepal, Pakistan

The study identified a number of major drivers and barriers to the capacity to adopt technological innovation:

- Cost and financing
- Laws and policies
- Social values, public opinion, and politics
- Infrastructure
- Privacy concerns
- Use of resources and environmental health
- R&D investment
- Education and literacy
- Population and demographics

Clearly, therefore, countries need to do more than providing support for the pursuit of S&T if they wish to reap the economic and development benefits. Investments in science, including chemistry, must be coupled with national policies on the applications of science and creation of national environments that foster innovation. Brazil provides an example of a country that is pursuing this approach (Box 1.5) [120–125]. Areas now receiving considerable attention, which have a chemistry linkage, include bio-fuels and pharmaceuticals.

Among the emerging economies, China has demonstrated an extremely high growth rate in recent years, coupled with accelerated investments in R&D (Figure 1.6). An important aspect of China's development has been the capacity to take a long-term approach to investment and planning. It is remarkable that, in 2008, the Chinese Academy of Sciences published a 50-year science strategy as an extension to the Mid-to-Long-Term Plan for Development of Science and Technology (2006–2020) issued by the State Council of China [126].

While India's overall investment in S&T has lagged behind those of other emerging economies such as Brazil, China and South Africa, nevertheless, it

Box 1.5 Brazil's experience of promoting of science, technology and innovation (ST&I)

Brazil began systematically investing in S&T in the early 1950s, establishing national councils for research and for postgraduate education. During the military rule period (1964–1985), the funding system was consolidated with the creation of agencies for innovation and S&T; and a 1988 law required Brazil's states to create S&T funds. Following a serious health crisis in the 1980s, much greater attention began to be given to S&T to solve national problems and increased investment was soon followed by increasing scientific output.

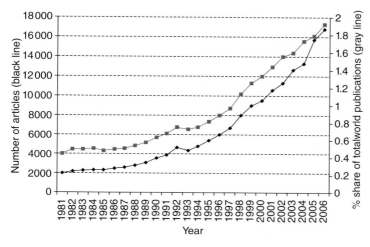

Brazilian SCIE-indexed publications, total and share of world output, 1981–2006.

However, the environment for innovation remained unfavorable with laws that restricted the ability of university-based researchers to develop their discoveries. As a result, Brazil lagged behind other economies with a more open approach.

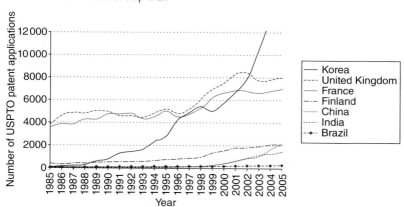

US Patent and Trademark Office patent applications, Brazil and selected countries, 1985–2005.

Continued

Brazil has taken some key steps to improve the environment for S&T and innovation:

- Sixteen Sectoral Funds were created between 1998 and 2008. They direct a fraction of the taxation of key industries to R&D projects to focus sector-specific research collaborations between enterprise, universities and research institutions, as well as to ensure the redistribution of research resources to help build capacity in less developed regions of the country.

- A new Innovation Law was adopted in 2004, to stimulate research innovation and remove barriers making it difficult for public sector researchers and private companies to collaborate.

- Brazil has significantly increased its investment in R&D in recent years, aiming to raise it from a level of around 1% of GDP to 1.5% by 2010 and with a prospect of increasing to 2% by 2020.

- Very importantly, the improved legal frameworks, investments and economic policies have been complemented in recent years by strong political will that recognizes that ST&I are at the center of economic development and social transformation.

See [120].

---

established a strong position in a number of chemistry-related fields. In particular, prior to accession to the World Trade Organization (WTO) in 2005 [127], India's non-recognition of product patents and innovative use of "reverse engineering" enabled it to establish itself as the "world's pharmacy", becoming the largest source of generic pharmaceuticals for other LMICs [128]. Having joined the WTO, the India pharmaceutical industry is being strongly encouraged to innovate and create its own intellectual property.

1.7
Capacity Building: Some Key Requirements for Chemistry's Role in Development

1.7.1
Evolution of Capacity Building Approaches in LMICs

Over the last few decades, the approach to capacity building in LMICs has evolved through several distinct stages [129]. Initially, the focus was on training of individual chemists – mainly by supporting students from LMICs to attend universities in HICs to obtain higher degrees and research experience. However, the graduates often found it difficult to pursue science careers on returning home and were frustrated by the lack of suitable laboratories, chemicals and equipment to follow their research interests. Some were assisted by maintaining close relationships with the HIC institutions where they had trained and by developing new North–South and South–South networks. Later, a more systematic effort emerged to develop high quality centers of advanced teaching and research in LMICs, creating their own cohorts of masters and PhD graduates and providing support and positions for those returning from advanced training abroad. However, chemistry and other academic disciplines have generally continued to have low esteem in many LMICs and to lack rewarding career pathways able to retain the brightest people. The importance of basic sciences like chemistry has been the subject of major emphasis in a number of conferences attended by scientists, policy makers and donor representatives [130, 131].

In the light of this mixed history of progress, but reflecting the strengthening recognition of the important roles that chemistry and other sciences need to play in the process of development, attention has turned, more recently, to addressing the entire system of teaching and research in a more comprehensive way.

Chemistry is taught within the overall framework of secondary and higher education. In many countries there are fragmented systems, sometimes involving separate ministries. Public sector teachers and lecturers in LMICs are often employed directly by the state as civil servants and subjected to a wide range of government regulations affecting salaries and terms and conditions of employment. Research is typically funded through a diverse and complex array of channels, including general university funding bodies, science councils and specific ministry programs and may involve federal, state and city-based sources and earmarked taxes. The private sector has developed some role in teaching, with many LMICs having private universities, but relatively little academic research is funded from private sources in these countries and few LMIC private universities have significant research capacity. On the other hand, overseas funding for research in LMICs may seem extremely large relative to national sources. The complex interplay of all these factors makes it difficult for a country to plan the growth and development of disciplines like chemistry and to enhance the roles that they play in addressing national priorities.

A further issue is that many scientists are resistant to the very notion of national planning, fearing that the identification of national priority areas for research will lead to the elimination of "blue skies research" and a corresponding loss of "academic freedom". An open public debate is needed which avoids taking an all-or-nothing extreme view about the direction of resources, but focuses on what are the appropriate proportions of national resources that should be apportioned to priority-focused versus undirected research and appropriate and evidence-informed mechanisms for selecting the priorities. Scientists in LMICs need to work more closely with their political leaders to show the importance of investment in long-term research in science. They are often very quick to blame their leaders for not allocating funds for research, especially when they wish to appeal for funding from external agencies.

1.7.2
National Policies for S&T

The evidence from emerging economies such as those in Brazil and China indicates that science and scientists fare well and make positive contributions to development when there are in place:

- National policies for science, technology & innovation, addressing
    - S&T education and training
    - financing national R&D
    - stimulation of innovation
    - exploitation of natural resources
    - environmental protection
    - regulation of medicines
    - regulation of intellectual property
- Key infrastructures for
    - education
    - R&D
    - innovation

Taiwan provides an example of a country which transformed its economy during the second half of the 20th century, with national planning and investment in chemistry capacity playing a key role (Box 1.6) [132].

1.7.3
Responsibilities

The government needs to take responsibility for instituting comprehensive policies for S&T, increasing levels of investment of public funds in R&D and fostering an environment that values knowledge and evidence and that promotes innovation. The scientists and academics have a corresponding set of responsibilities to be sensitive to national priority problems and to the need to communicate effectively with policy makers. The development of mutual trust and improved understanding between policy makers and researchers is now the subject of considerable attention – especially in areas such as health and the environment – and chemists must also engage in order to influence and benefit from the process.

It is also vital for international development partners to accept and orient their behavior towards the systems-based approach. Development assistance has undergone a revolution in recent years, with the preferred modality shifting away from project-based bilateral programs. Many such programs are now perceived to have had limited success and poor sustainability, being too donor-driven and failing to build country capacities and local support. In their place have come new multilateral arrangements based on sector-wide programs or general budget support, enabling the government to be in the driving seat in terms of national policies and fostering the building of government capacities for policy development, implementation and accountability. This new model has been formalized in the Paris Declaration on Aid Effectiveness and Accra Agenda for Action (Box 1.7) [133].

However, while the Paris/Accra principles are increasingly being applied to broad areas of development assistance to social and economic sectors and overall government finances, the entire field of research has been relatively neglected. Northern research institutions, research funding agencies and development assistance partners often still indulge in project-type approaches that engage individuals

Box 1.6   Taiwan's experience in national planning and investment in chemistry capacity

Taiwan's "economic miracle"

Taiwan's per capita GNP rose from US$919 in the 1950s, through US$1671 in the 1960s, US$3626 in the 1970s and US$6501 in the 1980s to US$7358 in 1990, as the agrarian economy was transformed into an export-oriented industrial one. In 1990, the total value of industrial production was US$165.3 billion, giving Taiwan foreign reserves of US$80 billion (first or second in the world).

By the early 1990s, the chemical industry was the largest industrial sector, contributing 24.2% of the total production value of US$165.3 billion, but only 8.5% directly to export sales of US$95.6 billion. This demonstrates the strategic importance of the chemical industry, as a supplier of materials and chemicals, in underpinning other export industries, including electrical/electronic goods and textiles.

The development of Taiwan's chemical industry can be divided into a number of phases:

1913–1943   Manufacturing of basic chemicals (e.g., fertilizer, chloralkali) (Japanese colony)

1944–1953   Production of substitutes for imported consumer goods (e.g., consumer commodities, agricultural products)

1954–1967   Development of light industries (emphasizing paper/food/textile products, etc)

1968–1975   Beginning of backward integration of petrochemical industry (boosting the export of textile products)

1976–1988   No. 3 and No. 4 naphtha crackers started with the fastest growth of the petrochemical industry

1989   Restructured strategy to shift away from commodities to higher value chemical products and advanced materials

Four main factors contributed to Taiwan's success in becoming one of the world's leading producers of a number of plastics and synthetic fibers by the 1990s:

1)   Establishing an integrated chemical industry – for example, integrating backwards from a garment industry dependent on cheap labor and raw materials by successively developing capabilities for the synthesis of earlier intermediates (terylene: ethylene glycol, terephthalic acid: ethylene, xylene); similarly the shoe industry was strengthened by development of the plastics industry.

2)   Development of a "debottle-necking" capacity – creating a cadre of skilled engineers and technicians able to de-bug and even improve on technology (e.g., for vinyl chloride production) originally imported from elsewhere.

3)   Cooperation between up/mid/downstream operators for example, in the pricing of chemical raw materials, intermediates and products to ensure competitiveness.

4)   Strong support by the government – including tax and investment incentives, well planned industrial zones, government-owned low-profit raw material and intermediate manufacturers and production/procurement agreement between these and petrochemical suppliers, custom/tariff protection and export incentives.

Information from [132]

Box 1.7 Paris Declaration on Aid Effectiveness and Accra Agenda for Action

The Paris Declaration, endorsed on 2 March 2005, is an international agreement to which over 100 Ministers, Heads of Agencies and other Senior Officials adhered and committed their countries and organizations to continue to increase efforts in harmonization, alignment and managing aid for results with a set of monitorable actions and indicators. Key principles are:

Ownership Partner countries set their own strategies for poverty reduction, improve their institutions and tackle corruption.

Alignment Donor countries align behind these objectives and use local systems.

Harmonization Donor countries coordinate, simplify procedures and share information to avoid duplication.

Results Partner countries and donors shift focus to development results and results get measured.

Mutual accountability Donors and partners are accountable for development results

The Accra Agenda for Action was drawn up in 2008 and builds on the commitments agreed in the Paris Declaration, focusing on:

Predictability Donors will provide 3–5 year forward information on their planned aid to partner countries.

Country systems Partner country systems will be used to deliver aid as the first option, rather than donor systems.

Conditionality Donors will switch from reliance on prescriptive conditions about how and when aid money is spent to conditions based on the partner country's own development objectives.

Untying Donors will relax restrictions that prevent partner countries from buying the goods and services they need from whomever and wherever they can get the best quality at the lowest price.

Based on [133].

or institutions in specific research or capacity-building programs. These not only ignore but often effectively undermine any existing national policies and programs, drawing scarce research resources into externally-driven activities. The Australian Centre for International Agricultural Research provides an example of an international research program that is shaped for congruence with the Paris/Accra principles [134]

Chemists in LMICs can help to counter these fragmentary approaches by engaging – among themselves, with science colleagues and with policy makers – in serious analysis and debate to promote the establishment of clear national policies on S&T and to support their governments to encourage HIC collaborators and development assistance partners to harmonize their approaches and work through and in alignment with the national policies.

1.7.4
Professional Associations and Cooperative Networks for Chemistry and Development

Professional associations such as national societies for chemistry or its sub-branches played an important role in facilitating the development of the subject and the profession in countries industrializing in the 19th and early 20th centuries. Through their impact on training, professional qualifications, the dissemination of information, encouraging good standards of practice, and the popularization of science, they contributed to strengthening the roles and contributions of chemistry in enhancing knowledge, wealth and health in these countries.

The world's largest professional society for chemistry is the American Chemical Society, founded in 1876 and now having over 160 000 members [135]. The Royal Society of Chemistry (RSC) [136] the largest organization in Europe for advancing the chemical sciences, was formed in 1980 by amalgamation of the Chemical Society (founded 1841); the Society for Analytical Chemistry (founded 1874, initially as the Society of Public Analysts); the Royal Institute of Chemistry (founded 1877, initially as the Institute of Chemistry of Great Britain) and the Faraday Society (founded 1903). Other early chemical societies include Australia (1917) [137], Austria (1897) [138], Brazil 1922 [139], Egypt (1928) [140], France (1901) [141], Germany (1867) [142], Japan (1878) [143], Netherlands (1903) [144], Norway (1893) [145], Portugal (1911) [146], Russia (1868) [147], South Africa (1912) [66], Spain (1903) [148], Sweden (1883) [149] and Switzerland (1901) [150].

The International Union of Pure and Applied Chemistry (IUPAC), founded in 1919, includes many national chemical societies among its members and provides global networking opportunities through its conferences and symposia [151]. Recently, international groupings of chemical societies have taken on regional networking and capacity building roles. For example:

- European Association for Chemical and Molecular Sciences – multiple society members from 37 European countries [152]

- Federation of African Societies of Chemistry (FASC) – established in 2006 with assistance from the Royal Society of Chemistry, it has 8 member societies, one of which is the West African Chemical Society (representing members from Benin, Burkina Faso, Côte d'Ivoire, Guinea, Mali, Niger, Senegal, Togo) [153].

- Federation of Asian Chemical Societies – 28 member chemical societies in the Asia-Pacific region [154].

In an example of inter-regional collaboration, FASC participated in launching the Pan African Chemistry Network (PACN). This is a program for Africa launched in November 2007 by the Royal Society of Chemistry with support from Syngenta. The PACN has established initial hubs in Nairobi and Addis Ababa and aims to help African countries to integrate into regional, national and international scientific networks [155].

One of the important ways that chemists can contribute to sustainable capacity development and utilization in LMICs is by developing and participating in South–South and North–South–South cooperation networks. During the course of the last century, advances in science generally have moved from being the work of highly gifted individuals (e.g., Galileo, Newton, Darwin, Einstein) to involving the work of localized groups of collaborators (e.g., Watson and Crick) and then to international networks of scientists. Such networks are not only a feature of "big" science such as particle physics or the human genome project, where dozens or hundreds of collaborators may be involved at a large number of centers, but also extend to a range of projects which are substantial scientific challenges that require a critical mass of workers tackling different aspects of a problem and/or that cross disciplinary boundaries [156, 157].

In addition to these types of networks that are driven by the demands of tackling large, complex challenges, there are also networks whose purpose is to provide support and capacity building for scientists working in settings with limited resources.

Both types of networks are of relevance to many chemists working in LMICs, since they provide opportunities for being associated with large, cutting-edge global programs as well as enabling research to be conducted in areas of local interest or relevance and reducing researchers' isolation [158]. There are recent trends showing the increasing engagement of chemists in networks aiming to foster research, capacity building and development, as summarized below.

The International Science Programme (ISP) at Uppsala University supports networks on a long-term basis. It aims at assisting developing countries to strengthen their domestic research capacity within the chemical, physical and mathematical sciences. Support focuses on regional networks and on research groups that are primarily in least developed countries targeted by the Swedish government for long-term cooperation [159].

The International Foundation for Science (IFS) receives funding from governmental and nongovernmental sources, as well as national and international organizations and has an annual budget of approximately US$ 5 million. A primary form of support provided is in the form of an IFS Research Grant to young scientists at the beginning of their research careers, which amounts to US$ 12000 and may be renewed twice. It is intended for the purchase of the basic tools needed to conduct a research project: equipment, expendable supplies, and literature. Since 1974 there have been 3500 IFS Grantees in Africa, Asia and the Pacific, and Latin America and the Caribbean. Of these 22% are women. IFS also acts as both enabler of existing and emerging networks and convener of new ones. Involvement is especially in the initial stages, with IFS providing seed money, co-operative and administrative assistance and funding to workshops, training courses, exchange visits and fellowship programs [160]. Examples of IFS-and ISP-supported networks active in chemistry include:

- AFASSA (Co-ordination of Networks for Research on Biological Resources in Africa, Asia and South America) [161]

    AFASSA was set up as a result of an international symposium on natural product research held in Montevideo, Uruguay, in 1999, with participating scientists from Africa, Asia and South America. Members to date are:
    - African Laboratory of Natural Products
    - Asian Network of Research on Anti-diabetic Plants
    - Latin American Network for Research on Bioactive Natural Compounds
    - Network for Analytical and Bio-assay Services in Africa
    - Natural Products Research Network for Eastern and Central Africa

  − Southern African Regional Co-operation in Biochemistry, Molecular Biology and Biotechnology

- ANCAP (African Network for the Chemical Analysis of Pesticides) [162]

  ANCAP was initiated in 2001 under IFS auspices. Member countries are to date Ethiopia, Kenya, Tanzania and Uganda.

- NABSA (Network for Analytical and Bio-assay Services in Africa) [163]

  See below. IFS and NABSA work on the issue of maintaining proper function of scientific equipment.

- NAPRECA (Natural Products Research Network for Eastern and Central Africa) [164]

  NAPRECA was initiated in 1984. Among the founding African scientists were several IFS grantees. In 1987, NAPRECA became affiliated to UNESCO as one of UNESCO's network programs. IFS continues to provide funding for specific projects. NAPRECA members are mainly chemists, but also biologists and pharmacologists, working on the chemistry, botany, biological activities and economic exploitation of natural products. The NAPRECA headquarters is located in Dar es Salaam, Tanzania, and there are local branches in nine countries (Botswana, DR Congo, Ethiopia, Kenya, Madagascar, Rwanda, Sudan, Uganda, Zimbabwe).

- NITUB (Network of Instrument Technical Personnel and User Scientists of Bangladesh) [165]

  NITUB was formed in 1994 with seed money from IFS. The main objective of NITUB is to improve the competence of scientists and technical personnel to operate, maintain, and repair scientific instruments. NITUB maintains an inventory of scientific instruments at Bangladeshi institutions, and aims to create a stock of spare parts. The General Secretary is located in the Department of Chemistry, University of Dhaka.

- NUSESA (Network of Users of Scientific Equipment in Southern and Eastern Africa) [166]

  NUSESA was initiated in 1989 with the aim to provide a forum for information and discussion on issues related to the purchase, use, and maintenance of scientific equipment in southern Africa. IFS passed on the coordination and administration of NUSESA activities when the NUSESA Secretariat was established in 1996 in Harare, Zimbabwe. Since NUSESA has secured funding from government agencies, IFS–NUSESA joint collaboration has evolved to policy-making and consultation. Local NUSESA chapters have been set up in 16 countries (Botswana, Eritrea, Ethiopia, Kenya, Lesotho, Madagascar, Malawi, Mauritius, Mozambique, Namibia, South Africa, Swaziland, Tanzania, Uganda, Zambia and Zimbabwe). The NUSESA Secretariat is located at University of Western Cape in South Africa.

- WANNPRES (Western Africa Network of Natural Products Research Scientists) [167]

WANNPRES was established in 2002 by COSTED (Committee on Science and Technology in Developing Countries) on the initiative of a group of scientists from universities and research institutes in West Africa. The objective with WANNPRES activities is to enhance research and capacity building in the conservation and effective use of natural resources in Africa. IFS involvement dates back to the founding assembly. The WANNPRES Secretariat is located at the Department of Chemistry at the University of Ghana.

In the area of providing support and capacity building for scientists working in settings with limited resources, IOCD [102] began a program in the 1980s to provide analytical services for chemists in LMICs. This was initially a North–South network, with chemists at City University, London (Stephen Matlin), the University of Missouri (Michael Tempesta) and the Universidad Nacional Autónoma de México (Carlos Rius) receiving samples from chemists in a range of countries in Africa, Asia and Latin America and providing, free of charge, infrared, ultraviolet, NMR and mass spectra. In some cases, at the invitation of the submitting group, assistance was provided with the interpretation of spectra and the elucidation of structures of synthetic and natural products.

In 1992, IOCD's Walter Benson participated in a meeting in Gaborone, Botswana which led to the launch of a new activity, the Network for Analytical and Bioassay Services in Africa (NABSA) and IOCD contributed launching funds. There has been strong support from the University of Botswana and additional funding for NABSA has come from a variety of international sources including USAID, UNESCO, International Programme in the Chemical Sciences (Uppsala) and TWAS.

Coordinated by Berhanu Abegaz at the University of Botswana [168], NABSA's objectives are:

- To promote the development of scientific activities in Africa by offering analytical, bioassay and literature support services to chemists.

- To cooperate with active scientists in a joint short-term intensive-research undertaking by inviting them to the reasonably well equipped laboratory in Botswana.

- To promote the professional development of young scientists by arranging sub-regional symposia.

NABSA collaborating centers contribute a range of spectroscopic facilities including 200–600 MHz NMR (Botswana, Ethiopia, Kenya, Lesotho, South Africa and Zimbabwe) (Figure 1.11). In the period 1998–2009, NABSA provided over 11 000 NMR spectra and over 2000 mass spectra to African scientists. Recipient countries of NABSA services have included Cameroon, Democratic Republic of Congo, Egypt, Ethiopia, Ghana, Kenya, Nigeria, Sierra Leone, South Africa, Sudan, Tanzania and Zimbabwe. More than 30 short-term visits to the University of Gabarone were arranged for chemists from different African countries [163].

From 2005, NABSA's focus shifted to research cooperation with research groups in selected countries and institutions, particularly in Cameroon, Ethiopia, Nigeria,

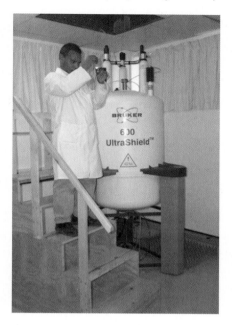

Figure 1.11 600 MHz NMR forming part of the NABSA analytical service at the University of Gabarone, Botswana. Photo from B.M. Abegaz.

South Africa, Tanzania and Zimbabwe, in order to help build and strengthen capacities and increase the overall impact of the collaboration. The NABSA centre itself has been involved in at least 63 publications in peer-reviewed international and regional journals Productive NABSA collaborations include a range of phytochemistry studies aimed at structure elucidation and the identification of bioactive natural products, such as anthraquinones [169]; Studies of Lathyrism, a disorder produced by toxic non-protein amino acids from the grass pea (*Lathyrus* species, *Leguminosae*) [170]; identification of dimeric sesquiterpenoid lactones from a south African plant (*Dicoma anomala*) possessing anti-plasmodial properties [171], leading to patenting; and identification of insect antifeedant bichalcones from *Rhus pyroides* and development of a general synthetic method for C–C linked bichalcones [172, 173].

NABSA has been an important source of energy and momentum for strengthening Africa's capacity for natural products research. In 2006 a new aspect was initiated with the establishment at the University of Botswana of the Centre for Scientific Research, Indigenous Knowledge and Innovation (CESRIKI), which has facilitated the Research Visitors' Programme of NABSA and also introduced a range of bioassay capabilities. In 2009, a consultative meeting in Gabarone led to an initiative for the establishment of a Pan-African Natural Products Library (p-ANPL) as a repository for compounds that have been isolated, identified and screened. The Board is drawn from several African countries.

Other networks operating in the African region include:

- SEANAC (Southern and Eastern African Network for Analytical Chemists). Established as a result of a SIDA-funded workshop in Gabarone in 2002, SEANAC's objectives are to (i) promote analytical chemistry in the region through collaboration, research, research training, teaching and information sharing; (ii) facilitate inventory, access, operation, maintenance and repairs of analytical equipment; and (iii) collaborate with organizations with similar aims [174].

- AAPAC (African Association of Pure and Applied Chemistry). AAPAC's objectives are to (i) provide a forum for the exchange among scientists and development agents of scientific information on the state of the chemical sciences in Africa; (ii) foster research in the chemical sciences; (iii) cooperate with other international bodies which pursue aims and objectives similar to those of AAPAC; (iv) promote mutually beneficial interdependent linkages between industry and other entrepreneurial bodies on the one hand and research institutes, including universities, on the other [175].

1.7.5
National Funding for Research

While funding from public and private international sources is very important, the availability of regular, national sources of public funding is essential if a critical mass of researchers is to be established and maintained [176].

In the 18th and 19th centuries, as the pace of technological progress increased before and during the industrial revolution, most scientific and technological research was carried out by individual inventors using their own funds [177]. Historically, the development of national channels for funding scientific research in Europe and North America is relatively recent and lagged considerably behind these first waves of technological progress. Apart from military research, they were driven by specific challenges such as public health problems and by broader concerns about international competitiveness. Some examples to illustrate this include:

- The Royal Society of England, founded in 1660 as a learned society, received a government grant in 1850 to assist scientists in their research and to buy equipment [178]. But it was in the early the 20th century that the challenge of tuberculosis led to the establishment in 1913 of the Medical Research Committee and the realization that Britain was falling behind its competitors led to the establishment in 1916 of the Department of Scientific and Industrial Research to fund applied research and technological innovation [179].

- Germany's Kaiser Wilhelm Gesellschaft zur Förderung von Wissenschaft und Forschung (KWG), now the Max-Planck-Gesellschaft, was created in 1911 [180]. The Deutsche Forschungsgemeinschaft (DFG, German Research Foundation) traces its origins back to its forerunner, the Notgemeinschaft der Deutschen Wissenschaft, which was established in 1920 on the initiative of Fritz Haber [181].

- Following Belgium's early industrialization and development of chemical industries (Box 1.2), it established the Fonds National de la Recherche Scientifique (FNRS) in 1928 [182].

- The Centre National de la Recherche Scientifique (CNRS) was created in 1939 and is the largest governmental research organization in France and the largest fundamental science agency in Europe [183, 184].

- In the USA, the first National Institute of Health was established in 1930 for medical research, but the main science-funding agency, the National Science Foundation, was not created until 1950 [185].

1.7.6
Gender Issues

Across the world, there is a substantial under-representation of women in science. In 121 countries with available data, women represent on average 29% of researchers. In 37% of these countries, they represent less than one-third. Only about 15% of countries have achieved gender parity, and only a handful of others have more women researchers than men. The proportion of women researchers is low in Africa (33%) and particularly low in Asia (18%) (Figure 1.12) [55].

Equality for women in all fields of human activity, including in science, is first and foremost a fundamental human right. In addition, as noted by the World Economic Forum [186] "Countries that do not capitalize on the full potential of one half of their societies are misallocating their human resources and undermining their competitive potential".

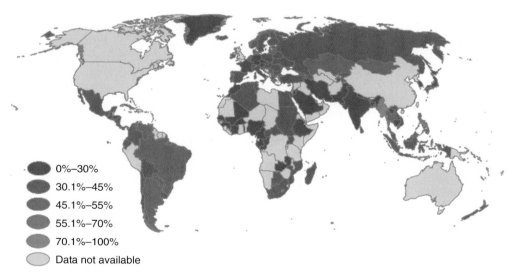

0%–30%
30.1%–45%
45.1%–55%
55.1%–70%
70.1%–100%
Data not available

Figure 1.12 The gender gap in science. Women as a share of total researchers, 2007 or latest available year [55]. (Please find a color version of this figure in the color plates.)

To ensure that women achieve equality of opportunity in the field of science, countries need to institute a range of measures including ensuring equal access and treatment in school science programs and helping girls to build confidence in their science abilities; eliminating biases in university entrance procedures; and enacting anti-discrimination measures in employment [187].

1.7.7
Open Access

Access to published information is critical for scientists in order to enable them to assemble, understand and build on prior knowledge [188]. In chemistry and other sciences, there has been a rapid proliferation in the numbers and costs of journals over recent decades and libraries have found it difficult or impossible to keep pace. This problem has been especially acute for libraries in LMICs. As access to the internet has gradually extended to lower-income countries and some existing and newly established journals have introduced open access policies, there has been some improvement in access for those working in resource-poor settings. However, the situation is still far from satisfactory and there remains an unresolved tension between those who favor a policy of completely open, free-for-all access to scientific publications, those promoting special schemes which provide discounts and subsidies to scientists in certain LMIC-based institutions, and those wishing to preserve the commercial basis of publications. Learned societies in HICs are often finding this a challenging issue, since many produce leading journals in their subject areas and have become dependent on the income from journal subscriptions to maintain their financial viability [189–191].

Chemists need to work with their learned societies, professional bodies and funding agencies to seek innovative solutions that will maximize access to published material for those working in resource-poor settings.

1.7.8
Technology Transfer

The term "technology transfer" is used in two contexts, both of which are important to the subject of chemistry for development.

1) **Transfer between academia and industry:** In HICs, many universities have established "technology transfer offices" or "development offices" whose role is to assist academic researchers to find commercial applications for their discoveries. These offices may facilitate patenting, identifying commercial partners, start-up costs for small-medium size businesses, and so on. A number of technologically strong universities have established their own business parks as a further way of promoting commercialization. To date, relatively few LMIC academic institutions have adopted such models, so that their chemists and other scientists find it much more difficult to develop their ideas and may find it necessary to become involved in overseas partnerships instead.

LMIC governments and academic institutions can benefit from encouragement and assistance with establishing appropriate frameworks and building human resource capacities to be able to institute academic–industrial technology transfer arrangements as part of an overall enhancement of national innovation. There are instructive examples to be found in some successful initiatives in LMICs [192–194].

2) **Transfer of technology between countries:** The diffusion of technologies from one part of the world to another is not a new phenomenon. Prominent historic examples in the field of chemistry include a range of processes for chemical manufacture adopted from the Islamic world by European countries over a period of several centuries [195]. Interest now centers on the acquisition by LMICs of processes for manufacturing chemicals, pharmaceuticals and advanced materials. Tanzania provides an example (Box 1.8) of a country that has benefitted from giving greater attention to S&T and technology transfer [196, 197]).

Sources of technology for transfer may include the private sector [198, 199], international organizations [200] and public–private partnerships [201]. For LMICs,

---

Box 1.8   Science and technology (S&T) and the transfer of technology in Tanzania

S&T has made major contributions to Tanzania's economic development, including:

- Improvement of overall productivity by instituting structural changes in the models and methods of production which lead to greater efficiency and innovativeness in economic activities.

- Boosting exports by improving the quality of products, reducing the costs of production.

- Improvement of food security by supplying technical information to agri-producers and providing reliable markets for their products.

- Spreading income-earning opportunities by making technical information available to small entrepreneurs in rural and urban Tanzania.

- Upgrading the technical skills altitudes and productivity of the labor force through science and technology education and popularization.

Transfer of Technology

In the past, Tanzania had no deliberate strategies or plans for appropriate selection, acquisition and transfer of technology for effective integration of imported technologies with local capacity for R & D. However, currently, deliberate efforts have been put into place in order to make sure that the speed of technology transfer is effective and sustainable for example, establishment of the Tanzania Commission for Science and Technology in 1986 and the Centre for the Development and Transfer of Technology in 1994, to institute a workable mechanism for the coordination of capacity building in the selection assessment, negotiation, adoption, R&D, information exchange and extension services.

Policy/Regulatory Framework

The first National S&T policy was enacted in 1985 and revised in 1995, its major thrust being to establish a prioritized program for generating new knowledge and to determine strategies for

Continued

the application of S&T development in Tanzania. The broad objectives of the Tanzania S&T Policy are to:

- Promote science and technology as tools for economic development, the improvement of human, physical and social well being and for the protection of national sovereignty.

- Promote scientific and technological self-reliance in support of economic activities through the upgrading of R&D capabilities.

- Promote and encourage the public and private productive sectors in developing S&T.

- Promote active participation of women in S&T.

- Establish and/or strengthen national S&T institutions.

Legal Framework and Technology Policy Instruments

A scientific and technical advisory committee on S&T has been established in order to advise the President in addition to the Inter-Ministerial Technical Committee of the Cabinet.

A legal framework was laid down through the (Investment Promotion and Protection) Act of 1990, and the Tanzania Commission for Science and Technology Act No. 7 of 1986 which spells out the establishment of a National Centre for Development and Transfer of Technology. This Centre is charged with powers to establish rules and regulations for rationalizing the acquisition evaluation, choice coordination and development of technology.

In order to achieve the national goal of steady economic growth, maximum utilization of local resources and technology, expansion of technical education through development of local research units in enterprises, and long-term comprehensive technological policies integrated within the overall national development plans have been adopted by the Tanzania government.

Extracts from Tanzania *National Website [197]*

---

technology transfer may be the entry point for developing high value-added industries. In recent times, the growth of the chemical industries in Brazil, China and India have reflected a combination of technology transfer, adaptation and, in some cases, "reverse engineering" to devise new processes for making known high-value products.

The challenge for LMICs is to develop the combination of policy and economic conditions that favor inward investment, guarantee orderly labor relations and access to markets and that ensure the supply of well qualified scientists, managers and administrators.

1.8
Chemistry and Future Challenges to Health, Wealth and Wellbeing

1.8.1
"Glocal" – Thinking and Acting from Global to Local

Some have argued that the pursuit of S&T capability is not at all essential for lower income countries, but is an expensive luxury, while essential knowledge and technologies are best acquired by transfer from the richer nations that can afford to create them. But experience has shown that this view is not correct. Countries substantially accelerate their own development, enhance their competitiveness and strengthen their international negotiating positions by increasing their S&T capabilities. They need to be able to adapt and localize technologies even if these are created elsewhere; and they need the capacity to address their own problems, which are not always shared with higher-income countries (e.g., tropical diseases, local agricultural sustainability, specific water pollutants). Moreover, all countries need a basic capacity in fundamental areas like analytical chemistry in order to be able to monitor what is happening, identify problems and develop and apply solutions – for example, in relation to the quality of the environment or the quality and authenticity of pharmaceuticals in local supply.

These considerations are encouraging a growing realization that many problems have a dual character, involving a global dimension which requires joint international action on the one hand, but a local adaptation, application or focus on the other. Thus, as globalization progresses, the slogan "think globally, act locally" is evolving into a "glocal" or "glolocal" principle which requires everyone to "think and act globally and locally" and which has been adopted in the private sector [202, 203].

The following discussion of some major challenges provides a number of examples of the importance of this global–local duality of approach.

1.8.2
Agriculture, Food and Nutrition

The term "green revolution" describes a series of research, development, and technology transfer initiatives involving the development of high-yielding varieties of cereal grains, expansion of irrigation infrastructure, and distribution of hybridized seeds, synthetic fertilizers, and pesticides to farmers. They enabled many LMICs to increase their industrialized agriculture production and thereby to help meet the food and nutrition needs of growing populations (Box 1.9) [204]. Supported by the Rockefeller Foundation, the Nobel Laureate Norman Borlaug (who was also one of IOCD's early advisers) played a leading role in the green revolution, developing new, high-yield dwarf wheat that resisted a variety of plant pests and diseases and yielded two to three times more grain than traditional varieties.

Overall, raised productivity in three of the world's main staple food crops – rice, wheat, and corn – had substantial impact in a number of LMICs, including Mexico, Pakistan, India, and China. While this approach worked in Asia and other places where rice and wheat are the staple crops, it provided little benefit in Africa, where sorghum, millet, and cassava were consumed by the poor and where many countries also suffered from poor soil and uncertain rainfall, a shortage of trained agriculturalists and lack of technology. Borlaug was convinced that, while traditional plant breeding methods remained important, agricultural biotechnology and herbicide-resistant crops had a vital role to play in places like Africa [205, 206].

Box 1.9   The green revolution

"Record yields, harvests of unprecedented size and crops now in the ground demonstrate that throughout much of the developing world – and particularly in Asia – we are on the verge of an agricultural revolution.

- In May 1967 Pakistan harvested 600 000 acres to new high-yielding wheat seed. This spring (1968) the farmers of Pakistan will harvest the new wheats from an estimated 3.5 million acres. They will bring in a total wheat crop of 7.5 to 8 million tons – a new record. Pakistan has an excellent chance of achieving self-sufficiency in food grains in another year.

- In 1967 the new high-yielding wheats were harvested from 700 000 acres in India. This year they will be planted to 6 million acres. Another 10 million acres will be planted to high-yield varieties of rice, sorghum, and millet. India will harvest more than 95 million tons in food grains this year – again a record crop. She hopes to achieve self-sufficiency in food grains in another three or four years. She has the capability to do so.

- Turkey has demonstrated that she can raise yields by two and three times with the new wheats. Last year's Turkish wheat crop set a new record. In 1968 Turkey will plant the new seed to one-third of its coastal wheat growing area. Total production this year may be nearly one-third higher than in 1965.

- The Philippines have harvested a record rice crop with only 14% of their rice fields planted to new high-yielding seeds. This year more land will be planted to the new varieties. The Philippines are clearly about to achieve self-sufficiency in rice.

These and other developments in the field of agriculture contain the makings of a new revolution. It is not a violent Red Revolution like that of the Soviets, nor is it a White Revolution like that of the Shah of Iran. I call it the Green Revolution.

This new revolution can be as significant and as beneficial to mankind as the industrial revolution of a century and a half ago. To accelerate it, to spread it, and to make it permanent, we need to understand how it started and what forces are driving it forward. Good luck – good monsoons – helped bring in the recent record harvests. But hard work, good management, and sound agricultural policies in the developing countries and foreign aid were also very much involved."

Extract from speech by William Gaud, 6 March 1968 [204]

With the world's population expected to increase by 50% in the first half of the 21st century, there is now seen to be a need for another green revolution [49, 207] – one which provides adequate amounts of nutritious and affordable food-stuffs in high yield and through agricultural practices that do not require large increases in the global amount of land under cultivation and do not cause pollution of earth, water or air.

Chemistry has many roles to play in meeting these challenges, from soil chemistry to pollution monitoring, from creation of better methods of plant crop protection to helping develop new, more productive and more robust varieties.

1.8.3
Climate Change

There is now incontrovertible evidence that human activities during the last couple of centuries are responsible for a significant rise in average global temperature and a concomitant increase in extreme weather events such as floods and droughts and heat waves and severe winters. If continuing unchecked, this global warming and shifts in weather patterns threaten the lives and livelihoods of many millions of people, due to drought, desertification, floods, inundation of major tracts of land and even entire small island states, disruption of agriculture and the spread of water- and vector-borne diseases [208].

While there has been a temptation to see the impact of climate change as being something that will be felt some decades in the future, there is much evidence that the adverse consequences are already being seen, Three broad categories of health impacts are associated with climatic conditions (Figure 1.13) [209] and current excess mortality due to climate change may now be a few hundred thousand extra deaths per year [209, 210].

Furthermore, it is of special concern that the most severe impact of climate change is being felt by vulnerable populations who have contributed least to the problem. The risk of death or disability and economic loss due to the adverse impacts of climate change is increasing globally and is concentrated in poorer countries [48, 209].

Needed contributions from chemistry towards the international response to climate change involve a combination of measures to mitigate its extent and to adapt to the unavoidable consequences. LMICs will require substantial assistance, including technology transfer [211]. Chemistry has already made innumerable contributions to identifying climate-related problems (e.g., through environmental analysis of ozone depletion and greenhouse gas emission); understanding their underlying causes and contributing solutions (e.g., through studies of the roles of

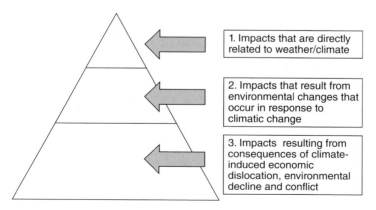

Figure 1.13   Health impacts associated with climatic conditions. Reproduced from [209].

gases like carbon dioxide and methane and the atmospheric photochemistry of fluorocarbon refrigerants and aerosol propellants).

Much more will be required of chemistry in the future, including better, cheaper and more robust field methods for environmental monitoring and impact assessment and developing new processes for energy generation, industrial production, materials recycling, and so on.

1.8.4
Energy

Intimately linked to the issue of climate change is the challenge of generating energy in an environmentally appropriate and sustainable manner.

World energy use (measured in kg of oil equivalent per capita) rose from 1338 kg per capita in 1971 to 1819 kg per capita in 2007, while world population rose from 3.8 billion to 6.6 billion during that period, representing an absolute increase in energy consumption of 236% [212]. With 80–90% of world energy production coming from the combustion of fossil fuels, there is an extremely urgent need to develop environmentally sustainable methods of energy generation, as well as methods for increasing the efficiency of energy use.

Among the many roles that chemistry is playing are:

1) devising ways of extracting and fixing carbon dioxide and other pollutants generated in burning coal and oil and reducing the emission of particulates that damage the respiratory system;

2) contributing to the improvement of efficiency of combustion engines, thereby reducing consumption;

3) developing new bio-fuels which reduce the net difference between current absorption and emission of greenhouse gases;

4) innovating new materials for the capture of sunlight and its transformation into energy;

5) creating new, environmentally clean processes for energy production, such as fuel cells.

1.8.5
Environment and Sustainable Development

Human activities such as the clearing of land for agriculture; diversion of water courses and extraction of water from underground water tables; extraction and transformation of raw materials to provide physical products and energy for power appliances; and the creation of domestic and industrial waste materials, all contribute to changes in the natural environment.

The pace of these changes has accelerated as the population of Earth has grown, especially in the last couple of centuries, but it is only within the last few decades that the extent of the problem has been recognized internationally [213] and a concerted global effort has begun to address the issues which have come to be expressed in the objective of "sustainable development". This term encapsulates the need to ensure that human beings can live in an environment free from pollutants and health hazards and that the sum of human activities does not cause degradation of the physical and biological environments of the planet. In particular, the Rio Declaration emerging from the UN Conference on Environment and Development, held in Rio de Janeiro in 1992, provided the fundamental principles and a program of action ("Agenda 21") for achieving sustainable development, establishing linkages among economic and social development and environmental protection [57]. This was followed up a decade later by the World Summit on Sustainable Development held in Johannesburg in 2002 [58].

Sustainable chemistry is understood as the contribution of chemistry to the implementation of the Rio Declaration and Agenda 21 and of follow-on processes such as the Johannesburg Declaration. The focus of sustainable development concerns in areas such as water, energy, nutrition, human habitation and the quality of the environment mean that chemistry has many central roles to play—in monitoring the environment, in understanding, preventing and mitigating adverse chemical/biological impacts of human activities, and in devising new, cleaner and more environmentally sustainable process. Strengthening capacity for analytical chemistry in LMICs is indispensable to achieving these objectives [214].

1.8.6
Health

The nature of health challenges faced in every part of the world is changing, as a result of shifting patterns of disease, the globalization of health threats, changes in the environment and in human behavior. Some of the key health challenges and roles for chemistry in meeting them are highlighted below.

Non-communicable diseases (NCDs—e.g., cancer, diabetes, heart disease, stroke) are becoming the most prevalent causes of ill-health and death everywhere—but in many LMICs they are appearing alongside still-prevalent *infectious diseases*—including some that are global challenges (e.g., HIV/AIDS and tuberculosis) and some specific to tropical regions (e.g., malaria, African and South American trypanosomiasis, visceral and cutaneous lieshmaniasis, schistosomiasis). Those diseases that are prevalent globally, like NCDs and HIV/AIDS, have attracted considerable R&D investment and drugs to prevent or treat them have become available, but often these are too expensive for use in LMICs and/or require functioning health systems able to support patients with chronic conditions—which many LMICs lack. Many of the tropical infectious diseases have been neglected by the global pharmaceutical industry and new or improved drugs are still needed [215]. Chemists have a central role to play in the discovery of new drugs for these communicable and chronic diseases—drugs that are safe, effective, affordable and suitable for use in resource-poor settings where there may be a dearth of cold chains, laboratory diagnostic and clinical analysis facilities and specialist medical facilities and personnel.

The *quality of pharmaceuticals in circulation* in developing countries has become an issue of very serious concern in recent years. Drugs that are sub-standard, illegal imitations and non-effective fakes are widely available in many LMICs, as well as authentic drugs that are out of date or have deteriorated due to poor conditions of transport and storage [216]. Every country needs mechanisms to identify such pharmaceuticals – requiring the establishment and maintenance of well-equipped and staffed national analytical laboratories able to conduct reliable and speedy analyses to internationally-recognized standards. Recently, there has been demand from some LMICs to be able to localize the production of pharmaceuticals essential to their health needs, which increases the need for local capacities for drug regulation and quality control.

Pandemics such as those caused by viruses responsible for severe respiratory diseases (e.g., SARS, avian flu, swine flu) have emerged as a serious global threat in recent years, with the potential to cause ill-health, death and economic disruption on a very large scale [217]. The importance of the contributions of clinical chemistry to diagnosis and of medicinal chemistry to developing preventions and treatments cannot be underestimated.

Demographic changes are occurring on an unprecedented scale of speed and scope. The world's population grew from a level of about 1 billion in 1800 to 2 billion around 1920 and then leapt to 6 billion in 2000. It is predicted to reach around 9 billion by 2050. The birth rate in HICs declined to around replacement level during the 20th century, so that most of the 50% increase anticipated in the 21st century will be in LMICs. While average population ages are rapidly increasing in HICs, creating serious challenges in the worker/dependent ratios, many LMICs currently have very high proportions of young people [218]. These demographic shifts have major implications for patterns of consumption and demands for physical and energy resources – and also for health. For example, the largest cohort of adolescents that the world has ever seen requires greatly increased attention to sexual and reproductive health services, including safe, effective and acceptable means of family planning, while the growing numbers of aging people will present growing challenges in the management of chronic diseases, including disabling conditions such as arthritis. Chemistry can make a major contribution, not only to the development of new drugs, diagnostics and medical devices appropriate to these changing populations but also to the creation of new materials that enhance their quality of life.

In 2007, for the first time the world's population living in urban settings was as large as the rural population and this *urbanization* trend is continuing. As most of the world's increase in population is happening in LMICs, a substantial proportion of the new citizens of the planet in the next half century will be living in cities in less wealthy countries, where currently many people live in very unhealthy conditions in crowded urban slums and informal settlements with poor access to clean water and sanitation [219]. Chemistry and allied sciences such as chemical engineering and materials, food, energy and sewage treatment sciences have much to offer in helping to ensure the availability of salubrious living conditions including healthy dwellings, clean water, sanitation and safe foodstuffs.

1.8.7
Intellectual Property

About US$ 1.1 trillion was invested globally in R&D in 2007, up from about US$ 525 billion in 1996. This intensification of investment in the creation of new knowledge has, in turn, fueled demand for rights over knowledge – that is, for the protection of intellectual property (IP). In 2007, 1.85 million patent applications, 3.3 million trademark applications and 621 000 industrial design applications were filed around the world. This creation of new knowledge is increasingly taking place in emerging economies and involving researchers in LMICs. As one indicator of the globalization of research, 21.9% of scientific articles in 2007 were internationally coauthored, three times as many as in 1985. Further scope for participation of LMIC scientists in the creation of IP is being provided by the increasing trend of "open innovation" – the tendency for firms to look outside themselves to satisfy their innovation needs, whether through traditional means, such as licensing, sub-contracting, R&D contracts or joint ventures, or through newer means, such as the use of problem-solvers on the Internet or open source cooperation [39].

Two aspects of IP issues are of particular relevance in the context of chemistry for development:

1) **Protection of IP rights of LMIC citizens and institutions:** This has been of particular concern in relation to the discovery or invention of useful processes and products by researchers in LMICs, and the protection of indigenous knowledge in areas like traditional agriculture and medicine.

With regard to this aspect, it is particularly important for LMIC governments to ensure that strong IP policies and legislation are developed and implemented and that governments and institutions in LMICs strengthen their capacities for the management of IP at the national levels and for the negotiation of IP issues at the global level.

- The World Intellectual Property Organization (WIPO) is a specialized UN agency established in 1967, dedicated to developing a balanced and accessible international IP system which rewards creativity, stimulates innovation and contributes to economic development while safeguarding the public interest [220]. WIPO's new Medium Term Strategic Plan [39] includes multiple technical assistance and capacity building activities.

- Among the tools available to assist in capacity building, the handbook of best practices in intellectual property management in health and agricultural innovation is an outstandingly useful resource [221].

2) **Protection of IP rights of LMICs to flexibilities under the rules of the World Trade Organization (WTO):** WTO deals globally with the rules of trade between nations, providing standards, enforcement and dispute settlement under the Agreement on Trade-Related Aspects of Intellectual Property Rights (TRIPS) which came into effect on 1 January 1995 [222]. The TRIPS Agreement includes flexibilities in rules concerning the observance of patent rights

on medicines and allows an exception that Members may exclude from patentability diagnostic, therapeutic and surgical methods for the treatment of humans or animals. In some cases, this can allow member states to import or synthesize drugs considered essential to their national health needs. However, the rules also make provision for agreements among parties which introduce additional restrictions. These "TRIPS Plus" variations have been extremely controversial in cases where they have the effect of reducing the ability of LMICs to protect the public interest. One notorious example has been the dispute over the rights of LMICs to access drugs for the treatment of HIV/AIDS without having to pay the extremely high prices that were being charged by multinational pharmaceutical companies [223–225].

Following the work of a Commission on Intellectual Property Right, Innovation and Public Health [226], an inter-governmental negotiating process under the aegis of the World Health Organization led to agreement on the WHO Global Strategy and Plan of Action on Public Health, Innovation and Intellectual Property [227]. In this document, Member States endorsed by consensus a strategy designed to promote new thinking in innovation and access to medicines, which would encourage needs-driven research rather than purely market-driven research to target diseases which disproportionately affect people in developing countries. The challenge now is to achieve consistency in the ways that member states approach IP issues in their dealings with WIPO, WTO and WHO.

1.8.8
Natural Resources Exploitation

Exploitation of the Earth's physical and biological resources has always been a feature of human activities and the pace and extent of this exploitation have increased markedly in the last two centuries – both driven by and feeding technology advances, economic growth and population expansion.

Many LMICs which have been mainly the source of raw materials such as minerals and primary agriculture products now wish to reap the economic and developmental benefits of increasing production and adding value to the materials through processing. At the same time, there is pressure on these countries to conserve their natural resources, engage in sustainable development and not follow the historic pathways set by HICs which have led to pollution, exhaustion of resources and loss of biodiversity.

Countries everywhere are now being faced with the challenges of dwindling stocks of many key natural resources such as minerals, oil, and gas, as well as a global reduction of biological diversity due to factors such as deforestation, overfishing, hunting, and excessive use of monocultures in agriculture.

Some examples of key contributions of chemistry to these challenges include developing cleaner, more efficient, less energy-intensive and less polluting extraction and refining methods for minerals; methods for the recycling of inorganic and organic materials, and new substitute materials.

Chemical studies of natural products have made exceptionally valuable contributions to human health and wealth. As an example of "chemical prospecting", Eisner [228] cites the example of ivermectin, a fungal metabolite of highly complex structure which has been marketed for treatment of parasitic worm infections in animals and human beings. As well as generating US$ 1 billion in sales, its donation by Merck to the World Health Organization laid the basis for the treatment and prospective eradication of river blindness (onchocerciasis) in Africa.

The exploitation of biological resources has become an area of particular concern. Conservation of biodiversity is considered vital for long-term human survival because plants, animals and bacteria can be the source of new nutrients, genes conferring resistance to crop pests, and drugs for combating diseases. This implies that studies are undertaken globally to uncover these valuable assets, but exploitation needs to conserve their stocks as well as ensuring appropriate rewards for their owners. Countries which have some of the most valuable and diverse and least studied biological resources – often LMICs – have sometimes experienced "biopiracy" in which samples of plants or knowledge about their uses have been taken abroad and exploited without benefit to the country of origin or to the local inhabitants whose indigenous knowledge has been the key.

Valuable lessons have been learned from the experience of LMICs that have developed ways to meet these challenges. One very instructive example has been that of Costa Rica, a tiny Central American country which covers 0.04% of the world's total land area, yet is believed to harbor about 4–5% of the estimated terrestrial biodiversity of the Earth. In 1989, Costa Rica founded a national institution to gather knowledge on the country's biological diversity, its conservation and its sustainable use. The Instituto Nacional de Biodiversdad (INBio) was established as a non-profit, public interest NGO with a high degree of autonomy and with initial financial assistance from the Swedish Cooperation Agency (SIDA) and the MacArthur Foundation [229]. In 1991, INBio instituted an innovative agreement with a multinational pharmaceutical company, in which Merck was granted the right to evaluate the commercial prospects of up to 10 000 plant, insect, and microbial samples collected in Costa Rica. In return for these "bioprospecting" rights, Merck paid INBio US$ 1 million over two years, and provided equipment for processing samples and scientific training. Merck also agreed to pay a royalty – to be shared equally by INBio and the Costa Rican Ministry of Environment and Energy – on the profits of any future pharmaceutical product or agricultural compound that was isolated or developed from an INBio sample. Subsequently, INBio negotiated several further bioprospecting contracts involving other partners than Merck, including Eli Lilly, with the result that income from INBio's bioprospecting activities rose to about US$ 1 million per year [230].

The Costa Rica example demonstrates the possibility of conducting research to identify new medicinal products from natural sources in an LMIC, in a way that preserves property rights, generates a financial return and encourages capacity building. The contrasting experience of Eli Lilly's efforts to work in Cameroon highlights the importance of clear government policies and laws that establish the legal basis for conducting bioprospecting [231, 232].

Many of the lessons were taken forward in the International Cooperative Biodiversity Groups Program, initiated in 1992 to make multi-disciplinary, multi-institutional awards to foster work on the three interdependent issues of drug discovery, biodiversity conservation, and sustainable economic growth [233, 234]. On a more modest scale, as a result of Eisner's encouragement, in 1996 IOCD established a program to facilitate and catalyze ethical bioprospecting. This has included work in South Africa, Kenya and Uganda – in the latter case, most recently assisting policy makers in the development of draft legislation [235].

1.8.9
Water

Unsafe water, coupled with a lack of basic sanitation, kills at least 1.6 million children under the age of five every year. The target under MDG Goal 7 aims to halve, by 2015, the proportion of people without sustainable access to safe drinking water and basic sanitation. There has been some progress: between 1990 and 2006, over 1.5 billion people gained access to improved drinking water sources (usage of drinking water from improved sources rising to 87%, as compared to 77% in 1990) and over 1.1 billion gained access to improved sanitation (usage of improved sanitation facilities rising to 62%, as compared to 54% in 1990) (Figure 1.14). However, while the world is on track to meet the MDG drinking-water supply target by 2015 at the global level, many countries in sub-Saharan Africa and in Oceania are currently projected to miss MDG country targets, leaving significant portions of the population without access to improved drinking-water supplies. Moreover, the world is not on track to meet the MDG sanitation target by 2015. About 340 million Africans lack access to safe drinking water and almost 500 million lack access to adequate sanitation [236–239].

Meeting the challenge requires coordinated action by international agencies, governments and civil society partners, with a strong focus on sanitation [240] and major inputs from science and technology. Efforts among the various UN agencies with an interest in water and/or sanitation are coordinated by UN-Water [241].

The role of the chemical sciences is crucial, as highlighted in the report [242] of the Pan African Chemistry Network's "Sustainable Water Conference 2009". One of the key messages from this conference was that increasing Africa's capacity in analytical chemistry is imperative in order to support chemical monitoring and water management activities.

1.9
Conclusions

Chemistry has demonstrated its capacity to contribute substantially to increasing health, wellbeing and economic growth [243] and has great potential to assist in the development of LMICs. But what is needed for the future is not simply more of the same. The world has changed substantially in the last two centuries and

(a)

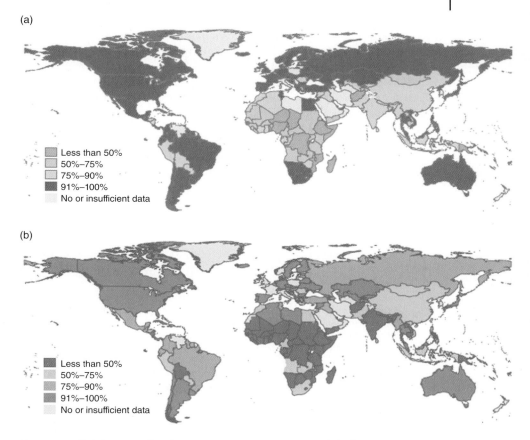

Less than 50%
50%–75%
75%–90%
91%–100%
No or insufficient data

(b)

Less than 50%
50%–75%
75%–90%
91%–100%
No or insufficient data

Figure 1.14 Global water and sanitation coverage [236]. (a) Improved drinking-water coverage, 2006. (b) Improved sanitation coverage, 2006. (Please find a color version of this figure in the color plates.)

many of the challenges it now faces are not only bigger but in some cases fundamentally different from those in the past. Globalization, urbanization, aging and population increases, threats of new diseases, pandemics, irreversible environmental damage and climate change, the revolutions in information technology, genomics and nanotechnology – all of these, individually and collectively, present new playing fields and new opportunities in an increasingly competitive and interconnected world.

Chemistry is a platform science. Its potential is, therefore, not only to help improve the human condition through its own direct contributions in fields such as the analysis and transformation of chemical entities, but also through the underpinning it provides to such diverse fields as medicine, genetics, biotechnology, materials and energy.

To take advantage of the new opportunities and to ensure that chemistry fulfils its potential, LMICs must invest in their own future by strengthening their science base and ensuring that the conditions exist for science, technology and innovation to flourish. This requires establishing sound policies, adequate funding mechanisms and environments where research is valued and its products utilized. It can be assisted by development partners, both public and private, who support the country's own policies and help build individual, institutional and country capacities to conduct, manage and exploit the fruits of science and technology [244].

Acknowledgments

The authors thank members of the Board of IOCD and in particular its President, Jean-Marie Lehn, and Executive Director, Alain Krief, for helpful discussions.

References

1 Gardner, C.A., Acharya, T., and Yach, D. (2007) Technological and social innovation: a unifying new paradigm for global health. *Health Affairs*, **26** (4), 1052–1061. http://content.healthaffairs.org/cgi/content/abstract/26/4/1052 (accessed 6 January 2011).

2 Buchmann, I. (2001) *Batteries in a Portable World*, 2nd edn, Cadex, Vancouver, http://www.buchmann.ca (accessed 6 January 2011).

3 Burns, R. (1951) Industrial and engineering chemistry – "electrochemical industry". *Ind. Eng. Chem.*, **43** (2), 301–304. http://pubs.acs.org/doi/abs/10.1021/ie50494a603 (accessed 6 January 2011).

4 Chemical Heritage Foundation (2011) Herbert Henry Dow. http://www.chemheritage.org/discover/chemistry-in-history/themes/electrochemistry/dow.aspx (accessed 6 January 2011).

5 Beck, T.R. (2008) Electrolytic production of aluminium. Electrochemistry Encyclopedia, http://electrochem.cwru.edu/encycl/art-a01-al-prod.htm (accessed 6 January 2011).

6 Yess, M. (2008) The electrochemical society: the first hundred years, 1902–2002. Electrochemistry Encyclopedia, http://electrochem.cwru.edu/encycl/art-e04-echem-soc.htm (accessed 6 January 2011).

7 Morris, P.J.T. and Travis, A.S. (1992) A history of the international dyestuff industry. *Am. Dyestuff Rep.*, **81** (11). Reproduced in: http://www.colorantshistory.org/HistoryInternationalDyeIndustry.html (accessed 6 January 2011).

8 Institute Charles Gerhardt (2010) *Charles Gerhardt 1816–1856*, Institute Charles Gerhardt, Montpellier, http://www.icgm.fr/spip.php?article306 (accessed 6 January 2011).

9 Wikipedia (2010) Louis Pasteur http://en.wikipedia.org/wiki/Louis_Pasteur (accessed 6 January 2011).

10 Gilman, A.G., Goodman, L.S., Rall, T.W., and Murad, F. (1985) *The Pharmacological Basis of Therapeutics*, 7th edn, Macmillan, New York.

11 Papac, R.J. (2001) Origins of cancer therapy. *Yale J. Biol. Med.*, **74**, 391–398. http://www.ncbi.nlm.nih.gov/pmc/articles/PMC2588755/pdf/yjbm00015-0028.pdf (accessed 6 January 2011).

12 Association of the British Pharmaceutical Industry (2010) Facts & Statistics from the Pharmaceutical Industry: Pharmaceuticals and the UK Economy. ABPI, http://www.abpi.org.uk (accessed 6 January 2011).

13 Griminger, P. (1972) Casimir Funk, a biographical sketch (1884–1967). *J. Nutr.*, **102**, 1105–1114.

http://jn.nutrition.org/cgi/reprint/102/9/1105.pdf (accessed 6 January 2011).

14 Rosenfeld, L. (1997) Vitamine–vitamin. The early years of discovery. *Clin. Chem.*, **43** (4), 680–685. http://www.clinchem.org/cgi/reprint/43/4/680 (accessed 6 January 2011).

15 Pauling, L., Itano, H., Singer, S.J., and Wells, I. (1949) *Science*, **110** (2865), 543–548. http://www.sciencemag.org/cgi/content/citation/110/2865/543 (accessed 6 January 2011).

16 Crick, F., Watson, J., and Wilkins, M. (2010) The Nobel Prize in Physiology or Medicine 1962. Nobelprize.org. http://nobelprize.org/nobel_prizes/medicine/laureates/1962/index.html (accessed 6 January 2011).

17 Sanger, F. (2010) The Nobel Prize in Chemistry 1958. Nobelprize.org. http://nobelprize.org/nobel_prizes/chemistry/laureates/1958/ (accessed 6 January 2011).

18 Perutz, M.F. and Kendrew, J.C. (2010) The Nobel Prize in Chemistry 1962. Nobelprize.org. http://nobelprize.org/nobel_prizes/chemistry/laureates/1962/kendrew-bio.html (accessed 6 January 2011).

19 Ferracane, J.L. (2001) *Materials in Dentistry: Principles and Applications*, 2nd edn, Lippincott Williams and Wilkins, New York.

20 Smil, V. (2001) *Enriching the Earth: Fritz Haber, Carl Bosch, and the Transformation of World Food Production*, MIT.

21 Hager, T. (2008) *The Alchemy of Air*, Harmony Books, New York.

22 Wikipedia (2010) Haber process http://en.wikipedia.org/wiki/Haber_process#cite_note-13.

23 Centres for Disease Control and Prevention (2011) *The History of Malaria, An Ancient Disease*, Centres for Disease Control, Atlanta, http://www.cdc.gov/malaria/history/index.htm (accessed 6 January 2011).

24 Müller, P. (2010) The Nobel Prize in Physiology or Medicine, 1948. Nobelprize.org. http://nobelprize.org/nobel_prizes/medicine/laureates/1948 (accessed 6 January 2011).

25 Hartley, D. and Kidd, H. (1983) *The Agrochemicals Handbook*, Royal Society of Chemistry, Nottingham, UK.

26 Senchenkova, E.M. (1976) Tsvet (or Tswett), Mikhail Semenovich (1872–1919), in *Dictionary of Scientific Biography*, vol. 13 (ed. C.C. Gillispie), American Council of Learned Societies, Charles Scribner & Sons, New York, pp. 486–488.

27 Wikipedia (2010) William Herschel http://en.wikipedia.org/wiki/William_Herschel (accessed 6 January 2011).

28 Ingle, J.D.J. and Crouch, S.R. (1988) *Spectrochemical Analysis*, Prentice Hall, Englewood Cliffs, NJ.

29 Röntgen, W.C. (2010) The Nobel Prize in Physics, 1901. Nobelprize.org. http://nobelprize.org/nobel_prizes/physics/laureates/1901/rontgen-bio.html (accessed 6 January 2011).

30 Thomson, J.J. (2010) The Nobel Prize in Physics, 1906. Nobelprize.org. http://nobelprize.org/nobel_prizes/physics/laureates/1906/thomson.html (accessed 6 January 2011).

31 Rabi, I.I. (2010) The Nobel Prize in Physics, 1944. Nobelprize.org. http://nobelprize.org/nobel_prizes/physics/laureates/1944 (accessed 6 January 2011).

32 The Charles Goodyear Story. (2010) Goodyear Company, http://www.goodyear.com/corporate/history/history_story.html (accessed 6 January 2011).

33 Kauffman, G.B. (2011) Rubber, *Chemistry Explained*. http://www.chemistryexplained.com/Ru-Sp/Rubber.html (accessed 6 January 2011).

34 Biography of Alexander Parkes (1813–1890). (2010) yourdictionary.com, http://www.yourdictionary.com/biography/alexander-parkes (accessed 6 January 2011).

35 Leo Hendrik Baekeland, 1863–1944. (2010) Chemical Achievers: The Human Face of the Chemical Sciences. The Chemical Heritage Foundation, http://www.chemheritage.org/classroom/chemach/plastics/baekeland.html (accessed 6 January 2011).

36 Wallace Carothers. (2010) National Historic Chemical Landmaks. American Chemical Society, http://acswebcontent.acs.org/landmarks/nylon/carothers.html (accessed 6 January 2011).

37 Hamilton, J. (2004) *A Life of Discovery: Michael Faraday, Giant of the Scientific Revolution*, Random House, New York.

38 Maddison, A. (ed.) (2001) HS–8: the world economy, 1–2001 AD, in *The World Economy: A Millennial Perspective*, OECD, Paris, pp. 241–263, http://www.ggdc.net/maddison/ (accessed 6 January 2011).

39 World Intellectual Property Organization (WIPO) (2010) Medium-Term Strategic Plan 2010–15: Consultation Paper. http://www.wipo.int/export/sites/www/about-wipo/en/pdf/mtsp.pdf (accessed 6 January 2011).

40 Dye, C. (2010) Is wealth good for your health? *Gresham Lecture Notes*, http://www.gresham.ac.uk/uploads/Dye%20 7%20Wealth-Health.ppt (accessed 6 January 2011).

41 Easterlin, R. (1999) How benevolent is the market? A look at the modem history of mortality. *Eur. Rev. Econ. Hist.*, **3**, 257–294.

42 Jamison, D., Sandbu, M.E., and Wang, J. (2004) Why has infant mortality decreased at such different rates in different countries? Disease Control Priorities Project, Working Paper No. 21.

43 Jamison, D., Breman, J.G., Measham, A.R., Alleyne, G., Claeson, M., Evans, D.B., Jha, P., Mills, A., and Musgrove, P. (eds) (2006) *Disease Control Priorities in Developing Countries (DCP2)*, 2nd edn, World Bank, Washington, DC, http://files.dcp2.org/pdf/DCP/DCPFM.pdf (accessed 6 January 2011).

44 United Nations (2000) Millennium Summit, New York, http://www.un.org/millennium/summit.htm (accessed 6 January 2011).

45 UNDP (2010) Millennium Development Goals. http://www.undp.org/mdg/basics.shtml (accessed 6 January 2011).

46 World Bank (2010) *What is poverty?* World Bank, Washington, DC, http://www.thequietworld.com/ahealthyworld/index.php?page=poverty (accessed 6 January 2011).

47 Juma, C. and Lee, Y.-C. (2005) *Task Force on Science, Technology and Innovation. Innovation: Applying Knowledge in Development. UN Millennium Project*, Earthscan, London, http://www.unmillenniumproject.org/documents/Science-complete.pdf (accessed 6 January 2011).

48 United Nations (2010) *Millennium Development Goals Report 2010*, United Nations, New York, http://bit.ly/bmwAhK (accessed 6 January 2011).

49 Nature Editorial (2010) How to feed a hungry world, *Nature*, **466**, 531–532. http://www.nature.com/nature/journal/v466/n7306/full/466531a.html (accessed 6 January 2011).

50 UNESCO Institute for Statistics (2005) *Children out of School: Measuring Exclusion from Primary Education*, UIS, Montreal, http://www.uis.unesco.org/template/pdf/educgeneral/OOSC_EN_WEB_FINAL.pdf (accessed 6 January 2011).

51 Universal Declaration of Human Rights (1948) adopted by UN General Assembly, http://www.un.org/en/documents/udhr/index.shtml (accessed 6 January 2011).

52 Convention on the Elimination of All Forms of Discrimination against Women (CEDAW) (1979), adopted by UN General Assembly, http://www.un.org/womenwatch/daw/cedaw/cedaw.htm (accessed 6 January 2011).

53 *Beijing Declaration and Platform for Action* (1995) Fourth World Conference on Women, Beijing, http://www.un.org/womenwatch/daw/beijing/platform (accessed 6 January 2011).

54 International Institute for Sustainable Development (1994) Programme of action. International Conference on Population and Development Cairo, 5–13 September, http://www.iisd.ca/cairo.html (accessed 6 January 2011).

55 UNESCO Institute for Statistics (2009) A Global Perspective on Research and Development. UIS Fact Sheet, No. 2. http://www.uis.unesco.org/template/pdf/S&T/Factsheet_No2_ST_2009_EN.pdf (accessed 6 January 2011).

56 Hogan, M.C. *et al.* (2010) *Building Momentum: Global Progress Toward Reducing Maternal and Child Mortality*, IHME, Seattle, http://www.healthmetricsandevaluation.org/resources/policyreports/2010/building_momentum_0610.html (accessed 6 January 2011).

57 Report of the United Nations Conference on Environment and Development, (1992) Rio de Janeiro, http://www.un.org/esa/sustdev (accessed 6 January 2011).

58 *Johannesburg Declaration*, (2002) World Summit on Sustainable Development, Johannesburg, http://www.un-documents.net/jburgdec.htm#fn1 (accessed 6 January 2011).

59 IPCC (2007) Fourth Assessment Report. Intergovernmental Panel on Climate Change, http://www.ipcc.ch (accessed 6 January 2011).

60 United Nations Human Settlements Programme (2009) *State of the World's Cities 2006/7*, Earthscan, London, http://www.unhabitat.org/pmss/listItemDetails.aspx?publicationID=2101 (accessed 6 January 2011).

61 ELDIS (2010) *Access to Medicines and International Issues*, Institute of Development Studies, Sussex, http://www.eldis.org/go/topics/resource-guides/health-systems/access-to-medicines-and-international-issues (accessed 6 January 2011).

62 Solvay Company: History. (2010) http://www.solvay.com (accessed 6 January 2011).

63 Essenscia (2010) Belgium: A World Champion for Chemicals and Plastics. Belgian Federation for Chemistry and Life Sciences Industries, http://www.essenscia.be/01/MyDocuments/WORLD_CHAMPION_BAN_030310.pdf (accessed 6 January 2011).

64 Institut Pasteur International Network. (2010) http://www.pasteur-international.org/ip/easysite/pasteur-international-en/institut-pasteur-international-network/the-network (accessed 6 January 2011).

65 Indian Institute of Science (2009) Celebrating 100 Years of Achievement, http://www.iisc.ernet.in/ (accessed 6 January 2011).

66 South African Chemical Institute (2010) http://www.saci.co.za/ (accessed 6 January 2011).

67 Oswaldo Cruz Foundation (Fiocruz), Brazil (2010) http://www.fiocruz.br/cgi/cgilua.exe/sys/start.htm?UserActiveTemplate=template%5Fingles&infoid=2297&sid=281 (accessed 6 January 2011).

68 Fiocruz: Farmanguinhos Medicines and Drugs Technology Institute (2010) http://www.fiocruz.br/cgi/cgilua.exe/sys/start.htm?infoid=820&sid=122&UserActiveTemplate=template_ingles (accessed 6 January 2011).

69 Rubber Research Institute of Sri Lanka (2008) http://www.rrisl.lk/sub_pags/aboutus_home.html (accessed 6 January 2011).

70 Rubber Research Institute of Nigeria (2010) http://www.icpsr.org.ma/?Page=showInstitute&InstituteID=RRIN123&CountryID=Nigeria (accessed 6 January 2011).

71 Rubber Research Institute of Malaya (2010) http://sejarahmalaysia.pnm.my/portalBI/detail.php?section=sm05&spesifik_id=47&ttl_id=15 (accessed 6 January 2011).

72 Rubber Research Institute of India (2005) http://www.irikerala.org/ (accessed 6 January 2011).

73 UN Agenda for Development, (2005) adopted by the General Assembly, Science and Technology: Paras 72–75. http://www.un.org/Docs/SG/ecodev.htm (accessed 6 January 2011).

74 UNESCO (1945) Constitution of the United Nations Educational, Scientific, and Cultural Organization (UNESCO), adopted 16 November 1945, http://www.icomos.org/unesco/unesco_constitution.html (accessed 6 January 2011).

75 Dixon, D. (2005) World Bank Puts Science Back on the Agenda. http://www.scidev.net/en/news/world-bank-puts-science-back-on-the-agenda.html (accessed 6 January 2011).

76 Swedish International Development Cooperation Agency (SIDA) (2010) Secretariat for Research Cooperation. http://www.sida.se/English/Partners/Universities-and-research/From-funding-research-to-fighting-poverty/About-FORSKSEK/ (accessed 6 January 2011).

77 United Nations Economic and Social Commission for Asia and the Pacific (2010) Statistical Yearbook for Asia and the Pacific 2009. Chapter 1: Research

and Development, 95–97. http://www.
unescap.org/stat/data/syb2009/
15-Research-and-development.asp
(accessed 6 January 2011).

78 Leadbeater, C. and Wilsdon, J. (2007) *The
Atlas of Ideas: How Asian Innovation Can
Benefit Us All*, Demos, London, http://
www.demos.co.uk/files/Overview_
Final.pdf (accessed 6 January 2011).

79 African Union/New Partnership for
Africa's Development (NEPAD) (2005)
Africa's Science and Technology
Consolidated Plan of Action. http://
www.africa-union.org/root/AU/
Conferences/2010/March/Amcost/docs/
Africa's%20Consolidated%20Plan%20
of%20Action.pdf (accessed 6 January
2011).

80 UNESCO Natural Sciences Sector
(2010) *Science for Sustainable
Development*, UNESCO, Paris, http://
www.unesco.org/science/psd/
publications/rep_usa_eur_05.shtml
(accessed 6 January 2011).

81 United Nations Educational, Scientific
and Cultural Organization (2010)
International Science Prizes, UNESCO,
Paris, http://www.unesco.org/science/
intern_prizes.shtml (accessed 6 January
2011).

82 UNDP (2007) Globalization and the
least developed countries: issues in
technology. United Nations Ministerial
Conference of the Least Developed
Countries: Making Globalization Work
for the LDCs. Istanbul, 9–11 July,
http://www.un.int/turkey/1.pdf
(accessed 6 January 2011).

83 Academy of Sciences for the Developing
World (TWAS) (2010) http://
twas.ictp.it/ (accessed 6 January 2011).

84 Organization of Women in Science for
the Developing World (2011) formerly
Third World Organization for Women
in Science (TWOWS), http://twows.ictp.
it/ (accessed 6 January 2011).

85 International Association of Science
and Technology for Development
(IASTED) (2011) http://www.iasted.org
(accessed 6 January 2011).

86 *United Nations Millennium Declaration.*
(2000) UN General Assembly, New
York, http://www.un.org/millennium/
declaration/ares552e.pdf (accessed 6
January 2011).

87 Commission on Macroeconomics and
Health (2001) *Macroeconomics and
Health: Investing in Health for Economic
Development*, WHO, Geneva, http://
whqlibdoc.who.int/
publications/2001/924154550x.pdf
(accessed 6 January 2011).

88 Juma, C. (ed.) (2005) *Going for Growth:
Science, Technology and Innovation in
Africa*, The Smith Institute, London.

89 Juma, C. (2005) We need to reinvent
the African university. SciDevNet, 14
June, http://www.scidev.net/en/
opinions/we-need-to-reinvent-the-
african-university.html (accessed 6
January 2011).

90 Coober, D.I., Langer, S.S., and Pratt,
J.M. (eds) (1992) *Chemistry and
Developing Countries*, Commonwealth
Science Council/Royal Society of
Chemistry, London.

91 Chemical Research Applied to World
Needs (CHEMRAWN) (2010) http://
www.iupac.org/web/ins/021 (accessed 6
January 2011).

92 Malin, J.M. (2006) History and
Effectiveness of CHEMRAWN
Conferences, 1978–2006. IUPAC,
http://old.iupac.org/standing/
chemrawn/CR_History_061027.pdf
(accessed 6 January 2011).

93 Crabbé, P. and Cardyn, L. (1983) *The
Time for Another World*, University
Printing Services, Columbia, MO.

94 Crabbé, P. (1983) A new challenge for
the university. *Interciencia*, **8**, 279.

95 Crabbé, P., Archer, S., Benagiano, G.,
Diczfalusy, E., Djerassi, C., Fried, J.,
and Higuchi, T. (1983) Long-acting
contraceptive agents: design of the
WHO Chemical Synthesis Programme.
Steroids, **41**, 243–253. http://www.ncbi.
nlm.nih.gov/pubmed/6658872 (accessed
6 January 2011).

96 Seaborg, G.T. (1984) An international
effort in chemical science. *Science*, **223**,
9. http://www.sciencemag.org/cgi/pdf_
extract/223/4631/9 (accessed 6 January
2011).

97 Matlin, S.A., Chan, L., Hadjigeogiou, P.,
Prazeres, M.A., Mehani, S., and Roshdi,
S. (1983) Long-acting contraceptive
agents: analysis and purification of
steroid esters. *Steroids*, **41**, 361–367.
http://www.ncbi.nlm.nih.gov/

pubmed/6419409 (accessed 6 January 2011).

98 Matlin, S.A., Chan, L., Prazeres, M.A., Mehani, S., and Cass, Q.B. (1987) Long-acting androgens: analytical and preparative HPLC of testosterone esters. *J. High Resolut. Chromatogr. Chromatogr. Commun.*, **10**, 186–190.

99 Waites, G.M.H. (2003) Development of methods of male contraception: impact of the World Health Organization Task Force. *Fertility and Sterility*, **80**, 1–15. http://www.fertstert.org/article/S0015-0282(03)00577-6/abstract (accessed 6 January 2011).

100 Crabbé, P., Diczfalusy, E., and Djerassi, C. (1980) Injectable contraceptive synthesis: an example of international cooperation. *Science*, **209**, 992–994. http://www.ncbi.nlm.nih.gov/pubmed/7403868 (accessed 6 January 2011).

101 History of the International Organization for Chemical Sciences in Development (2007) http://www.iocd.org/ourhistory.shtml (accessed 6 January 2011).

102 Inter-Academy Council Study Panel (2003) Inventing a better future. IAC Report, http://www.interacademycouncil.net/CMS/Reports/9866/9430.aspx?returnID=9866 (accessed 6 January 2011).

103 Matlin, S.A., Landriault, E., and Monot, J.-J. (2009) The 2009 Report Card on financing research and development for health, in *Monitoring Financial Flows for Health Research 2009: Behind the Global Numbers* (eds E. Landriault and S.A. Matlin), Global Forum for Health Research, Geneva, pp. 153–193, http://www.globalforumhealth.org/Media-Publications/Publications/Monitoring-Financial-Flows-for-Health-Research-2009-Behind-the-Global-Numbers (accessed 6 January 2011).

104 Eurostat (2010) R & D Expenditure. Eurostat Statistics: Main Tables. European Commission-Eurostat, http://epp.eurostat.ec.europa.eu/portal/page/portal/science_technology_innovation/data/main_tables (accessed 6 January 2011).

105 Nordling, L. (2010) Big spending on science promised for East Africa. SciDevNet, 11 June, http://www.scidev.net/en/news/big-spending-on-science-promised-for-east-africa.html (accessed 6 January 2011).

106 National Advisory Council on Innovation Indicators Reference Group (2008) South African Science and Technology Indicators 2008, South Africa, http://www.nacinnovation.biz/south-african-science-and-technology-indicators-2008/ (accessed 6 January 2011).

107 New Partnership for Africa's Development (NEPAD) (2010) African Science, Technology & Innovation Indicators Initiative (ASTII), http://www.nepadst.org/astii/index.shtml (accessed 6 January 2011).

108 Obama, B. (2009) Speech to US National Academy of Sciences, 27 April, http://www.whitehouse.gov/the_press_office/Remarks-by-the-President-at-the-National-Academy-of-Sciences-Annual-Meeting/ (accessed 6 January 2011).

109 Council for Scientific and Industrial Research (2005) *The Tshwane Consensus on Science and Development* CSIR, Pretoria. Media release, 19 October, http://ntww1.csir.co.za/plsql/ptl0002/PTL0002_PGE013_MEDIA_REL?MEDIA_RELEASE_NO=7323626 (accessed 6 January 2011).

110 Esteves, B. (2006) Brazil to boost health research capacity in Angola. SciDevNet, 7 August, http://www.scidev.net/News/index.cfm?fuseaction=readNews&itemid=3028&language=1 (accessed 6 January 2011).

111 Worldmapper maps (2006) http://www.worldmapper.org (accessed 6 January 2011).

112 UNESCO Institute for Statistics (2005) What do Bibliometric Indicators Tell Us about World Scientific Output? UIS Bulletin on Science and Technology Statistics. UIS, Issue No. 2. http://www.uis.unesco.org/template/pdf/S&T/BulletinNo2EN.pdf (accessed 6 January 2011).

113 Pendlebury, D.A. (2008) *White Paper: Using Bibliometrics in Evaluating Research*, Thomson Reuters, Philadelphia, http://wokinfo.com/media/pdf/UsingBibliometricsinEval_WP.pdf (accessed 6 January 2011).

114 Bibliometrics Special Interest Group, Universiti Technologi MARA, Shah

Alam (2004) *Science and Technology Knowledge Productivity in Malysia 2003*, Malaysian Science and Technology Information Centre, Putrajaya, http://myais.fsktm.um.edu.my/590/ (accessed 6 January 2011).

115 Dumont, J.C. and Lemaître, G. (2005) Counting Immigrants and Expatriates: A New Perspective. OECD, Social Employment and Migration Working papers. OECD, http://www.oecd.org/dataoecd/27/5/33868740.pdf (accessed 6 January 2011).

116 Hunt, J. and Gauthier-Loiselle, M. (2008) How Much Does Immigration Boost Innovation? Working Paper No. 14312. National Bureau of Economic Research, Cambridge, http://www.nber.org/papers/w14312.pdf (accessed 6 January 2011).

117 UNDP (2009) *Human Development Report 2009. Overcoming Barriers: Human Mobility and Development*, Palgrave Macmillan, New York, http://hdr.undp.org/en/media/HDR_2009_EN_Complete.pdf (accessed 6 January 2011).

118 Chu, R. and Pugatch, M. (2010) From test tube to patient – national innovation strategies for the biomedical field. Stockholm Network, http://www.stockholm-network.org/downloads/publications/From_Test_Tube_to_Patient_Final.pdf (accessed 6 January 2011).

119 Silberglitt, R., Antón, P.S., Howell, D.R., and Wong, A. (2006) The Global Technology Revolution 2020: In-Depth Analyses. RAND Corporation, http://www.rand.org/pubs/technical_reports/2006/RAND_TR303.pdf (accessed 6 January 2011).

120 Bound, K. (2008) *Brazil: The Natural Knowledge Economy*, Demos, London, http://www.demos.co.uk/publications/brazil (accessed 6 January 2011).

121 Páscoa, M.B.A. (2005) *In Search of An Innovative Environment – the New Brazilian Innovation Law*, World Intellectual Property Organization, Geneva, http://www.wipo.int/sme/en/documents/brazil_innovation.htm (accessed 6 January 2011).

122 Veneu, F. (2004) Brazil adopts innovation law. SciDevNet, 20 December, http://www.scidev.net/en/news/brazil-adopts-innovation-law.html (accessed 6 January 2011).

123 Massarani, L. (2006) Editorial. Brazil's innovation law: lessons for Latin America. SciDevNet, 3 August, http://www.scidev.net/en/editorials/brazils-innovation-law-lessons-for-latin-america.html (accessed 6 January 2011).

124 Petherick, A. (2010) High hopes for Brazilian science. *Nature*, **465**, 674–675. http://www.nature.com/news/2010/100609/full/465674a.html (accessed 6 January 2011).

125 Ryan, M.P. (2010) Patent Incentives, Technology Markets, and Public-Private Bio-Medical Innovation Networks in Brazil. Creative and Innovative Economy Center Research Paper, http://www.law.gwu.edu/Academics/research_centers/ciec/Documents/research%20papers/GW%20CIEC%20Brazil%20patents%20bio-medical%20technology%20networks.pdf (accessed 6 January 2011).

126 Peng, K. (2009) China issues 50-year science strategy. SciDevNet, 6 July, http://www.scidev.net/en/news/china-issues-50-year-science-strategy-.html (accessed 6 January 2011).

127 World Trade Organization (2010) Understanding the WTO: Membership, Alliances and Bureaucracy. WTO, http://www.wto.org/english/thewto_e/whatis_e/tif_e/org3_e.htm (accessed 6 January 2011).

128 Shetty, P. (2010) BioMed analysis: India's patent catch-22. SciDevNet, 20 May, http://www.scidev.net/en/opinions/biomed-analysis-india-s-patent-catch-22.html (accessed 6 January 2011).

129 Nuyens, Y. (2005) *No Development without Research: A Challenge for Research Capacity Strengthening*, Global Forum for Health Research, Geneva, http://www.globalforumhealth.org/Media-Publications/Publications/No-Development-Without-Research-A-challenge-for-research-capacity-strengthening (accessed 6 January 2011).

130 Abegaz, B.M. (1998) The Status of basic sciences in Africa, in *Basic Sciences and Development: Rethinking Donor Policy* (eds M.J. Garett and C.G. Granqvist), Ashgate, Aldershot, pp. 73–102.

131 Swedish International Development Cooperation Agency (2009) International Conference on the Strengthening of Research and Higher Education in Basic Sciences by Regional and Interregional Cooperation; Relevance for Developing Countries. Addis Ababa, 1–4 September 2009, http://www2.math.uu.se/~leifab/addis/programme_pdf.pdf (accessed 6 January 2011).

132 Wu, T.K. (1992) Planning a chemical industry – Taiwan ROC: a case study, in *Chemistry and Developing Countries* (eds D.I. Coober, S.S. Langer, and J.M. Pratt), Commonwealth Science Council/Royal Society of Chemistry, London, pp. 141–146.

133 Development Cooperation Directorate (OECD-DAC) (2010) *The Paris Declaration and Accra Agenda for Action*, OECD, Paris, http://www.oecd.org/document/18/0,3343,en_2649_3236398_35401554_1_1_1_1,00.html (accessed 6 January 2011), http://www.oecd.org/dataoecd/11/41/34428351.pdf. (accessed 6 January 2011).

134 Australian Centre for International Agricultural Research (2010) *ACIAR and the Paris Declaration on Aid Effectiveness*, Australian Centre for International Agricultural Research, Canberra, http://aciar.gov.au/node/2405 (accessed 6 January 2011).

135 American Chemical Society (2011) http://www.acs.org (accessed 6 January 2011).

136 Royal Society of Chemistry (2011) http://www.rsc.org (accessed 6 January 2011).

137 Royal Australian Chemical Institute Inc. (2011) http://www.raci.org.au/ (accessed 6 January 2011).

138 Gesellschaft Österreichischer Chemiker (2011) http://www.goech.at/info_english.shtml (accessed 6 January 2011).

139 Brazilian Chemical Society (2011) http://www.sbq.org.br/ingles/history.php (accessed 6 January 2011).

140 Egyptian Chemical Society (ECS) (2009) http://www.egy-chem-soc.org/ (accessed 6 January 2011).

141 Société Chimique de France (2011) http://www.societechimiquedefrance.fr/ (accessed 6 January 2011).

142 Gesellschaft Deutscher Chemiker (2011) http://www.gdch.de/gdch__e.htm (accessed 6 January 2011).

143 The Chemical Society of Japan (2011) http://www.chemistry.or.jp/ (accessed 6 January 2011).

144 Koninklijke Nederlandse Chemische Vereniging (2011) http://www.kncv.nl/ (accessed 6 January 2011).

145 Norwegian Chemical Society (NKS) (2011) http://www.kjemi.no/english/ (accessed 6 January 2011).

146 Sociedade Portuguesa de Quimica (2009) http://www.spq.pt/ (accessed 6 January 2011).

147 European Association for Chemical and Molecular Sciences (2011) Mendeleev Russian Chemical Society, http://www.euchems.org/MemberSocieties/Mendeleev/ (accessed 6 January 2011).

148 Real Sociedad Española de Química (2011) http://www.rseq.org/ (accessed 6 January 2011).

149 Svenska Kemistsamfundet (2011) http://www.chemsoc.se/ (accessed 6 January 2011).

150 Swiss Chemical Society (SCS), Schweizerische Chemische Gesellschaft (SCG), Société Suisse de Chimie (SSC) (2011) http://www.swiss-chem-soc.ch/org (accessed 6 January 2011).

151 International Union of Pure and Applied Chemistry (IUPAC) (2011) http://www.iupac.org/ (accessed 6 January 2011).

152 European Association for Chemical and Molecular Sciences (EuCheMS) (2010) http://www.euchems.org/ (accessed 6 January 2011).

153 Federation of African Societies of Chemistry (2011) http://www.faschem.org (accessed 6 January 2011).

154 Federation of Asian Chemical Societies (FACS) (2011) http://www.facs-as.org/index.php?page=facs (accessed 6 January 2011).

155 Pan African Chemistry Network (2011) http://www.rsc.org/Membership/Networking/InternationalActivities/

PanAfrica/index.asp (accessed 6 January 2011).

156 Barabási, A.-L. (2005) Network theory – the emergence of the creative enterprise. *Science*, **308**, 639–641. http://www.sciencemag.org/cgi/content/short/308/5722/639 (accessed 6 January 2011).

157 Esparza, J. and Yamada, T. (2007) The discovery value of "big science". *J. Exp. Med.*, **204** (4), 701–704. http://ukpmc.ac.uk/backend/ptpmcrender.cgi?accid=PMC2118535&blobtype=pdf (accessed 6 January 2011).

158 Nordling, L. (2010) Africa analysis: professional societies need networking. SciDevNet, 29 January, http://www.scidev.net/en/opinions/africa-analysis-professional-societies-need-networking-1.html (accessed 6 January 2011).

159 International Science Programme (2010) http://www.isp.uu.se/ (accessed 6 January 2011).

160 International Foundation for Science (2011) http://www.ifs.se/Partners/networks.asp (accessed 6 January 2011).

161 AFASSA (Co-ordination of Networks for research on Biological Resources in Africa, Asia and South America) (2009) http://www.afassa.org (accessed 6 January 2011).

162 ANCAP (African Network for the Chemical Analysis of Pesticides) (2010) http://www.ancap.org/ (accessed 6 January 2011).

163 NABSA (Network for Analytical and Bio-assay Services in Africa) (2010) http://www.nabsaonline.org/ (accessed 6 January 2011).

164 NAPRECA (Natural Products Research Network for Eastern and Central Africa) (2011) http://www.napreca.net/ (accessed 6 January 2011).

165 NITUB (Network of Instrument Technical Personnel and User Scientists of Bangladesh) (2005) http://www.nitub.org/ (accessed 6 January 2011).

166 NUSESA (Network of Users of Scientific Equipment in Southern and Eastern Africa) (2008) http://www.ansti.org/new/index.php?option=com_content&task=view&id=313&Itemid=40 (accessed 6 January 2011).

167 WANNPRES (Western Africa Network of Natural Products Research Scientists) (2010) http://www.ifs.se/Partners/networks.asp (accessed 6 January 2011).

168 Abegaz, B.M. (2001) Promoting research and research-training in Eastern and Southern Africa through science-focused networks, societies and academies: an overview of progress and potentials. Paper presented at the Ford Foundation Conference on "Innovations in African Higher Education" October 1–3, 2001, Nairobi, Kenya.

169 Bringmann, G., Mutanyatta-Comar, J., Knauer, M., and Abegaz, B.M. Knipholone and related 4-phenylanthraquinones: structurally, pharmacologically, and biosynthetically remarkable natural products. *Nat. Prod. Rep.*, 2008, **25**, 696–718. http://pubs.rsc.org/en/Content/ArticleLanding/2008/NP/b803784c (accessed 6 January 2011).

170 Abegaz, B.M., Tekle-Haimanot, R., Palmer, V.S., and Spencer, P.S. (eds) (1994) *Nutrition, Neurotoxins and Lathyrism: The ODAP Challenge*, Third World Medical Research Foundation, New York.

171 Abegaz, B.M. Personal communication. Data from paper presented at Conference on Regional and Interregional Cooperation to Strengthen Basic Sciences in Developing Countries. Addis Ababa, 1–4 September 2009. http://www.math.uu.se/~leifab/addis/programme_pdf.pdf (accessed 6 January 2011); http://www.math.uu.se/~leifab/addis/abstracts/Abegaz%20IPICS%20NABSA.pdf (accessed 6 January 2011).

172 Masesane, I.B., Yeboah, S.O., Liebscher, J., Muegge, C., and Abegaz, B.M. (1999) A bichalcone from the twigs of *Rhus pyroides*. *Phytochemistry*, **53**, 1005–1008.

173 Mdee, L.K., Yeboah, S.O., and Abegaz, B.M. (2003) Rhuschalcones II-VI, five new bichalcones from the root bark of *Rhus pyroides*. *J. Nat. Prod.*, **66** (5), 599–604.

174 Southern and Eastern African Network for Analytical Chemists (2007) http://www.seanac.org (accessed 6 January 2011).

175 African Association of Pure and Applied Chemistry (2001) http://old.iupac.org/links/ao/aapac.html (accessed 6 January 2011).

176 Becker, E., Ober, C., and Henry, B. (2009) Chemistry research funding. *Chem. Int.*, **31**, 23–24. http://www. iupac.org/publications/ci/2009/3104/ july09.pdf (accessed 6 January 2011).

177 Wikipedia (2010) Research funding http://en.wikipedia.org/wiki/Research_ funding (accessed 6 January 2011).

178 History of the Royal Society (2011) http://royalsociety.org/History-of-the-Royal-Society/ (accessed 6 January 2011).

179 Royal Society of Chemistry (2006) *Funding Science and Technology: Who Pays? Who Benefits? Report of a Seminar Organised by the Royal Society of Chemistry at the House of Commons, London*, Royal Society of Chemistry, London. http://www.rsc.org/images/ funding%20science%20report_tcm18-91112.pdf (accessed 6 January 2011).

180 Die Kaiser Wilhelm Gesellschaf (2010) http://www.dhm.de/lemo/html/ kaiserreich/wissenschaft/kwg/ index.html (accessed 6 January 2011).

181 Deutsche Forschungsgemeinschaft (2010) http://www.dfg.de (accessed 6 January 2011).

182 Fonds National de la Recherche Scientifique (2010) Notre histoire et nos statuts. FNRS. http://www2.frs-fnrs.be/ index.php?option=com_content&view=a rticle&id=76&Itemid=4&lang=fr (accessed 6 January 2011).

183 Centre National de la Recherche Scientifique (2011) http://www.cnrs.fr (accessed 6 January 2011).

184 Aux origines du CNRS (1989) *Colloque sur l'Histoire du CNRS des 23 et 24 octobre 1989*, CNRS, Paris, http:// www.histcnrs.fr/origines_cnrs_1.html (accessed 6 January 2011).

185 Standler, R.B. (2009) Funding of Research in Basic Science in the USA. 1–13. http://www.rbs0.com/funding.pdf (accessed 6 January 2011).

186 Lopez-Claros, A. and Zahidi, S. (2005) Women's Empowerment: Measuring the Global Gender Gap. World Economic Forum, http://www. weforum.org/pdf/Global_ Competitiveness_Reports/Reports/ gender_gap.pdf (accessed 6 January 2011).

187 Jenkins, E.W. (1997) Gender and science & technology education. *Educ.*

Newsl., **23** (1). UNESCO, Paris, http:// www.unesco.org/education/educprog/ ste/newsletter/eng_n1/gender.html (accessed 6 January 2011).

188 Inter-Academy Council Study Panel (2003) Inventing a Better Future. Ch. 3: Expanding human resources. IAC Report, 43–60. http://www. interacademycouncil.net/Object.File/ Master/6/742/Chapter%203.pdf (accessed 6 January 2011).

189 Directory of Open Access Journals. (2010) Lund University Libraries, http://www.doaj.org/ doaj?func=loadTempl&templ=about (accessed 6 January 2011).

190 Royal Society of Chemistry (2006) *Chemistry in the Developing World*, RSC, London, http://www.rsc.org/ ScienceAndTechnology/Policy/ Bulletins/Issue3/ChemDeveloping World.asp (accessed 6 January 2011).

191 Trager, R. (2007) Chemistry's Open Access Dilemma. Chemistry World, http://www.rsc.org/chemistryworld/ Issues/2007/December/ ChemistrysOpenAccessDilemma.asp (accessed 6 January 2011).

192 Bueno, R. (2009) The inova success story – technology transfer in Brazil. WIPO Magazine, November. http:// www.wipo.int/wipo_magazine/en/2009/ 06/article_0009.html (accessed 6 January 2011).

193 Teng, H. (2010) University-industry technology transfer: framework and constraints. *J. Sust. Dev.*, **3**, 296–299. http://www.ccsenet.org/journal/ index.php/jsd/article/viewFile/6345/ 5102 (accessed 6 January 2011).

194 Ganguli, P. (2005) Industry-Academic Interaction in Technology Transfer and IPR. The Indian Scene – An Overview. WIPO, http://www.wipo.int/export/ sites/www/uipc/en/partnership/pdf/ ui_partnership_in.pdf (accessed 6 January 2011).

195 al-Hassan, A.Y. (2005) *Transfer of Islamic Technology to the West. Part III: Technology Transfer in the Chemical Industries – Transmission of Practical Chemistry*, IRCICA, Istanbul, http:// www.history-science-technology.com/ Articles/articles%2072.htm#_edn1 (accessed 6 January 2011).

196 Okoso-Amaa, K. and Mapima, C. (1995) Technology transfer and acquisition of managerial capability in Tanzania, in *Technology Policy and Practice in Africa* (eds O.M. Ogbu, B.O. Oyeyinka, and H.M. Mlawa), IDRC, Ottawa, Chapter 7. http://www.idrc.ca/en/ev-9301-201-1-DO_TOPIC.html (accessed 6 January 2011).

197 Government of Tanzania (2010) Science and technology. Tanzania National Website, http://www.tanzania.go.tz/science_technology.html (accessed 6 January 2011).

198 International Federation of Pharmaceutical Manufacturers & Associations (2010) *Technology Transfer and Anti-Retroviral Licensing in Developing Countries*, IFPMA, Geneva, https://www.ifpma.org/index.php?id=3878 (accessed 6 January 2011).

199 Glaxo SmithKline (2007) Technology transfer, capacity building and the developing world Glaxo SmithKline, http://www.gsk.com/policies/GSK-on-technology-transfer-capacity-building.pdf (accessed 6 January 2011).

200 United Nations Industrial Development Organization (UNIDO) (2011) http://www.unido.org (accessed 6 January 2011).

201 Diamant, R., Davison, H., and Pugatch, M.P. (2007) Promoting technology transfer in developing countries: lessons from public-private partnerships in the field of pharmaceuticals. Stockholm Network, http://www.stockholm-network.org/downloads/publications/Promoting_Technology_Transfer_1.pdf (accessed 6 January 2011).

202 Wikipedia (2010) Glocalisation – definition http://en.wikipedia.org/wiki/Glocalisation (accessed 6 January 2011).

203 See, for example, *AMF Americas Network* website: http://www.amfamericasnetwork.com/index.html (accessed 6 January 2011).

204 Gaud, W.S. (1968) The Green Revolution: Accomplishments and Apprehensions. Speech to the Society for International Development, 8 March, http://www.agbioworld.org/biotech-info/topics/borlaug/borlaug-green.html (accessed 6 January 2011).

205 Norman Borlaug: The Nobel Peace Prize 1970. (2010) Nobelprize.org, http://nobelprize.org/nobel_prizes/peace/laureates/1970/ (accessed 6 January 2011).

206 Borlaug, N. (2002) Biotechnology and the green revolution. An ActionBioscience.org original interview, http://www.actionbioscience.org/biotech/borlaug.html (accessed 6 January 2011).

207 Strengthening Food Security: Alliance for a Green Revolution in Africa (AGRA) (2010) Rockefeller Foundation, http://www.rockefellerfoundation.org/what-we-do/current-work/strengthening-food-security-alliance/ (accessed 6 January 2011).

208 Pachauri, R.K. and Reisinger, A. (eds) (2007) *Climate Change 2007: Synthesis Report. Contribution of Working Groups I, II and III to the Fourth Assessment Report of the Intergovernmental Panel on Climate Change*, IPCC, Geneva, http://www.ipcc.ch/publications_and_data/ar4/syr/en/contents.html (accessed 6 January 2011).

209 Shahab, S., Ghaffar, A., Stearns, B.P., and Woodward, A. (2008) *Strengthening the Base: Preparing Health Research for Climate Change*, Global Forum for Health Research, Geneva, http://www.globalforumhealth.org/Media-Publications/Publications/Strengthening-the-base-preparing-health-research-for-climate-change (accessed 6 January 2011).

210 Campbell-Lendrum, D. and Woodruff, R. (2006) Comparative risk assessment of the burden of disease from climate change. *Environ. Health Perspect.*, **114** (12), 1935–1941. http://www.bvsde.paho.org/bvsacd/cd67/comparative.pdf (accessed 6 January 2011).

211 Third World Network (2008) Developing countries call for new technology transfer mechanism. TWN Bonn News Update 4, 6 June. http://www.twnside.org.sg/title2/climate/news/TWNbonnupdate4.doc (accessed 6 January 2011).

212 Data from World Bank Development Indicators (2010). http://data.worldbank.org/indicator (accessed 6 January 2011).

213 *Report of the United Nations Conference on the Human Environment, Stockholm, 5–16 June* (1972). United Nations Publication, Sales No. E.73.II.A.14 and corrigendum.

214 Solomon, T., Åkerblom, M., and Thulstrup, E.W. (2003) Chemistry in the developing world. *Anal. Chem.*, **75** (5), 107A–113A. http://pubs.acs.org/doi/pdfplus/10.1021/ac0312458 (accessed 6 January 2011).

215 Matlin, S.A. (2006) The changing scene in *Monitoring Financial Flows for Health Research 2006: The Changing Landscape of Health Research for Development* (eds A. de Francisco and S.A. Matlin), Global Forum for Health Research, Geneva, pp. 3–32, http://www.globalforumhealth.org/Media-Publications/Publications/Monitoring-Financial-Flows-for-Health-Research-2006-The-changing-landscape-of-health-research-for-development (accessed 6 January 2011).

216 International Pharmaceutical Federation (2003) Statement of Policy on Counterfeit Medicines. IFPMA, http://www.fip.org/www/uploads/database_file.php?id=164&table_id= (accessed 6 January 2011).

217 Wikipedia (2010) Pandemic http://en.wikipedia.org/wiki/Pandemic (accessed 6 January 2011).

218 United Nations Population Funds (2009) *State of World Population 2009*, UNFPA, New York, http://www.unfpa.org/public/publications/search_pubs/swpreports (accessed 6 January 2011).

219 United Nations Human Settlements Program (2010) *State of the World's Cities 2010/2011 – Cities for All: Bridging the Urban Divide*, UN-Habitat, Nairobi, http://www.unhabitat.org/pmss/listItemDetails.aspx?publicationID=2917 (accessed 6 January 2011).

220 World Intellectual Property Organization (2011) http://www.wipo.int (accessed 6 January 2011).

221 Krattiger, A. *et al.* (ed.) (2009) *Intellectual Property Management in Health and Agricultural Innovation: Handbook of Best Practices. Executive Guide*, 2nd edn, MIHR, PIPRA, Oswaldo Cruz Foundation, and bioDevelopments-International Institute, http://www.iphandbook.org/handbook/about/attachment_files/ipHandbook%20Executive%20Guide%20Final.pdf (accessed 6 January 2011).

222 World Trade Organization (WTO) Overview: The TRIPS Agreement, http://www.wto.org/english/tratop_e/trips_e/intel2_e.htm (accessed 6 January 2011).

223 Musungu, S.F. and Dutfield, G. (2003) *Multilateral Agreements and A TRIPS-Plus World: The World Intellectual Property Organisation (WIPO)*, Quaker United Nations Office (QUNO), Geneva, http://www.quno.org/geneva/pdf/economic/Issues/Multilateral-Agreements-in-TRIPS-plus-English.pdf (accessed 6 January 2011).

224 Berger, J. and Prabhala, A. (2005) *Assessing the Impact of TRIPS-Plus Patent Rules in the Proposed US-SACU Free Trade Agreement*, WHO, Geneva, http://www.who.int/hiv/amds/capacity/tza2_oxfamreport_pricing_financing.pdf (accessed 6 January 2011).

225 Collins-Chase, C.T. (2008) The case against TRIPS-Plus protection in developing countries facing aids epidemics. *University of Pennsylvania Law School J*, **29**, 763–802. http://www.law.upenn.edu/journals/jil/articles/volume29/issue3/CollinsChase29U.Pa.J.Int'lL.763(2008).pdf (accessed 6 January 2011).

226 Commission on Intellectual Property Rights, Innovation and Public Health (2006) *Report of the Commission on Intellectual Property Rights, Innovation and Public Health*, World Health Organization, Geneva, http://www.who.int/intellectualproperty/en/ (accessed 6 January 2011).

227 World Health Organization (2008) *WHO Global Strategy and Plan of Action on Public Health, Innovation and Intellectual Property*, Health Organization, Geneva, http://www.who.int/phi/en/ (accessed 6 January 2011).

228 Eisner, T. (1994) Chemical prospecting: a global imperative. *Proc. Am. Philos. Soc.*, **138**, 385–395. http://www.jstor.org/pss/986744 (accessed 6 January 2011).

229 National Biodiversity Institute (INBio) of Costa Rica (2010) http://www.inbio.ac.cr/en/inbio/inb_queinbio.htm (accessed 6 January 2011).

230 Eberlee, J. (2000) Assessing the benefits of bioprospecting in Latin America. IDRC Reports Online, 21 January, http://www.idrc.ca/en/ev-5571-201-1-DO_TOPIC.html (accessed 6 January 2011).

231 Jenner, A. (2010) Experiences with bio-prospecting and access & benefit sharing. Presentation at the Geneva Pharma Forum, 6 May, http://www.ifpma.org/fileadmin/webnews/2010/pdfs/20100506_IFPMA_GPF_ABS_05052010_AJEN.pdf (accessed 6 January 2011).

232 Weiss, C. and Eisner, T. (1998) Partnerships for value-added through bioprospecting. *Technology in Society*, **20**, 481–498. http://www.sciencedirect.com/science?_ob=ArticleURL&_udi=B6V80-3VGSGFS-R&_user=10&_coverDate=11%2F30%2F1998&_rdoc=1&_fmt=high&_orig=search&_sort=d&_docanchor=&view=c&_acct=C000050221&_version=1&_urlVersion=0&_userid=10&md5=4036d4f2568177cd6839609f1435ccb1 (accessed 6 January 2011).

233 International Cooperative Biodiversity Groups Program (2011) http://www.icbg.org/ (accessed 6 January 2011).

234 Rosenthal, J. (1997) Integrating drug discovery, biodiversity conservation, and economic development: early lessons from the International Cooperative Biodiversity Groups, in *Biodiversity and Human Health* (eds F. Grifo and J. Rosenthal), Island Press, Washington DC, Ch. 13, http://www.icbg.org/pub/documents/lessons.pdf (accessed 6 January 2011).

235 IOCD: Working Group on Biotic Exploration (2009) http://www.iocd.org/bioticexpfund.shtml (accessed 6 January 2011).

236 World Health Organization (2008) *UN-Water Global Annual Assessment of Sanitation and Drinking-Water (GLAAS)*, WHO, Geneva, http://www.who.int/water_sanitation_health/glaas_2008_pilot_finalreport.pdf (accessed 6 January 2011).

237 United Nations (2008) *The Millennium Development Goals Report*, UN, New York, http://mdgs.un.org/unsd/mdg/Resources/Static/Products/Progress2008/MDG_Report_2008_En.pdf#page=42 (accessed 6 January 2011).

238 World Water Assessment Programme (2009) *Water in A Changing World. Third UN World Water Development Report*, UNESCO, Paris and Earthscan, London, http://www.unesco.org/water/wwap/wwdr/wwdr3/ (accessed 6 January 2011).

239 World Health Organization (2006) *Meeting the MDG Drinking-Water and Sanitation Target: the Urban and Rural Challenge of the Decade*, WHO, Geneva and UNICEF, Paris, http://www.who.int/water_sanitation_health/monitoring/jmpfinal.pdf (accessed 6 January 2011).

240 Botting, M.J., Porbeni, E.O., Joffres, M.R., Johnston, B.C., Black, R.E., and Mills, E.J. (2010) Water and sanitation infrastructure for health: the impact of foreign aid. *Global Health*, **6** (12), 1–8. http://www.globalizationandhealth.com/content/pdf/1744-8603-6-12.pdf (accessed 6 January 2011).

241 UN-WATER (2008) *Work Programme, 2008-9*, UN, New York, http://www.unwater.org/downloads/unw_workplan_2008.pdf (accessed 6 January 2011).

242 Pan African Chemistry Network (2010) *Africa's Water Quality – A Chemical Science Perspective*, PACN/Royal Society of Chemistry, London, http://www.rsc.org/Membership/Networking/InternationalActivities/PanAfrica/waterreport.asp (accessed 6 January 2011).

243 Alliance for Chemical Sciences and Technology in Europe (2010) Chemistry: Europe & the Future. AllChemE, http://www.cefic.org/allcheme/ (accessed 6 January 2011).

244 Conway, G. and Waage, J. (2010) *Science and Innovation for Development*, UK Collaborative on Development Sciences, London, http://www.ukcds.org.uk/publication-Science_and_Innovation_for_Develop%3E%20ment-172.html (accessed 6 January 2011).

2
The Role of Chemistry in Addressing Hunger and Food Security

Jessica Fanzo, Roseline Remans, and Pedro Sanchez

2.1
Chemistry is the Backbone of Food and Nutrition

Chemistry has provided the backbone in understanding the structure, organization and functions of living matter. Biochemistry, in particular, is composed of the structural chemistry of living matter, the metabolism or chemical reactions of those living matters, and the molecular genetics of heredity. The ability of plants to derive energy from sunlight and animals and humans to derive energy from food begins with chemistry and the principles of thermodynamics, and the basics of food itself are made of chemical and biological structures – amino acids, sugars, lipids, nucleotides, vitamins, minerals and hormones.

The chemical elements are key to understanding our modern day food and nutritional needs. In the late 18th century, many of the chemical elements had been defined, including nitrogen from ammonia, followed by the discovery of protein in egg albumin, inorganic elements and amino acids. The characterization of energy and calorimetry were also critical for the food and nutrition science world and could not have been understood without the use of physiological chemistry [1]. By studying persons engaged in labor and exercise and the amount of heat released, and understanding metabolics, the kilocalorie was defined – the energy needed to raise the temperature of 1 kg of water by 1 degree Celsius.

Starting in the 1800s, scientists worked backwards by characterizing disease states and, from defined foods, they discovered what was lacking or deficient, honing in on specific vitamins and minerals. For example, in the 1880s, deficiency of thiamin through fractionation of rice polishing was discovered to cause beriberi. In 1870, vitamin C deficiency was discovered to be the root cause of scurvy [1]. In the mid 1700s, British naval commander James Lind pleaded with the British Navy to make citrus foods available on all sea voyages. In a book he authored after an especially long journey with high mortality among the crew, he described miracle cures achieved with the use of lemon juice. Almost 60 years later, the British Navy did provide citrus foods when Captain Cook succeeded in avoiding scurvy altogether by giving his sailors lime juice on three successive voyages

The Chemical Element: Chemistry's Contribution to Our Global Future, First Edition.
Edited by Javier Garcia-Martinez, Elena Serrano-Torregrosa
© 2011 Wiley-VCH Verlag GmbH & Co. KGaA. Published 2011 by Wiley-VCH Verlag GmbH & Co. KGaA.

Table 2.1 The known 51 essential nutrients for sustaining human life (adapted from [3]).

Air, water and energy	Protein (amino acids)	Lipids-fat (fatty acids)	Macrominerals	Trace minerals	Vitamins
Oxygen	Histidine	Linoleic acid	Na	Fe	A
Water	Isoleucine	Linolenic acid	K	Zn	D
Carbohydrates	Leucine		Ca	Cu	E
	Lysine		Mg	Mn	K
	Methionine		S	I	C (Ascorbic acid)
	Phenylalanine		P	Fe	B_1 (Thiamine)
	Threonine		Cl	Se	B_2 (Riboflavin)
	Tryptophan			Si	B_3 (Niacin)
	Valine			Mo	B_5 (Pantothenic acid)
				Co (in B_{12})	B_6 (Pyroxidine)
				B[a]	B_7/H (Biotin)
				Ni[a]	B_9 (Folic acid, folacin)
				Cr[a]	B_{12} (Cobalamin)
				V[a]	
				As[a]	
				Li[a]	
				Sn[a]	

a) Not generally recognized as essential but some supporting evidence published.

(between 1768 and 1779) [2]. It was not until later that scientists established the definitive link between scurvy and vitamin C (ascorbic acid) deficiency. Chemistry not only identified these critical elements for human health but defined what is considered "essential" for sustaining life (see Table 2.1) [3].

Agriculture, the source of most food, became a science when Justus von Liebig (1803–1873) discovered the essential nutrient elements in plants. Fritz Haber (1868–1934) and Carl Bosch (1874–1940) invented ammonia synthesis that produced nitrogen fertilizers. Together with Gregor Mendel (1822–1884), the father of genetics and Martinus Beijerinck (1851–1900) discoverer of biological nitrogen fixation by legumes, the chemical foundations of agriculture were established. The International Union of Pure and Applied Chemistry devoted its second CHEMRAWN (Chemical Research Applied to World Needs) conference, held in Manila, Philippines in 1982, to chemistry and world food supplies [4]; followed by CHEMRAWN XII (Role of Chemistry in Sustainable Agriculture and Human Well-being in Africa) held at Stellenbosch, South Africa in 2007, reflecting the close relationship between chemistry and agriculture.

At the root of these chemical compounds is food, the backbone of human survival and evolution. Yet, today, we find that food remains an issue – either too much or not enough – in the continuing development of the human race. This chapter will explore how chemistry has historically influenced food security through food production, its access and nutrition. We will also explore how chemistry continues to provide and contribute to technology that defines the future of food and, hopefully, provides enough high quality food to sustainably nourish the estimated 9 billion people who will live on this planet by 2050.

2.2
Global Hunger and Malnutrition in the World Today

At the Millennium Summit in September 2000, the largest gathering of world leaders in history adopted the UN Millennium Declaration, committing their nations to a bold global partnership to reduce extreme poverty and to address a series of time-bound health and development targets [5]. These targets, the Millennium Development Goals (MDGs) and their indicators could be used to set benchmarks and monitor country-level progress. Among these MDGs is a commitment to reduce the proportion of people who suffer from hunger by half between 1990 and 2015 [6]. In 2010, many countries remain far from reaching this target, and ensuring global food security persists as one of the greatest challenges of our time. In the developing world, reductions in hunger witnessed during the 1990s have recently been eroded by the global food price and economic crises [7], which together added 105 million to the ranks of the hungry since 2008 [8].

There is clearly much progress to be made in addressing both hunger and undernutrition. As of 2009, 1 billion people are hungry, and 200 million children under five years of age are undernourished, with the majority of these children living in just 36 countries [9]. Vitamin A and zinc deficiency alone contribute to over half a million child deaths annually – both deficiencies are amenable to simple, effective and low-cost interventions [9a].

The MDG1 hunger target has two specific measures to track success: the prevalence of underweight children under five years of age, and the proportion of the population below a minimum level of dietary energy consumption. Obtaining an accurate measure of progress towards the MDG1 hunger target, and "food security" is challenging, and these two measures of the MDG1 are flawed.

2.2.1
Progress on the Proportion of Children Who are Underweight

Nonetheless, in the developing world, the proportion of children under five years of age who are underweight, declined from 31% to 26% between 1990 and 2008 [7, 10]. The progress made to reduce the number of children who are undernourished is insufficient to meet the goal of cutting underweight prevalence in half globally. When taking the 2008–2009 financial and economic crises into account, the task will be more difficult, but not unachievable in some countries (Figure 2.1) [9c].

Making progress
Insufficient progress
No progress

Figure 2.1 Country progress in meeting the MDG1 indicator for prevalence of children underweight (source [9c]). (Please find a color version of this figure in the color plates.)

2.2.2
Progress on the Proportion of the Population Who are Undernourished

The proportion of undernourished persons in developing countries, as measured by the proportion of the population below the minimum level of dietary energy consumption, decreased from 20% to 17% (a decrease in absolute numbers of 9 million) in the 1990s but both the proportion and absolute numbers have reversed course and increased in 2008 due to the food price crisis, which has severely impacted sub-Saharan Africa and Oceania regions [7]. Sub-Saharan Africa has the highest proportion of undernourished with 29% followed by Southern Asia, including India, at 22% [7].

2.3
Hunger, Nutrition, and the Food Security Mandate

What does it mean to be hungry? In its common usage, hunger describes the subjective feeling of discomfort that follows a period without eating [11]; however, even temporary periods of hunger can be debilitating to longer term human growth and development [12]. *Acute hunger* is when lack of food is short term and is often caused by shocks, whereas *chronic hunger* is a constant or recurrent lack of food [13]. The term *undernourishment* defines insufficient food intake to continuously meet dietary energy requirements [8] with FAO further defining hunger as the consumption of less than 1600–2000 calories per day.

What does it mean to have enough to eat?

The definition of food security set out at the 1996 World Food Summit stated that "food security exists when all people at all times have both physical and economic access to sufficient food to meet their dietary needs for a productive and healthy life".

Hunger often goes hand in hand with food security. The concept of food security goes beyond caloric intake and addresses both hunger and undernutrition [14]. Reducing levels of *hunger* places the emphasis on the quantity of food, and refers to ensuring a minimum caloric intake is met. Conversely, ensuring adequate *nutrition* refers to a diet's quality. A diet rich in proteins, essential fatty acids, and micronutrients has been proven to improve birth weight, growth, and cognitive development while leading to lower levels of child mortality [9a, 15]. A lack of these essential vitamins and minerals often results in "hidden hunger" where the signs of malnutrition and hunger are less visible in the immediate sense.

The achievement of food security depends upon three distinct but connected pillars. The first is *food availability*, which refers to ensuring sufficient quantity and diversity of food is available for consumption from the farm, the marketplace or elsewhere. Such food can be supplied through household production, other domestic output, commercial imports, or food assistance. The second, *food access*, refers to households having the physical and financial resources required to obtain appropriate foods for a nutritious diet. Access depends on income available to the household, on the distribution of income within the household, and on the price of food. The third, *food utilization*, implies the capacity and resources necessary to use food appropriately to support healthy diets, including sufficient energy and essential nutrients, potable water and adequate sanitation. Effective food utilization depends, in large measure, on knowledge within the household of food storage and processing techniques, basic principles of nutrition and proper child care, and illness management [14a]. Most precisely, the concept of "nutrition security" has been defined as "having adequate protein, energy, vitamins, and minerals for all household members at all times" [16].

For many years, food security was simply equated with enhancing the availability of food, and was linked to innovations in agricultural production. In many developing countries, agriculture remains the backbone of the rural economy. Increasing agricultural outputs impacts economic growth by enhancing farm productivity and food availability [17], while providing an economic and employment buffer during times of crisis [8]. In the 1970s and 80s, large investments in agriculture, technology, roads and irrigation led to major improvements in food production, particularly in Asia and Latin America. Chemistry was at the heart of some of these tools and technologies. During this period the proportion of official development assistance devoted to agriculture peaked at 15–20% [8]. Over the past decade, decreasing levels of agriculture aid and investment, particularly the dismantling of input, credit and market subsidies, reduced public support for research and extension, and declining infrastructure investments have been linked to rising numbers of people being undernourished [8]. The reverse relationship has also been suggested, with hunger and undernourishment carrying substantive economic and social costs with reduced labor productivity, investment in human capital, and escalating poverty [18].

While food availability is clearly important to achieving food security, having the means to effectively access and utilize food remains central to good nutrition. This wider focus is important. There is a growing recognition that food security must

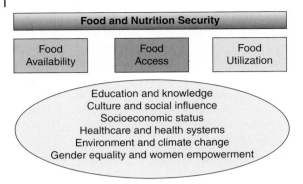

Figure 2.2   The determinants of food security (source [19]).

be viewed as inseparable from the other MDGs—and that addressing poverty, education, health and basic infrastructure are also critical. This interdependence is illustrated in Figure 2.2 [19] which makes the point that achieving sustainable gains in reducing hunger and undernutrition, and improving food security, will depend fully on concerted and synergistic efforts on a number of fronts.

2.4
Chemistry's Influence on the Pillars of Food Security

2.4.1
Food Availability

It is estimated that to meet the population's demands of 2050, a doubling of grain production will be needed, however, yield increases of the world's cereals have begun to stagnate [20] and yields in many regions of the world suffer from nutrient limitations and lack of access to irrigation. Future production will be further threatened by increased soil degradation, climate change, and the increased volatility of oil production and its impact on fertilizer prices [21].

In many poor rural settings, addressing hunger is inextricably linked to improving soil fertility and crop management [22]. Soil chemistry and applications play a critical role in developing soil and crop management practices through enhanced understanding of soil processes, plant nutrition, fertilizer production, development of improved crop varieties and methods for controlling pests and diseases.

2.4.2
Chemistry and the Green Revolution

The 1960s was a decade of despair with regard to the world's ability to cope with the food–population balance, particularly in developing countries. Most of the lands suitable for agriculture in Asian countries were cultivated while population growth rates accelerated, owing to the rapidly declining mortality rates that resulted from advances in modern medicine and health care. Massive starvation was predicted and international organizations and concerned professionals raised awareness of the ensuing food crisis and mobilized global resources to tackle the problem [19, 23].

Fortunately, large-scale famines and social and economic upheavals were averted, thanks largely to the marked increase in cereal grain yields in many Asian developing countries that began in the late 1960s [24]. This phenomenon – coined the "Green Revolution" – was due largely to the development and widespread adoption of chemical-based technologies. Key was the development and extension of genetically improved high-yielding varieties of cereal crops that were responsive to the application of advanced agronomic practices, including, most importantly, fertilizers and improved irrigation [23a, 24b]. Norman Borlaug, one of the fathers of the Green Revolution, summed up the role that nitrogen (N) fertilizer played in this grand agricultural transformation by using a memorable kinetic analogy: "If the high yielding dwarf wheat and rice varieties are the catalysts that have ignited the Green Revolution, then chemical fertilizer is the fuel that has powered its forward thrust..." [25]. The fuel for the transformation was made available by Haber's brilliant discovery of ammonia synthesis from its elements, in 1909, and the extraordinarily rapid commercialization of this invention, led by Bosch, that made large-scale production of ammonia possible by 1913 [26]. As shown in Figure 2.3, rapid post-1950 diffusion of N-fertilizer applications had increased

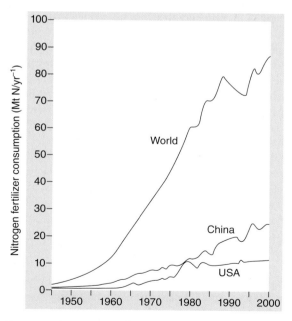

Figure 2.3 Consumption of nitrogenous fertilizers, 1950–1999 (source [27a]).

their worldwide use to nearly 80 million tonnes (Mt) by the late 1980s and, after a period of stagnation (related to the fall of the USSR), to 87 Mt N in the year 2000 [26, 27]. The projected demand for nitrogen from chemical fertilizer is estimated to increase to 236 MT in 2050 [27b].

The Green Revolution had a tremendous impact on food production and socio-economic conditions. Between 1966 and 2005, food production in South Asia increased by 240% [28]. Applying advanced technology to high-yielding varieties of cereals caused the marked achievements in world food production. The gradual replacement of traditional varieties of rice, maize and wheat–crops which account for almost 50% of calories in most diets–by improved varieties, and the associated improvement in farm management practices, had a great effect on the growth of rice, wheat and maize output, particularly in Asia. For example, the average rice yield in South Asia increased by 240% from 1966 to 2007. During the same period, daily caloric intake per capita improved on average by approximately 25% [28].

In addition to fertilizers, pesticides have also played an important role in increasing agricultural production during the Asian Green Revolution. Insect pests, diseases, weeds and rodents are serious constraints to agricultural production, especially in the humid tropics. Scientific efforts to remove these constraints have focused on the breeding of resistant varieties of crop plants, as well as on the development of pesticides, insecticides and herbicides and integrated pest management strategies. In developing countries, most of the pesticides are, however, applied to exported crops, such as cotton and tropical fruits, rather than to locally consumed food crops.

In this way, the Green Revolution was able to address food availability challenges. The widespread adoption of high-yielding varieties has helped many Asian and Latin American countries to meet their growing food needs from productive lands and has reduced the pressure to open up more fragile lands. Despite its achievements, the first Green Revolution did not solve all food and nutrition security issues, partly because the efforts emphasized the food availability component of food security over the food access, food utilization and sustainability components, resulting in the neglect of core nutrition elements and environmental challenges. Although massive efforts were taken to decrease hunger in India, 50% of children across South Asia continue to suffer from undernutrition [9c, 29]. Further, excessive use of fertilizers and pesticides, as well as the monoculture of a few crop cultivars, created serious environmental problems, including the breakdown of resistance and the degradation of soil fertility [24b, 30]. It is now critical that chemical science invests and takes a leading role in cross-disciplinary efforts to predict when and where the use of agrochemicals and chemical-based technologies are pushing food production systems over sustainable boundaries [31], and to develop innovative strategies that can enhance social, environmental and economic sustainability of food systems.

While Asia and Latin America dramatically increased their agricultural productivity over the past 40 years, Africa's agricultural growth stagnated due to high transport costs, poor infrastructure, low levels of fertilizer use and the dismantling of public agricultural institutions for research, extension, credit and marketing [11,

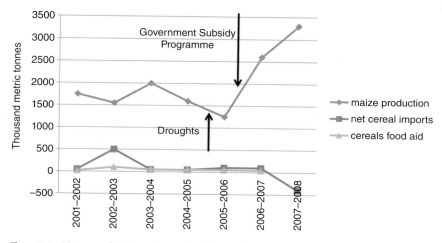

Figure 2.4 Maize production and cereal trade in Malawi (1990 to 2007) (source [34]).

32]. However, after decades of neglect of Africa's agricultural systems, a Green Revolution for Africa is emerging and there is now optimism about sub-Saharan Africa's ability to rapidly increase its agricultural productivity. This is partly due to some key successes – at the local and national levels – of policies that support smallholder farmers. In Malawi, because of a smart input subsidy program implemented by the government, maize harvests have greatly surpassed those of previous years, turning that country from a recipient of food aid into a food exporter and food aid donor to neighboring countries [33] (Figure 2.4).

Micronutrient fertilizers

Under certain soil conditions, the use of micronutrient fertilizers, in balanced combination with macronutrient fertilizers, has promising potential to increase production, disease resistance, stress tolerance, and the nutritional quality of crops. The increase in yield from the use of micronutrients deficient in the crops, notably zinc and boron, should compensate their cost and can also make the use of macronutrient fertilizers more cost effective as a package of balanced nutrients.

As more money and attention galvanizes much-needed action on the African Green Revolution, a vigorous debate is required to ensure that the mission of improving food security on the world's poorest continent is achieved in the most effective, comprehensive and inclusive manner possible. The African Green Revolution cannot be limited to increasing yields of staple crops but must be designed as a driver of sustainable development, which includes nutrition elements. Advances in chemistry can again play a pivotal role in this process. For example, our understanding of human N (protein) needs has undergone many revisions, and, although some uncertainties still remain, it is clear that average protein intakes are excessive in rich countries and inadequate for hundreds of millions of people in Asia, Africa, and Latin America. More dietary protein will be needed to eliminate these disparities but the future global use of N fertilizers can be moderated not just by better agronomic practices but also by higher feeding efficiencies and by gradual changes of prevailing diets. As a result, it could be possible to supply adequate nutrition to the world's growing population without any massive increases in N inputs. The addition of micronutrients to fertilizers is another area of interest [3, 35].

2.4.3
Genetically Engineered Crops and Food Production

A highly debated topic and example of a current technology at the crossroads of agriculture and chemistry is genetic engineering (GE), a modern technology for modifying crops and livestock. GE is one of several tools in the modern crop biotechnology kit and allows the introduction of genes from the same species or from any other species, including species that are beyond the normal reproductive range of the plant, into the plant or animal. The need to develop new crop varieties that are adapted to local conditions, conducive to sustainable agriculture, and remain high-yielding in the absence of irrigation or large inputs of petrochemicals, is an exceptionally tall and urgent order. Many plant scientists believe that GE can contribute significantly to achieving these goals [36]. However, there are a multitude of concerns about the effects of GE crops on human health, environment, social well-being and ethics which are fueling a polarized debate. One side perceives that excessive regulation is slowing the delivery of benefits [37]; the other is concerned that adoption is proceeding hastily and without adequate safeguards [38]. This is embedded in a multidimensional debate, including scientific, social, economic, political and ethical issues.

GE crops and foods have been commercially available in the US since 1995 and their adoption around the world followed, showing increases each year (Figure 2.5). In 2008, the global area of commercially grown GE crops was 125 million hectares, involving 25 countries [40]. The four primary GE crops in terms of land area are soybean, maize, cotton and canola (oilseed rape). In 2008, GE crops were being grown on 9% of the global arable land, 70% of soybean cropland was planted with a GE variety, for maize cropland this was 24%, for cotton 46% and for canola 20% [40], in total corresponding to 40% of the cropland of these main crops.

Nutritionally improved GE seeds

The first use of GE to alter nutritional quality was the introduction of three genes into rice to create the much publicized *Golden Rice* variety, enriched in vitamin A. Many more GE crops with enhanced nutritional value have followed, such as increased protein quality and levels in maize, increased calcium levels in potato and increased folate levels in tomato. However, none of these crops is commercially available yet.

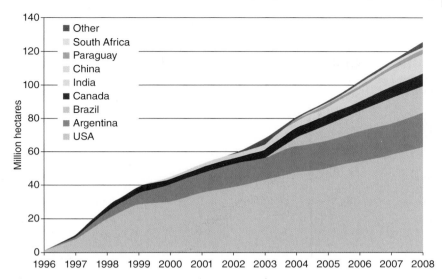

Figure 2.5 Million hectares of GE crops per country (source [39]). (Please find a color version of this figure in the color plates.)

Despite sizeable GE crop acreage, the current diversity of crop types and traits in commercial production is limited. Nearly all major-acreage, commercial releases of GE crops are at present based on pest protection via genes from *Bacillus thuringiensis* (Bt), a widespread soil bacterium that produces insecticidal proteins called Bt toxins, or herbicide tolerance (HT), or a combination of both [40, 41]. HT crops are tolerant to certain broad-spectrum herbicides such as glyphosate and glufosinate, which are more effective, less toxic, and usually cheaper than selective herbicides. By cutting the costs and labor of weed or insect control, these first generation commercial pest and/or herbicide tolerant GEs have been shown to provide a tangible economic benefit to farmers [37c, 41, 42], to result in time savings, increased ease of agricultural practices and reduction in yield losses and pesticide use [37c, 41–43].

The average reduction in pesticide use by using Bt crops has been shown to vary from 0% for Bt maize in Argentina to 77% for Bt cotton in Mexico [37c]. Average increases in *effective yield* – through reduction of yield losses – by using Bt crops vary between 0% for Bt cotton in Australia to 37% for Bt cotton in India [37c]. For HT crops, in most cases no increase in yield is observed compared to conventional crops [37c].

Although commercialized GE crops are limited in trait diversity, proof-of-concept for many other traits has been reported in laboratory experiments and small-scale field trials. While the first generation of commercialized crops focused largely on input agronomic traits, the coming generation of crop plants can be grouped into four broad categories of impact: *agronomic performance, environment, human health* [44] *and rural livelihood.*

2.4.4
Food Access

Food access involves whether households have the physical and financial resources required to obtain appropriate food for a nutritious diet. This access depends on income available to the household, on the distribution of income within the household, and on the price of food. Access also depends on what happens to the food after production, such as during post-harvest storage.

2.4.4.1   **Post-Harvest Treatment and Storage**
One of the main causes of food insecurity in Africa is the high prevalence of storage pests. Maize is an excellent food source and an ideal breeding site for storage pests [45]. Pests can be defined as those organisms that cause damage resulting in economic loss to maize and other plants in the field or in storage [46]. The Larger Grain Borer (LGB, Figure 2.6), which is sometimes referred to as the Greater Grain Borer (GGB), and given names like "Osama" is the single most serious pest of stored maize and dried cassava roots (chips). The primary host is maize, in particular maize on the cob, both before and after harvest. LGB destroys maize giving unusable flour (Figure 2.7). LGB also bores into non-food substances such as wood, bamboo, and even plastic, which poses a challenge to controlling the pest.

Figure 2.6   The Larger Grain Borer (source [45]). (Please find a color version of this figure in the color plates.)

Figure 2.7   Maize destroyed by the Large Grain Borer to flour (source [45]). (Please find a color version of this figure in the color plates.)

According to national maize experts, maize losses due to storage pests range from 30–60% in Malawi, Tanzania and Kenya, much of which is attributed to the presence of the Large Grain Borer and these figures far exceed what is currently recorded in the literature. Maize losses experienced by farmers are variable, but farmers in all countries of the study confirmed experiencing losses, even if they used inorganic or organic storage insecticides, and all confirmed that losses can be up to 100% if maize is not protected with insecticides before storing. Also, effectiveness of some insecticides is questionable. There are cases of purportedly effective insecticides bought direct from the importing company. It was even demonstrated by one of the company's experienced sales representatives, but the maize treated by the farmers and the maize treated for demonstration by the representative was destroyed by LGB [45].

It is apparent that maize storage is a crucial component of ensuring greater food security and should be included in efforts by research institutes, national governments and development partners, especially in countries where such efforts have yielded substantial returns in maize or other food crop productivity. Recommendations and management strategies form an integrated approach to the management of storage pests and include chemical-based technologies, particularly drying techniques, drying cribs and treatment of maize, before and during storage, with insecticides (Figure 2.8). However, due to lack of awareness and access to proper technologies, farmers end up selling their maize soon after harvest, only to buy it back from the same people at more than twice the price they sold it for just a few months after harvest, resulting in a continual poverty trap. If efforts to increase food security included storage, and farmers were able to store their maize properly, they would save between US$10 to US$20 per bag of maize needed for household consumption throughout the year. These may appear like small savings, but from analysis of family sizes in rural sub-Saharan Africa and the corresponding maize required to feed larger traditional families, these translate into huge savings per

Figure 2.8 Traditional granary (a) and improved granary/crib (b) in the Millennium Village Project Mwandama, Malawi (source: MDG Centre East and Southern Africa). (Please find a color version of this figure in the color plates.)

family. Furthermore, storage pests, in particular LGB, and in combination with other pests such as Maize weevils, cause substantial losses at national levels, which translates, approximately, to between US$150 and US$300 million, money which could be used to provide other essential in-country services [45].

A case study: millennium villages in sub-Saharan Africa

The Millennium Villages Project was initiated in 2005 to accelerate progress towards the MDG targets, including MDG 1 – to eradicate extreme poverty and hunger. The Millennium Villages are situated in "hunger hotspots", where at least 20% of children are malnourished and where severe poverty is endemic. The countries where Millennium Villages are located are Ethiopia, Ghana, Kenya, Malawi, Mali, Nigeria, Rwanda, Senegal, Tanzania, and Uganda. They were chosen to reflect a diversity of agro-ecological zones, representing the farming systems found in over 90% of sub-Saharan Africa and are demonstration and testing sites for the integrated delivery of science-based interventions in health, education, agriculture and infrastructure. Within the Project, hunger and undernutrition is being addressed with an integrated food- and livelihood-based model that delivers a comprehensive package of development interventions.

By supporting farmers with fertilizers, improved crop germplasm and intensive training on appropriate agronomic practices, average yields of $3\,t\,ha^{-1}$ were exceeded in all sites where maize is the major crop, compared to average cereal yields of less than 1 tons (t) hectare $(ha)^{-1}$ before intervention [47]. Households produced enough maize to meet basic caloric requirements, with the exception of farms smaller than 0.2 ha in Sauri, Kenya. Value to cost ratios of 2 and above show that the investment in seed and fertilizer is profitable, provided surplus harvests were stored and sold at peak prices [47]. Increased crop yields are the first step in the African Green Revolution, and must be followed by crop diversification for improving nutrition and generating income and a transition to market-based agriculture. A multi-sector approach that exploits the synergies among improved crop production, nutrition, health, and education is essential to achieving the MDGs.

Key Interventions in Food Production in the Millennium Villages include:

- **Soil rehabilitation techniques.** Replenish nutrients in the soil with mineral fertilizers, nitrogen-fixing legumes, and other organic materials, and, by returning crop residues to the soil, and soil conservation techniques, reduce run-off and erosion and maintain the investments in soil rehabilitation.

- **Access to improved seeds.** Provide farmers with access to and information about improved seeds for basic food crops, livestock, grain legumes, root and tuber crops, vegetable, tree, and fodder crops, as well as developing, where appropriate, the capacity of community members to produce their own seed or planting materials using seed multiplication plots, seed orchards, and nurseries.

- **Small-scale irrigation systems.** Promote efficient irrigation technologies for supplementation of rain-fed crops and increased and off-season production of cash crops. Train farmers and other groups in techniques such as rainwater harvesting and storage, gravity, and low-pressure irrigation systems, and improving existing irrigation systems; provide access to equipment required for these techniques.

- **Grain storage facilities.** Minimize post-harvest losses and store food beyond subsistence needs by training farmers and farmer groups in the construction of household and community grain storage structures. *Creation of cereal banks* by communities to store surplus for later sales at better prices.

- **Agricultural extension services.** Provide field training on land preparation, plant spacing, fertilizer placement, and integrated soil fertility management practices, including agro-forestry. Provide information to agricultural extension officers to ensure that they have the latest and most appropriate information on soil health, small-scale water management, improved seeds, livestock, agro-forestry, and other locally relevant agricultural techniques.

- **Crop diversification for income generation and nutritional security.** Promotion of crops that help improve household nutrition. Crops include vegetables, fruits, grain legumes, livestock, and dairy. This diversification strategy must include nutrition education programs around the various crops.

- **Farmer organizations.** Help establish and train village organizations to develop organized systems to sell products to more distant markets, engage with microfinance institutions to purchase farm inputs, and promote other skills required for developing commercial farming enterprises.

2.4.5
Food Utilization

To have food available and accessible to eat, whether purchased at the market or grown at home, is not the sole solution to food security. A person's body must be in good physical condition in order to properly use the food. This is termed food utilization – the ability to use food efficiently in order to live life to the fullest. Focusing on the individual level, food utilization also takes into consideration the biological utilization of food. Biological utilization refers to the ability of the human body to take food and convert it into energy, either used to undertake daily activities, or stored. Utilization requires not only an adequate diet, but also a healthy physical environment, including safe drinking water, adequate sanitation and hygiene, decreased burden of infectious disease, and the knowledge and understanding of proper care for oneself, for food preparation and safety.

2.4.5.1 Balanced Diets and Utilization of Nutrients: The Chemical Components

Chemistry has taught us that not only is it essential that our bodies absorb the nutrients from foods that we eat, but that the chemical composition of the foods consumed in specific combinations and the quality of the diet are critical in meeting dietary needs. The role of essential nutrients in human health and the synergies in their physiological functions are being increasingly recognized and support the notion that nutrient deficiencies rarely occur in isolation [48]. The challenge is to provide the adequate amount and diversity of nutrients required for a complete human diet. This urges a multidimensional approach. Optimizing for nutrient diversity can be presented schematically as maximizing the various

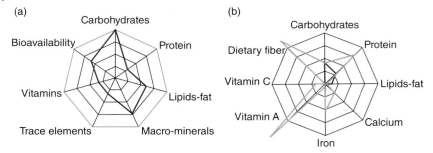

Figure 2.9 Ecological spider web presenting diversity requirements in a human diet. (a) Nutrient composition of an ideal diet that meets all nutritional needs is shown in pink. An example of nutrient composition of a diet that meets carbohydrate demand but lacks protein and micronutrients or trace elements is shown in blue. (b) Nutrient composition data of three food crops are shown as % of daily requirement (100%). The blue line represents one cup of white corn (166 g), the green line one cup of black beans (194 g), and the orange line one cup of pumpkin (116 g) (nutrition facts from http://www.nutrition-data.com). The spiderdiagram shows the complementarity between the three food crops for carbohydrates, proteins, dietary fiber, and vitamin A (source [49]). (Please find a color version of this figure in the color plates.)

arms of an ecological spider diagram, as illustrated in Figure 2.9. Figure 2.9a shows the nutrient composition of an ideal diet that meets all nutritional needs, shown in pink. An example of nutrient composition of a diet that meets carbohydrate demand but lacks protein and micronutrients or trace elements is shown in blue. In Figure 2.9b, nutrient composition data of three food crops are shown as % of daily requirement (100%). The blue line represents one cup of white corn (166 g), the green line one cup of black beans (194 g) and the orange line one cup of pumpkin (116 g). The spider diagram shows the complementarity between the three food crops for carbohydrates, proteins, dietary fiber, and vitamin A and the importance of the chemical composition of foods in meeting the nutritional requirements of the diet.

On a global basis, plants provide approximately 65% of the world supply of edible protein whereas animal products contribute 35% [50], much coming from cereal grains. Important differences among and between food products are the concentrations of proteins and the essential amino acids they contain. Eight amino acids are generally regarded as essential for humans: isoleucine, leucine, lysine, methionine, phenylalanine, threonine, tryptophan, and valine. In addition, arginine, cysteine, histidine and tyrosine are required by infants and growing children. These amino acids are considered essential because the body does not synthesize them, making it essential to include them in one's diet in order to obtain them.

Diets rich in cereals and of vegetable origin do not contain all the essential amino acids necessary for daily consumption and requirements. Instead, near-complete proteins are found in plant sources. In contrast, animal sources such as meat, poultry, eggs, fish, milk, and cheese provide all of the essential amino acids but are not consumed on a daily basis by the majority of the global population (particularly the developing world) due to cost and supply. Often in the developing

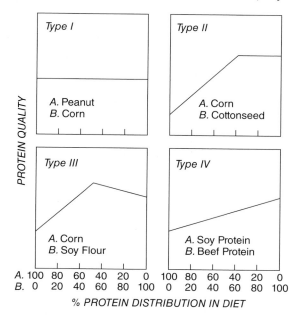

Figure 2.10   Protein complementation and four response types (source [50]).

world, the diet is not adequate in quality protein when consumed in the traditional sense as the diet is made up mainly of plant-based sources. This often leads to children with protein energy malnutrition and faltering growth, particularly among children aged 6 to 24 months [51]. Yet, despite this challenge, it is not necessary to consume animal sources containing complete proteins as long as a reasonably varied diet is maintained and other sources rich in proteins, such as legumes which contain essential amino acids, provide adequate full complementation of the essential amino acids and protein quality for adequate health and nutrition. By consuming a wide variety of plant foods, a full set of essential amino acids will be supplied and the human body can convert the amino acids into proteins. This is at the core of the chemical composition of diets.

   To consume a varied diet made up mainly of plant sources, it is important to think about chemical composition and combinations of food. When two dietary proteins are combined, different types of responses result [50]. Four types have been classified (Figure 2.10) through chemical bioassay studies. Type I indicates when no protein complementary effects occur. This can occur with peanuts mixed with corn diets in which both are deficient in essential amino acids. This is obviously not an optimal combination to provide protein needs for the day. Type II is when two sources of food such as corn and cottonseed have a limiting amino acid, in this case lysine. Some of the essential amino acids are met, but not completely. Type III demonstrates a true complementary effect working synergistically to meet the needs with corn and soybean. The sum of both meets the protein needs. Soy is considered a high quality protein source. Type IV occurs when both sources have a common amino acid deficiency and is not considered ideal [50].

2.4.5.2 **Antinutrients**

With plant foods being the predominant source of the diet in much of the poor world, antinutrients and promoters contained in these plant foods should be taken into consideration with regard to bioavailability of nutrients to humans. Most antinutrients in foods inhibit the absorption of micronutrients that are essential for growth and are often deficient in the developing world–predominantly iron and zinc. Antinutrients include phytic acid, fiber, tannins, oxalic acid, goitrogens and hemagglutinins [52]. Phytic acid or phytates, one of the greater concerns, are often found in whole legumes, and cereal grains–the staples of the diets in resource-poor communities.

Several traditional food processing and preparation methods, that work on the basics of chemistry, are often used at the household level to enhance the bioavailability of micronutrients, including mechanical processing, soaking, fermentation and germination or malting [53]. For example, boiling of tubers can induce moderate losses of phytic acid [54]. Fermentation can also induce phytate hydrolysis via the action of microbial phytase enzymes which hydrolyze phytate to lower inositol phosphates [55]. This has been done in maize, soy beans, sorghum, cassava, cocoyam, cowpeas and lima beans, all common foods in the developing world. Low-molecular weight organic acids such as citric acid can increase fermentation and enhance the absorption of zinc and iron [56].

Cassava is an important tropical root crop providing energy to approximately 500 million people (Figure 2.11). The presence of the two cyanogenic glycosides, linamarin and lotaustralin, in cassava is a major factor limiting its use as food and can be toxic. Traditional processing techniques practiced in cassava production are known to reduce the cyanide chemical in tubers and leaves. These including sun drying, soaking followed by boiling and fermentation, as used for traditional African cassava end products such as gari and fufu. The best processing method for the use of cassava leaves as human food is pounding the leaves and cooking the mash in water [57].

Figure 2.11 Cassava in Africa (source: Nestle). (Please find a color version of this figure in the color plates.)

2.4.5.3 **Fortification of Food Vehicles: One Chemical at a Time**

Food fortification is one of the food-based strategies for preventing micronutrient deficiencies. Currently, over 2 billion people globally are deficient in micronutrients, and many of these are women and children [9a, 9c]. In developing countries, fortification is increasingly recognized as an effective medium- and long-term and accessible approach to improve the micronutrient status of communities [58], and is considered cost effective. Fortification does not require changes in the dietary habits of the population, can often be implemented relatively quickly, and can be sustainable over a long period of time. It is considered one of the most cost-effective means of overcoming micronutrient malnutrition [18a].

Improving foods for malnourished children

The Global Alliance for Improved Nutrition (GAIN) is working with governments, public–private partnerships, local companies, non-governmental organizations, and not-for-profit venture capitalists to improve infant and young child feeding through commercializing nutritious *fortified* food products, including fortified complementary foods, micronutrient powders and lipid-based nutrient supplements. One study done by Doctors without Borders showed that short-term supplementation of these fortified complementary feeding foods given to children who were not malnourished reduced the incidence of acute malnutrition in Niger.

Fortification is the addition of nutrients to commonly eaten foods such as flour, sugar, salt and cooking oil, to increase the consumption of essential micronutrients for health. The food that carries the nutrient is the vehicle; the nutrient added is the fortifier. Fortification of foods is aimed to provide levels of the nutrient (30 to 50% of daily requirements) at normal consumption of a food vehicle [58]. For decades, fortification has been widespread throughout the world. In the US, flour has been fortified and almost one quarter of iron intake comes from fortified sources, much from flour [59].

The selection of the right vehicle and the appropriate micronutrients is a precondition to assure success in addressing micronutrient deficiencies, although at times it can be difficult. Potential vehicles have a pyramid-type priority (Figure 2.12) in that stable foods form the base of the pyramid and will result in a broader dissemination of targeting a large portion of the population. Basic foods and value-added foods are also critical to ensure that all common products within the food chain with the potential to be fortified are targeted.

Successful fortification vehicles for vitamin A, a common deficiency in the developing world, include sugar, margarine and oil. For example, in Guatemala, national sugar fortification with vitamin A has eliminated vitamin A deficiency [60]. Iodization of salt has become the most commonly accepted method of iodine deficiency prophylaxis in most countries of the world. Its advantages include uniformity of consumption, universal coverage, acceptability, simple technology and low cost [18a]. 60 to 70% of all salt is now iodized [58]. During the past few years, attention has been given to the possible high prevalence of zinc deficiency

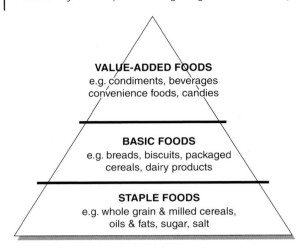

Figure 2.12 Food product pyramid for fortification (source [58]).

among children and its consequences. Results presented show that zinc has an impact on growth, especially in severely growth retarded and underweight children, and reduces morbidity. The role of seeds biofortified with zinc is being explored.

With this combination of technology and chemistry, food fortification is one of the most effective methods to eliminate micronutrient deficiencies, afflicting over 2 billion people worldwide. It has been shown to eliminate goiter, rickets, beriberi and pellagra from the western world [58] however, the focus should next be on the developing world where many remain hungry and undernourished.

2.4.5.4 Improving Utilization through Modern Medicine: The Contribution of Chemistry to Basic Medicines

Many of the determinants that impact food utilization are considered long-term poverty stricken determinants to poor nutrition. As shown in Figure 2.13, UNICEF's framework on the determinants of undernutrition, maternal and child care practices, which are associated with the situation of women in societies – education, knowledge, income generation, and reproductive practices [61] – are at the root of the problem.

Improving child feeding practices for young children is also a huge determinant of food utilization and child growth. This starts right at birth with exclusive breast-feeding and complementation of milk with food rich in energy and nutrients. Lastly, a robust primary health care systems approach must be in place to improve the nutritional situation and food utilization. Infectious diseases impede dietary intake and utilization, resulting in malnutrition. Consequently, one of the most important premises to improve nutrition is to control and prevent most common childhood infectious diseases by expanding immunization programs, providing diarrhea and malaria control and treatment programs, and decreasing parasitic

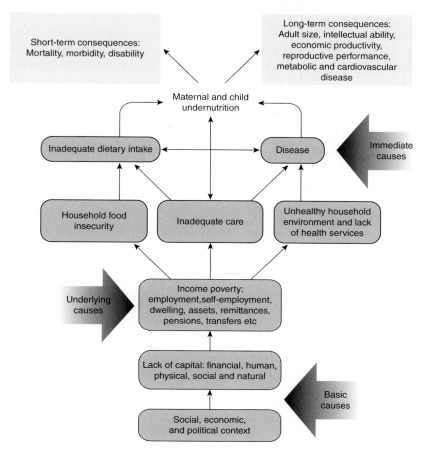

Figure 2.13 UNICEF's framework on undernutrition (source [61]).

burden. The backbone of some of these programs is water supply improvements and improving sanitation and hygiene in the home and schools.

One of the greatest contributions of chemistry to treating disease is through medicine. In the developing world, neglected tropical diseases such as worms, including ascaris roundworms, trichuris whipworms, and necator hookworms, in the intestines can contribute to anemia [62] and children suffer from deficits in physical growth, as well as reductions in intelligence, memory, and cognition [63]. The Global Burden of Disease caused by the three major intestinal nematodes is an estimated 22 million disability-adjusted life-years (DALYs) lost for hookworm, 10 million for Ascaris lumbricoides, 6 million for Trichuris trichiura, and 39 million for the three infections combined (as compared with malaria at 36 million) [63b]. Anorexia and perpetuated hunger, which can decrease intake of all nutrients in tropical populations on marginal diets, is likely to be the most important means by which intestinal nematodes inhibit growth and development.

In many cases, deworming once per year with a benimidazole anthelminthic drug such as albendazole or mebendazole, as discovered and made available by chemical companies, is sufficient [63b, 64]. These drugs are particularly effective for treating ascaris and trichuris worm infections [63b] and can be administered for as little as US$0.03 per person [65]. By treating preschool-age children and girls and women of childbearing age with these essential inexpensive medicines, morbidity and mortality can be prevented and the vicious intergenerational cycle of growth failure that entraps infants, children and girls and women of reproductive age in developing areas can be decreased [66].

2.5
Conclusion

Throughout the history of time, it is clear that chemistry has played a central role in the food and nutrition security agenda. Chemistry has been pivotal to food production from soil to seed, from pest control to human nutrition. Although food access is largely dependent on socioeconomic status, chemistry plays a role in improving the access to healthy foods through improved post-harvest storage loss. And lastly, chemistry is core to food utilization and combinations – by improved biofortification, food processing and essential medicines. Chemistry has contributed much to the food security agenda (Figure 2.14).

Evidence from the examples in this chapter suggests that food and nutrition security is complex, and requires efforts across a spectrum that includes enhancing food production while simultaneously increasing access and utilization with substantive political commitment to address the most vulnerable populations with an equitable, basic human rights approach. Chemistry plays a critical role in this spectrum and, in the future, requires a cross-sectoral approach.

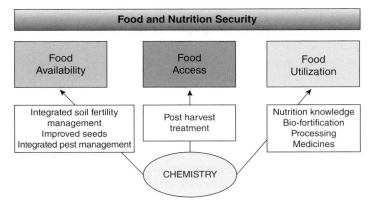

Figure 2.14 The role of chemistry in food and nutrition security.

Iyere [67] stated that "Sustainability and globalization therefore encompass addressing world hunger and poverty as well. Thus, there is a great need to gear chemistry contribution to mitigation of these problems to the chemistry of the past. In those times, agriculture was invigorated through the use of sensible chemistry and the development of the connections between chemistry, other disciplines, the environment, and daily life in such a way that interdisciplinary thinking and the relation of chemical concepts to societal issues became a way of life."

In the model of food security itself along with Iyere's thoughts on chemistry and sustainable development, addressing hunger requires a multi-disciplinary approach. Recent calls for greater attention to hunger and undernutrition highlight the importance of integrating technical interventions with broader approaches to address underlying causes of food insecurity – incorporating perspectives from agriculture, health, water and sanitation, infrastructure, gender and education – many rooted in the core science of chemistry. Such an approach would inherently build on the knowledge and capacities of local communities to transform and improve the quality of diets for better health and nutrition. Recent research has documented potential synergies between health and economic

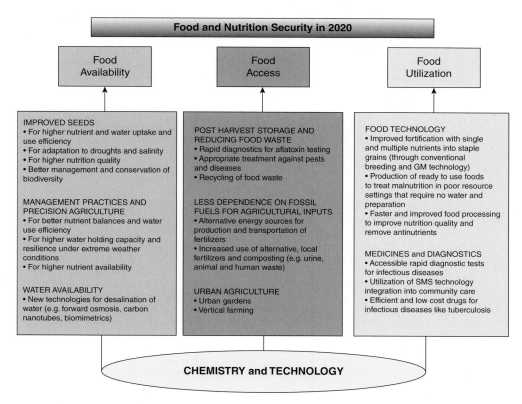

Figure 2.15 A 2020 vision for chemistry in achieving food security for all.

interventions, suggesting multi-sector approaches may generate a wider range of benefits than single sector approaches acting alone. The role of the chemical science and chemists will be challenged to work interdisciplinarily to address the global challenges that the world faces, and the hunger mandate calls for better tools and technologies to move forward.

In just 10 years from now, we envision that chemistry will become more and more important in all aspects of food security and nutrition (Figure 2.15). Although the numbers of those hungry and undernourished are staggering, the sciences such as chemistry can make huge strides to improve the situation. By 2020, much of the innovation and technology within chemistry can be earmarked and in motion to ensure that food security is achieved for all.

References

1 Carpenter, K. and Harper, A.E. (2006) *Evolution of Knowledge of Essential Nutrients*, 10th edn, Lippincott Williams & Wilkins, Philadelphia.

2 Carpenter, K. (1986) Captain Cook and the pneumatic chemistry (1770–1815), in *The History of Scurvy and Vitamin C*. Cambridge University Press, Cambridge, MA, pp. 75–97.

3 Graham, R.D., Welch, R.M., Saunders, D.A., IOrtiz-Monasterio, I., Bouis, H.E., Bonierbale, M., de Haan, S., Burgos, G., Thiele, G., Liria, R., Meisner, C.A., Beebe, S.E., Potts, M.J., Kadian, M., Hobbs, P.R., Gupta, R.K., and Twomlow, S. (2007) Nutritious subsistence food systems. *Adv. Agron.*, **92**, 1–72.

4 Bixler, G. and Shemilt, L.W. (1983) *Chemistry and World Food Supplies: The New Frontiers, CHEMRAWN II. Perspectives and Recommendations*, Pergamon Press, Manila, Philippines.

5 United Nations (2000) *United Nations Millennium Development*, United Nations, New York.

6 United Nations (2001) *Road Map Towards the Implementation of the United Nations Millennium Declaration*, UN General Assembly, New York.

7 United Nations (2009) *The Millennium Development Goals Report*, United Nations, New York.

8 Food and Agriculture Organization of the United Nations (2009) *Economic Crises – Impacts and Lessons Learned*, FAO, Rome.

9 (a) Black, R.E., Allen, L.H., Bhutta, Z.A., Caulfield, L.E., de Onis, M., Ezzati, M., *et al.* (2008) Maternal and child undernutrition: global and regional exposures and health consequences. *Lancet*, **371**, 243–260; (b) United Nations (2008) Goal 1: Eradicate Extreme Poverty and Hunger Fact Sheet; (c) UNICEF (2009) *Tracking Progress on Child and Maternal Nutrition*, UNICEF, New York.

10 UNICEF (2010) ChildInfo. http://www. childinfo.org/undernutrition_progress. html (accessed 15 January 2010).

11 UN Hunger Task Force (2004) *Halving Hunger: It Can Be Done*, UN Hunger Task Force, New York.

12 Narayan, S., Walker, J., and Trathen, K. (2009) *Who's Really Fighting Hunger?*, ActionAid, Johannesburg.

13 DFID (2009) *The Neglected Crisis of Undernutrition: Evidence for Action*, Department of International Development, Glasgow.

14 (a) USAID (1992) Policy Determination of 1992 for Definition of Food Security, http://www.usaid.gov/policy/ads/200/ pd19.pdf (accessed 1 January 2010); (b) Food and Agriculture Organization of the United Nations (1996) *Declaration on World Food Security*, Food and Agriculture Organization of the United Nations, Rome.

15 (a) Pelletier, D.L., Frongillo, E.A., and Habicht, J.P. (1993) Epidemiologic evidence for a potentiating effect of malnutrition on child mortality. *Am. J.*

Public Health, **83** (8), 1130–1133; (b) Ruel, M.T. (2002) *Is Dietary Diversity An Indicator of Food Security or Diet Quality? A Review of Measurement Issues and Research Needs*, International Food Policy Research Institute, Washington, DC; (c) Ruel, M.T. (2003) Operationalizing dietary diversity: a review of measurement issues and research priorities. *J Nutr*, **133** (11 Suppl. 2), 3911S–3926S; (d) Ruel, M.T. (2003) *Diet Quality and Diet Changs of the Poor: A Global Research Program to Improve Dietary Quality, Health and Nutrition*, A proposal for a global research program (GRP24), IFPRI, Washington, DC; (e) Habicht, J.P. (2008) Malnutrition kills directly, not indirectly. *Lancet*, **371** (9626), 1749–1750; author reply 1750.

16 Quisumbing, A.R., Haddad, L., and Peña, C. (1995) *Gender and Poverty: New Evidence from 10 Developing Countries*, International Food Policy Research Institute (IFPRI), Washington, DC.

17 Diao, J. (2007) *The Role of Agriculture in Development: Implications for Sub-Saharan Africa*, International Food Policy Research Institute, Washington, DC.

18 (a) World Bank (2006) *Repositioning Nutrition As Central for Development*, World Bank, Washington, DC; (b) Haddad, L. (2002) *Reducing Child Undernutrition: How Far Does Income Growth Take Us?* IFPRI, Washington, DC; (c) Bliss, C. and Stern, N. (1978) Productivity, wages and nutrition: parts I and II. *J. Dev. Econ.*, **5**, 331–397.

19 Negin, J., Remans, R., Karuti, S., and Fanzo, J. (2009) Integrating a broader notion of food security and gender empowerment into the African Green Revolution. *Food Secur.*, **1** (3), 351–360.

20 Cassman, K.G., Dobermann, A., Walters, D.T., and Yang, H. (2003) Meeting cereal demand while protecting natural resources and improving environmental quality. *Annu. Rev. Environ. Resour.*, **28**, 315–358.

21 Alston, J.M., Beddow, J.M., and Pardey, P.G. (2009) Agricultural research, productivity, and food prices in the long run. *Science*, **325**, 1209–1210.

22 (a) Sanchez, P.A. (2002) Ecology. Soil fertility and hunger in Africa. *Science*,

295 (5562), 2019–2020; (b) Sanchez, P.A. and Swaminathan, M.S. (2005) Public health. Cutting world hunger in half. *Science*, **307** (5708), 357–359.

23 (a) Khush, G.S. (2001) Green revolution: the way forward. *Nat. Rev. Genet.*, **2** (10), 815–822; (b) Paddock, W. and Paddock, P. (1967) *Famine 1975!* Little, Brown and Co., Boston, MA.

24 (a) Evenson, R.E. and Gollin, D. (2003) Assessing the impact of the green revolution, 1960 to 2000. *Science*, **300** (5620), 758–762; (b) Swaminathan, M.S. (2000) An evergreen revolution. *Biologist (London)*, **47** (2), 85–89.

25 Borlaug, N. (1970) The Green Revolution: Peace and Humanity. A speech on the occasion of the awarding of the *Nobel Peace Prize*, 1970 Nobel Peace Prize, Oslo Norway, Oslo Norwy.

26 Smil, V. (2001) *Enriching the Earth: Fritz Haber, Carl Bosch and the Transformation of World Food Production*, MIT Press, Cambridge, MA.

27 (a) Smil, V. (2002) Nitrogen and food production: proteins for human diets. *Ambio*, **31** (2), 126–131; (b) Tilman, D., Fargione, J., Wolff, B., D'Antonio, C., Dobson, A., Howarth, R., Schindler, D., Schlesinger, W.H., Simberloff, D., and Swackhamer, D. (2001) Forecasting agriculturally driven global environmental change. *Science*, **292** (5515), 281–284.

28 Food and Agriculture Organization of the United Nations (2010) FAOSTAT: Agriculture production data. http:// faostat.fao.org/ (accessed 6 February 2010).

29 Von Braun, J., Ruel, M., and Gulati, A. (2008) *Accelerating Progress Towards Reducing Malnutrition in India*, International Food Policy Research Institute, Washington, DC.

30 Gruber, N. and Galloway, J.N. (2008) An earth-systems perspective of the global nitrogen cycle. *Nature*, **451**, 293–296.

31 Rockstrom, J., Steffen, W., Noone, K., Persson, A., Chapin, F.S., 3rd, Lambin, E.F., Lenton, T.M., Scheffer, M., Folke, C., Schellnhuber, H.J., Nykvist, B., de Wit, C.A., Hughes, T., van der Leeuw, S., Rodhe, H., Sorlin, S., Snyder, P.K., Costanza, R., Svedin, U., Falkenmark,

M., Karlberg, L., Corell, R.W., Fabry, V.J., Hansen, J., Walker, B., Liverman, D., Richardson, K., Crutzen, P., and Foley, J.A. (2009) A safe operating space for humanity. *Nature*, **461** (7263), 472–475.

32 Denning, G., Kabambe, P., Sanchez, P., Malik, A., Flor, R., Harawa, R., Nkhoma, P., Zamba, C., Banda, C., Magombo, C., Keating, M., Wangila, J., and Sachs, J. (2009) Input subsidies to improve smallholder maize productivity in Malawi: toward an African green revolution. *PLoS Biol.*, **7** (1), e23.

33 Sanchez, P.A., Denning, G.L., and Nziguheba, G. (2009) The African green revolution moves forward. *Food Secur.*, **1**, 37–44.

34 Fanzo, J., Pronyk, P., Dasgupta, A., Towle, M., Menon, V., Denning, G., Zycherman, A., Flor, R., and Roth, G. (2010) *An Evaluation of Progress Toward the Millennium Development Goal One Hunger Target: A Country-Level, Food and Nutrition Security Perspective*, United Nations Development Group, New York.

35 Graham, R. (2008) Micronutrient deficiencies in crops and their global significance, in *Micronutrient Deficiencies in Global Crop Production* (ed. B.J. Alloway), Springer, Heidelberg, pp. 41–62.

36 Eckhardt, N., Cominelli, E., Galbiati, M., and Tonelli, C. (2009) The future of science: food and water for life. Meeting report. *Plant Cell*, **21**, 368–372.

37 (a) Bradford, K.J., Van Deynze, A., Gutterson, N., Parrott, W., and Strauss, S.H. (2005) Regulating transgenic crops sensibly: lessons from plant breeding, biotechnology and genomics. *Nat. Biotechnol*, **23**, 439–444; (b) Kalaitzandonakes, N., Alston, J.M., and Bradford, K.J. (2007) Compliance costs for regulatory approval of new biotech crops. *Nat. Biotechnol.*, **25** (5), 509–511; (c) Qaim, M. (2009) The economics of genetically modified crops. *Annu. Rev. Resour. Econ.*, **1**, 665–694.

38 Dona, A. and Arvanitoyannis, I.S. (2009) Health risks of genetically modified foods. *Crit. Rev. Food Sci. Nutr.*, **49** (2), 164–175.

39 Stein, A.J. and Rodríguez-Cerezo, E. (2010) International trade and the global pipeline of new GM crops. *Nat. Biotechnol.*, **28**, 23–25.

40 James, C. (2009) Global Status of Commercialized Biotech/GM Crops: 2008. International Service for the Acquisition of Agri-biotech application (ISAAA).

41 Lemaux, P.G. (2009) Genetically engineered plants and foods: a scientist's analysis of the issues (part II). *Annu. Rev. Plant Biol.*, **60**, 511–559.

42 (a) Brookes, G. and Barfoot, P. (2005) *GM Crops: The Global Socioeconomic and Environmental Impact—The First Nine Years*, PG Econ, Dorchester, IA; (b) Brookes, G. and Barfoot, P. (2008) *GM Crops: Global Socioeconomic and Environmental Impacts 1996–2008*, PG Econ, Dorchester.

43 (a) Fernandez-Cornejo, J. and Caswell, M. (2006) *The First Decade of Genetically Engineered Crops in the United States*, USDA, Washington, DC; (b) Huang, J., Hu, R., Rozelle, S., and Pray, C. (2005) Insect-resistant GM rice in farmers' fields: assessing productivity and health effects in China. *Science*, **308** (5722), 688–690.

44 (a) Newell-McGoughlin, M. (2008) Nutritionally improved agricultural crops. *Plant Physiol.*, **147**, 939–953; (b) Ye, X., Al-Babili, S., Kloti, A., Zhang, J., Lucca, P., Beyer, P., and Potrykus, I. (2000) Engineering the provitamin A (beta-carotene) biosynthetic pathway into (carotenoid-free) rice endosperm. *Science*, **287** (5451), 303–305.

45 Phiri, N.A. and Otieno, G. (2008) *Managing Pests of Stored Maize in Kenya, Malawi and Tanzania*, MDG Centre ESA, Nairobi, Kenya.

46 Haines, C. (1991) *Insects and Arachnids of Tropical Stored Products: Their Biology and Identification*, Nature Resources Institute.

47 Nziguheba, G., Palm, C.A., Berhe, T., Denning, G., Dicko, A., Diouf, O., Diru, W., Flor, R., Frimpong, F., Harawa, R., Kaya, B., Manumbu, E., McArthur, J., Mutuo, P., Ndiaye, M., Niang, A., Nkhoma, P., Nyadzi, G., Sachs, J., Sullivan, C., Teklu, G., Tobe, L., and Sanchez, P.A. (2010) The African Green Revolution: results from the Millennium Villages Project. *Adv. Agron.*, **109**, 75–115.

48 Frison, E.A., Smith, I.F., Johns, T., Cherfas, J., and Eyzaguirre, P.B. (2006) Agricultural biodiversity, nutrition, and health: making a difference to hunger and nutrition in the developing world. *Food Nutr. Bull.*, **27** (2), 167–179.

49 Remans, R., Fanzo, J.C., Palm, C.A., and DeClerk, F. (2010) Ecology and human nutrition, in *Integrating Ecology into Poverty Alleviation and International Development Efforts: A Practical Guide* (eds F. De Clerck, J.C. Ingram, and C. Rumbaitis del Rio), Springer, New York, In Press.

50 Young, V.R. and Pellett, P.L. (1994) Plant proteins in relation to human protein and amino acid nutrition. *Am. J. Clin. Nutr.*, **59** (5 Suppl.), 1203S–1212S.

51 Reeds, P.J. and Garlick, P.J. (2003) Protein and amino acid requirements and the composition of complementary foods. *J. Nutr.*, **133** (9), 2953S–2961S.

52 (a) Welch, R. and Graham, R. (2002) Breeding crops for enhanced micronutrient content. *Plant Soil*, **245**, 205–214; (b) Welch, R.M. (2002) Breeding strategies for biofortified staple plant foods to reduce micronutrient malnutrition globally. *J. Nutr.*, **132** (3), 495S–499S.

53 Hotz, C. and Gibson, R.S. (2007) Traditional food-processing and preparation practices to enhance the bioavailability of micronutrients in plant-based diets. *J. Nutr.*, **137** (4), 1097–1100.

54 Erdman, J.W. and Pneros-Schneier, A. (1994) Factors affecting nutritive value in processed foods, in *Modern Nutrition in Health and Disease* (eds M.E. Shils, J. Olsen, and M. Shile), Lea & Febiger, Philadelphia, pp. 1569–1578.

55 Hurrell, R.F. (2004) Phytic acid degradation as a means of improving iron absorption. *Int. J. Vitam. Nutr. Res.*, **74** (6), 445–452.

56 Teucher, B., Olivares, M., and Cori, H. (2004) Enhancers of iron absorption: ascorbic acid and other organic acids. *Int. J. Vitam. Nutr. Res.*, **74** (6), 403–419.

57 Padmaja, G. (1995) Cyanide detoxification in cassava for food and feed uses. *Crit. Rev. Food Sci. Nutr.*, **35** (4), 299–339.

58 Mannar, M.G. and Sankar, R. (2004) Micronutrient fortification of foods– rationale, application and impact. *Indian J. Pediatr.*, **71** (11), 997–1002.

59 Berner, L., Clydesdale, F.M., and Douglass, J.S. (2001) Fortification contributed greatly to vitamin and mineral intakes in the US. *J. Nutr.*, **131**, 2177–2183.

60 Krause, V.M., Delisle, H., and Solomons, N.W. (1998) Fortified foods contribute one half of recommended vitamin A intake in poor urban Guatemalan toddlers. *J. Nutr.*, **128** (5), 860–864.

61 UNICEF (1990) *Strategy for Improved Nutrition of Children and Women in Developing Countries*, UNICEF, New York.

62 Ezeamama, A.E., McGarvey, S.T., Acosta, L.P., Zierler, S., Manalo, D.L., Wu, H.W., Kurtis, J.D., Mor, V., Olveda, R.M., and Friedman, J.F. (2008) The synergistic effect of concomitant schistosomiasis, hookworm, and trichuris infections on children's anemia burden. *PLoS Negl. Trop. Dis.*, **2** (6), e245.

63 (a) Hotez, P. (2008) Hookworm and poverty. *Ann. N. Y. Acad. Sci.*, **1136**, 38–44; (b) Hotez, P.J. (2009) Mass drug administration and integrated control for the world's high-prevalence neglected tropical diseases. *Clin. Pharmacol. Ther.*, **85** (6), 659–664.

64 World Health Organization (2008) Progress report on number of children treated with anthelminthic drugs: an update towards the 2010 global target. *Wkly Epidemiol. Rec.*, **82**, 237–252.

65 Sinuon, M., Tsuyuoka, R., Socheat, D., Montresor, A., and Palmer, K. (2005) Financial costs of deworming children in all primary schools in Cambodia. *Trans. R. Soc. Trop. Med. Hyg.*, **99** (9), 664–668.

66 Stephenson, L.S., Latham, M.C., and Ottesen, E.A. (2000) Malnutrition and parasitic helminth infections. *Parasitology*, **121** (Suppl.), S23–S38.

67 Iyere, P. (2008) Chemistry in sustainable development and global environment. *J. Chem. Educ.*, **85** (12), 1604–1606.

3
Poverty

Mari-Carmen Gomez-Cabrera, Cecilia Martínez-Costa, and Juan Sastre

3.1
Contribution of Chemistry to Social and Economic Development

Chemistry is the science which deals with the properties of matter and the changes which they undergo, and hence it is a partner in business and also in our daily life [1]. Most products of the chemical industry serve as intermediates between the basic raw materials provided by nature and the final consumer goods of our life [2]. Chemical products have become so common and necessary that they are "invisible" to us and we do not value the work which is behind them.

The chemical industry was born in the middle of the 18th century in England and France out of the demands, created by other industries, for alkalis and sulfuric acid [2]. These chemical products were used for glassmaking and soapmaking, for dyes for textiles, and for bleach for raw cloth. At that time many natural dyes were obtained from India, Mexico or South America. Indigo was extracted from the leaves of a vegetable from Coromandel (India), whereas carmine red was extracted from female cochineal found on cacti in Mexico and Peru [2]. Red extracts from tinted wood were known as brazilwood, from which the country Brazil took its name [2]. In the 19th century natural dyes were substituted by synthetic dyes.

It was not until the first half of the 19th century that a relationship was established between plant growth and soil ingredients such as nitrogen, phosphorus, and potassium [2]. Well before the Spanish conquest, the Incas used guano as fertilizer. Guano deposits were made of bird excrement and were located in islands close to Peru's coastline. The use of guano was expanded in England and the United States, and it was later replace by Chile saltpeter (sodium nitrate) to improve agricultural production. Since Chilean nitrate deposits are almost exhausted, chemists are needed to provide enough supplies of nitrogen for agriculture by using the inexhaustible supply of nitrogen from the atmosphere [1].

The early inorganic or mineral industry evolved from dyes and explosives into the modern fine chemicals industry and pharmaceutical products. During the second half of the 19th century, great progress was accomplished by the German chemical industry, especially in pharmaceuticals, which was the result of long-standing efforts in the training of scientists [2]. Acetylsalicylic acid, known

The Chemical Element: Chemistry's Contribution to Our Global Future, First Edition.
Edited by Javier Garcia-Martinez, Elena Serrano-Torregrosa
© 2011 Wiley-VCH Verlag GmbH & Co. KGaA. Published 2011 by Wiley-VCH Verlag GmbH & Co. KGaA.

worldwide as aspirin, and chemotherapeutic agents were synthesized and manufactured during this period. The great work of the French chemist Louis Pasteur in this century needs to be highlighted. He discovered that harmful microorganisms were destroyed by heating above 45 °C, a process later called pasteurization and used worldwide in the food industry. He also halted the fatal silkworm epidemic that ruined the silkworm breeders and silk manufactures in 1865 [1, 2]. His studies also saved the French wine industry from the destructive ravages of phylloxera and halted chicken cholera and anthrax. And, most importantly, he was the first to vaccinate a man bitten by a mad dog, saving his life and many others along the years.

Breakthroughs in chemistry throughout the 19th century and during the beginning of the 20th century made possible the development of photography, insecticides, plastics, artificial fibers, cellophane, paints, pigments, varnishes, perfumes, fragrances, as well as a new science, that is, *biochemistry* (see Box 3.1). Kirchhoff discovered the conversion of starch into glucose by dilute acids and this observation led to the treatment of thousands of bushels of corn daily in corn products plants [1]. Some time ago cottonseed was a nuisance and the chemist converted it into a perennial source of wealth and the raw material for edible oils, soap stock, and cattle feeds [1]. The discovery by Schonbein in 1845 that cotton on exposure to nitric acid was converted into a new and highly explosive product was far reaching because it led to von Lenk and Abel to guncotton; Viele, Nobel, Abel and Dewar to smokeless powder; Hyatt to celluloid; Goodwin to photographic films; and du Chardoneet to artificial silk; and was the basis for the manufacture of patent leather, artificial leather, lacquers, nitrocellulose film, and many other products of daily use [1]. The development of synthetically produced nitrate explosives significantly affected the life of nations, for instance by boosting major public works and mining programs [2].

Box 3.1 Breakthroughs in chemistry in the 19th century and the first half of the 20th century

- The French chemist, Louis Pasteur, discovered that harmful microorganisms were destroyed by heating above 45 °C, a process later called pasteurization and used worldwide in the food industry. He also was the first to vaccinate a man bitten by a mad dog, saving his life and laying the basis for vaccination.

- Kirchhoff discovered the conversion of starch into glucose and this observation led to the conversion of cottonseed into edible oils, soap stock, and cattle feeds [1].

- The discovery by Schonbein in 1845 that cotton on exposure to nitric acid was converted into an explosive later led to guncotton, smokeless powder, celluloid, photographic films, and artificial silk [1].

- Following the discovery of penicillin by the physician Alexander Fleming in 1928, the chemists Ernst Boris Chain and Howard Walter Florey purified this antibiotic which enabled its synthesis and wide distribution to the population.

The discovery of penicillin and its industrial development during World War II led the pharmaceuticals industry increasingly to resort to biosynthesis for the preparation of its active principles [2]. It also promoted the development of biotechnology to obtain a variety of drugs and biological compounds.

Chemistry has also contributed greatly to transportation. Railroads operate on steel rails which are cheap because of the Bessemer process of making steel. Dudley's laboratory standardized the railroad practice and his guidelines also covered soaps, disinfectants, oils for lubricating, paints, steel in special forms for every use, car wheels, cement, and signal cord [1]. Nowadays, good roads are made of cement or bonded with asphaltic compounds. The automobile is in great part a chemical creation because chemistry supplies the alloy steel, the aluminum, artificial leather, plates the nickel, vulcanizes the rubber, provides lacquers, pigments, paints, and even the gasoline.

Chemistry serves the manufacturers and our daily life in so many ways that is has been essential for social and economic development all over the world. It is needed for control of the quality of raw materials, the scientific control of production processes, the control of formulas, temperatures, pressures, time and spacing, fineness of materials, moisture content, and many other factors that affect the quality and amount of our daily output [1].

The chemical industry contributes greatly to the economic output in many countries. The global chemical output was $3000 billion in 2004, and it was concentrated in Western Europe, North America and Japan. Recently, the chemical industry has contributed greatly to economic growth in Eastern Asia. Indeed, it is the third largest industry in China and accounts for 10% of its gross domestic product (GDP) [3]. Chemical industry also plays a significant role in the economies of some developing countries, such as India and South Africa, where it contributes to 3% and 3.7%, respectively, of the country's GDP [4]. The Chemical Industry comprises the companies that produce pharmaceuticals, agrochemicals, liquid fuels, polymers, plastics, and rubber, and forms the backbone of the industrial and agricultural development.

In recent years, the global growth of chemical industry has been around 4%. It is noteworthy that China and India exhibited the fastest growing chemical sectors in the world, with 13.9% and 15.8% growth, respectively, in 2004. Until 1991 India had a closed economy with protection and a highly regulated market for the domestic chemical industry, whereas over the last 15 years it has evolved into a mature industry within a liberalized economy. Chemical industry employs an important amount of workforce in Europe and many countries such as the United States, Japan, Australia, China, India, and South Africa. In 2007, in Germany alone, the chemical industry had a highly trained workforce of 441 100.

Many of the global challenges, like poverty eradication as the first Millennium Development Goal, need to be addressed with the aid of chemistry. Some of the most important areas in which chemistry will play an integral role in the future include medicine, food, sustainable energy, mobility, and clothing [5].

3.2
Concept and Historical Evolution of Poverty

Concern about poverty has a long tradition and has been taken into account in different religions from ancient time. Yet no consensus exists on what is poverty or how it should be measured. The most learned perspective equates poverty with the inability to participate in society with dignity. According to the classical economist Adam Smith, poverty is a lack of those necessities that the custom of the country renders it indecent for creditable people, even of the lowest order, to be without. According to the Nobel laureate Amartya Sen, the poor cannot participate adequately in communal activities, or be free of public shame from failure to satisfy conventions.

Poverty is often defined in absolute or relative terms [6]. Absolute poverty is defined in reference to a poverty line that has a fixed purchasing power, determined to cover needs that are physically and socially essential [7]. However, the poverty line may be multidimensional, comprised of two measures: an income-related poverty line (e.g., $1 a day) [8] and non-monetary measures (for other needs). Because basic needs are likely to evolve over time, absolute poverty lines do not need to be the same across countries or across long time intervals in a given country.

Alternatively to this absolute definition of poverty, which is in use in a large number of countries, according to the relative definition of poverty the poverty line is defined not in terms of some well defined basic needs, but as a fixed proportion of the mean income of the population. For instance, the European Union officially considers those whose economic resources are below 50% of the mean income in Member Countries as poor. Fixing the poverty line relative to average income can show rising poverty, even when the standard of living of the poor has in fact risen. While there is an increasing consensus among economists that relative deprivation is important, there is no such agreement that individual welfare depends only on one's relative position, and not at all on one's absolute standard of living as determined by income.

Sen [9] offers an alternative definition of poverty as capability deprivation, the inability to spread "economic opportunities through an adequately supportive social background, including high levels of literacy, numeracy and basic education; good general health care; completed land reforms and so on." Poverty as minimal income is, of course, closely related to poverty as capability deprivation, because enhanced capabilities would tend, typically, to expand a person's ability to be more productive and earn a higher income.

Different models of poverty imply different indicators. Money metric models require information on income or consumption; vulnerability models use indicators of wealth and exposure to risk, as well as income; models concerned with capability and functioning present indicators of life expectancy or educational achievement; models of well-being or social exclusion will include measures like the degree of social support. It is frequent practice, however, to present a wider set of indicators than is immediately required. For example, World Bank poverty assessments, concentrating on money metric measures, will also provide evidence on health, education, physical isolation, and other so-called correlates of poverty [10].

Some indicators are inherently more quantifiable than others, and more decomposable, in the sense that they can be subjected to statistical manipulation. Thus, income and consumption poverty are conventionally measured using measures proposed by Foster, Greer and Thorbecke [11]. These so-called F-G-T measures enable a calculation to be made both of the headcount, that is, the number of people below the poverty line, and the poverty gap, or shortfall of the poor below the poverty line. The latter provides a measure of the resources required to eliminate poverty. F-G-T calculations are usually based on representative sample data of income or consumption, and related to a poverty line. This, in turn, is established with reference to minimum consumption (in the case of absolute poverty). Additional problems arise in setting an international poverty line, in fact the current convention of using $US 1 per day has been questioned and different authors have increased this value to $US 1.5 per day. When poverty is considered multidimensional, problems arise about the weighting of different components and whether composite indicators add value to conventional measures [10].

From an historic perspective, thinking about poverty can be traced back at least to the codification of the poor laws in medieval England, through to the pioneering empirical studies, at the turn of the century, by Booth in London and by Rowntree in York. Rowntree's study, published in 1901, was the first to develop a poverty standard for individual families, based on estimates of nutritional and other requirements.

In the 1960s, the main focus was on the level of income, reflected in macroeconomic indicators like Gross National Product per head. This was associated with an emphasis on growth, for example in the work of the Pearson Commission, Partners in Development (1969).

In the 1970s, poverty became prominent, notably as a result of Robert MacNamara's celebrated speech to the World Bank Board of Governors in Nairobi in 1973, and the subsequent publication of "Redistribution with Growth". Two other factors played a part. First, was emphasis on relative deprivation, inspired by work in the UK by Runciman and Townsend. Townsend, in particular, helped redefine poverty: not just as a failure to meet minimum nutrition or subsistence levels, but rather as a failure to keep up with the standards prevalent in a given society.

The second shift was to broaden the concept of income-poverty, to a wider set of "basic needs", including those provided socially. Thus, following ILO's pioneering work in the mid-1970s, poverty came to be defined not just as lack of income, but also as lack of access to health, education and other services. The concept of basic needs inspired policies like integrated rural development. Its influence continues to be seen in current debates about human development.

New layers of complexity were added in the 1980s. The principal innovations were: (i) The incorporation of non-monetary aspects, particularly as a result of Robert Chambers' work on powerlessness and isolation. This helped to inspire greater attention to participation. (ii) A new interest in vulnerability, and its counterpart, security, associated with better understanding of seasonality and of the impact of shocks, notably drought. This pointed to the importance of assets as buffers, and also to social relations (the moral economy, social capital). It led to new work on coping strategies. (iii) A broadening of the concept of poverty to a wider construct, livelihood. This was adopted by the Brundtland Commission on Sustainability and the Environment, which popularized the term sustainable livelihood. (iv) Theoretical work by Amartya Sen, who had earlier contributed the notion of food entitlement, or access, emphasized that income was only valuable in so far as it increased the "capabilities" of individuals and thereby permitted "functionings" in society. (v) Finally, the 1980s was characterized by a rapid increase in the study of gender. The debate moved from a focus on women alone, to wider gender relations. Policies followed to empower women and find ways to underpin autonomy, or agency.

The 1990s saw further development of the poverty concept. Inspired by Sen, UNDP developed the idea of human development: "the denial of opportunities and choices to lead a long, healthy, creative life and to enjoy a decent standard of living, freedom, dignity, self-esteem and the respect of others...".

3.3
Asymmetry of Poverty in the World

While developed countries face health problems related to over-consumption, such as obesity and metabolic syndrome – with prevalence more than 60% among people over 60 years of age [12], more than a billion people in developing countries suffer from hunger and from lack of access to basic needs. Among the 6.7 billion people in the world more than a billion live in extreme poverty and lack clean water, and 1.6 billion lack electricity [13]. In addition, the global economic crisis has pushed tens of millions of people into extreme poverty, reversing the progress in the fight against poverty. Thus, between 1990 and 2005 the number of people living on less $1.25 a day decreased from 1.8 billion to 1.4 billion, but 55–90 million more people were living in extreme poverty in 2009. In the least developed countries, the percentage of people in extreme poverty was 63% in 1990 and this percentage went down to 53% in 2005, although it may rise in 2009 due to the economic crisis. In developing regions, this percentage was 45.5% in 1990 and was significantly lower in 2005 (26.6%), but later it is expected to fall at a much slower pace than before the downturn [14]. In North America, Europe and Australia the percentage of people in extreme poverty is lower than 2%, and in most developed countries it is less than 0.5%.

Progress in poverty eradication can also be assessed by the poverty gap ratio, which reflects the depth of poverty and its incidence. This ratio is obtained by multiplying the percentage of people below the poverty line by the difference between the poverty line and their average consumption. In developing regions, the poverty gap ratio was 15.6% in 1990 and 8% in 2005. In the least developed countries it was 27.5% in 1990 and 20% in 2005. Therefore, the reduction in this index of poverty depth and incidence has been significant during the last decades.

It is worth noting that poverty exhibits a gender dimension because the majority of the poor are women. Households headed by women face more obstacles to equal income and employment opportunities. Women generally earn less than men, perform more unpaid work and have lower access to resources [15].

Labor productivity (output per person employed in thousand US dollars), which is a key measure of economic performance, was 11 for developing regions and 71 for developed regions in 2008 [14]. The employment-to-population ratio has not changed significantly in the world during the last two decades, but the percentage of employed people living below $1.25 per day has markedly decreased from 1991 to 2008 (43.3% vs. 18%). These latter results show that important progress has been achieved in the reduction of poverty. This improvement has been huge in Eastern Asia, where the employed people living in extreme poverty have been markedly reduced from 69.5% in 1991 to 9.3% in 2008. Although Asia – particularly China – and the Pacific have made great progress in regional poverty reduction, they still have the largest number of poor people, accounting for slightly less than 2/3 of the world's poor [15].

The region with highest percentage of people in extreme poverty is Sub-Saharan Africa, with 57% in 1990 and 51% in 2005 (Figure 3.1). In fact, the number of

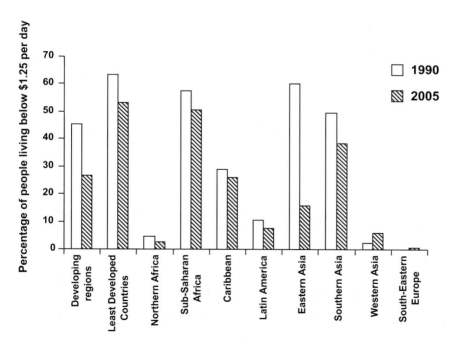

Figure 3.1 Percentage of people living in extreme poverty in developing countries. It shows the percentage of people living below $1.25 purchasing power parity per day. In the case of South-Eastern Europe, the percentage corresponds to transition countries in this European Region. Data from The Millennium Development Goals Report 2009, United Nations.

extremely poor people increased by 100 million in Sub-Saharan Africa between 1990 and 2005 [15]. Africa includes almost 30 of the least developed countries. Africa accounts for 13% of world population, produces 7% of the world's commercial energy, but only contributes 2% of the world's GDP [16]. In Africa, two thirds of the population does not have access to electricity. Africa is the lowest per capita consumer of modern energy of all regions of the world [16]. In East Africa, less than 3% of the rural population and 32% of its urban population is connected to electricity grids. In West Africa, only 12% of the 260 million inhabitants have access to electricity [17]. In both regions, under 10% of the rural population has access to electricity, and in some of the least-developed countries, such as Chad, this percentage falls to 1% or less.

The urban poor, and especially slum-dwellers, pay more for their cooking, water, and electricity than wealthier people connected to service networks. The very low incomes are not enough to procure energy services for basic needs such as cooking, transport, power pumps for potable water, sterile medical equipment, and heating. Severe social inequalities exist in many cities throughout developing countries, with around a billion impoverished inhabitants forced to subsist on heavily polluting sources such as wood and coal. The World Health Organization estimates that around 1.6 million deaths per year are associated with indoor air pollution; most of these are women and children from developing countries.

3.4
Causes of Poverty

Recently, it has been reported that all developing country regions have shown marked improvement in key indicators of poverty, health, economy, and food, except for sub-Saharan Africa [6] (see Figures 3.1 and 3.2). The percentage of extremely poor fell from 40% to 18%. However, in sub-Saharan Africa, the numbers almost doubled from 288 million (in 1980) to 516 million (in 2001), and the percentage stayed almost constant from 42% to 41% [8]. Regarding health, the life expectancy at birth in sub-Saharan Africa peaked in 1990 at 50 years but since then it has declined to 46 years, while steadily rising in all developing country regions to an average of 65 years [18]. Over the period 1960–2000, sub-Saharan Africa's per capita measure of annual economic growth (gross domestic product) was a mere 0.1%, whereas other developing country regions experienced accelerated growth averaging 3.6% [19]. Food production per capita grew by 2.3% per year between 1980 and 2000 in Asia, and by 0.9% in Latin America, but declined by 0.01% in tropical Africa [6]. Understanding African exceptionalism and contributing to its reduction is one of the grand challenges of sustainability science. Understanding the causes of poverty is essential to find strategies to fight against it. In this section we are going to focus on the main causes of poverty, especially taking into account the African continent as a reference.

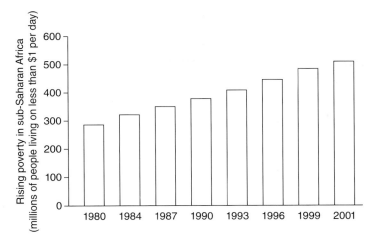

Figure 3.2 Rising poverty in sub-Saharan Africa. Millions of persons living on less than $1 per day.

3.4.1
Geopolitics

Africans and many people from other developing regions seem to be trapped in a vicious circle of little savings, leading to little capital investment, few skills, and truncated education [6], leading to conflict and maintaining high levels of corruption. The heritage of colonialism has been a favored explanation. It left little in the way of infrastructure, economy, health, and education, and much that would lead to future problems with conflict-laden borders and many small landlocked nations [6].

3.4.2
Geography

From a geographical point of view, the extensive aridity, environmental degradation [20], poor soils, distance to public resources [21], and endemic diseases, as well as the scattered populations in many small landlocked nations, are also considered relevant factors involved in the poverty of the African continent and other developing regions [6]. In rural areas where ~70% of the population live, poverty prevents farmers from self-financing or getting credit for needed farm inputs (fertilizer, improved seeds), and soils become depleted of nutrients after repeated crop cycles without sufficient replenishment.

3.4.3
Lack of Economic Growth

Different specialists argue that the cause of worldwide poverty is the lack of economic growth [19]. To explain the exceptional failures of certain economies to grow over the last quarter century, extensive sets of statistical regressions of national growth have been performed. Both physical and human features have been studied (resource scarcity and wealth, coastal and landlocked locations) and human (small national populations and ethnic diversity). In varying combinations, these features cause problems globally for many developing economies, but they are exceptionally prevalent in sub-Saharan Africa.

3.4.4
Deficient Governance

There is a widespread attribution of African exceptionalism to poor governance [22]. Although it has been considered for years that improved governance would lead to poverty reduction, this assumption has been questioned recently. Most political scientists believe the economy shapes governance institutions more than the reverse. More important is the finding that distinctive African socioeconomic and political conditions combine, so that poor people in Africa seek to meet their needs outside the "system" through an "economy of affection" and are less influenced either by the institutions of governance or by formal markets [23].

3.4.5
Deficient Health

Poverty and environmental degradation interact with a health crisis, particularly hunger, tuberculosis, malaria, AIDS, and neglected diseases such as amebiasis, Buruli ulcer, Chagas disease, Chlamydia infections, leprosy, leismaniashis, schistosomiasis, leptospirosis, and treponematosis. The significant increase in mortality rates in the poor countries generates a spiral of poverty. The resulting high child mortality blocks the demographic transition to low fertility rates. Rapid population growth and large families exacerbate poverty [6].

3.4.6
Failures of Effective and Sufficient Development Aid

The failure of international governance to provide sufficient public sector investment and aid and to make global markets accessible contributes as well. Mabogunje [24], in his urban poverty experiment, adopts the Sachs view of a poverty trap [25] but adds to the mix of governance considerations of the failure of African governments to enhance the capabilities of their own people and the desirable, but insufficient, efforts of civil society institutions to take up the slack. Poverty has been long seen as a non-issue, or as a transitory problem that would disappear with growth. However, it is a huge multifactorial problem influenced by geopolitics, governance, geography, health, and insufficient developmental aid.

3.5
Poverty, Malnutrition, and Life Expectancy

Poverty and the associated shortage in the access to food may lead to *protein-energy malnutrition (PEM)*, which is the major cause of mortality in children under five years of age. PEM can be the consequence of two clearly different situations: primary, exogenous or environmental PEM is the prevalent form in developing countries due to lack of sufficient amount and/or quality of food; secondary or endogenous PEM is caused by diseases that either impair nutritional and physiological processes, such as intake, absorption, or metabolism, or enhance nutritional requirements. The secondary form is prevalent in developed countries due to the survival achieved in most of these diseases.

Primary PEM is globally the prevalent form in the world, it is caused by poverty and is aggravated by the natural disasters and emergencies frequently associated with developing countries. According to WHO, the malnourished population lives mainly in Asia, Sub-Saharan Africa, and Latin America, and it is still frequent in countries suffering wars, and in refugees.

The circumstances that may lead to PEM can occur in children as well as in adults. However, PEM prevalence and severity are higher in children, especially during the first years of life, due to their higher growth rate and the corresponding high nutritional requirements.

When children suffer under nutrition, they pass first the stage of underweight, and, when it is aggravated, suffer from *severe wasting* (body mass index and weight for-height z-scores less than -3 SD applying World Health Organization (WHO) reference). This condition affects between 2 and 20% of children in developing countries (see Table 3.1). When this situation is long lasting or chronic, in addition to depleting their nutritional reserves, their growth is stopped, triggering the *stunting state*. This latter condition is still a great health problem reaching percentages between 15 and 59 in some countries (see Table 3.1).

According to the nutrients involved, under nutrition may evolve in two clinically well differentiated forms. If the deficit affects the global energetic supply *marasmus* will develop, whereas protein deficiency will lead to *kwashiorkor*, the latter with edema and intense immunodeficiency. Kwashiorkor is the most frequent form of malnutrition among the poor due to the lack of access to food that contains proteins of high biological value.

About 10 million children under five die each year [26] and most of them due to poverty related diseases. The major diseases responsible for the high *under-5 mortality* are neonatal disorders (mostly prematurity and low birth weight) followed by respiratory and digestive infections, malaria, measles, and AIDS. In all of them, poverty associated with undernutrition is the underlying cause in 50% of cases. Undernutrition refers to childhood malnutrition resulting in stunting and wasting, together with micronutrient deficiencies (iron, iodine, vitamin A and zinc).

In many developing countries the under-5 mortality rate is still very high and exhibits an inverse relationship to *life expectancy at birth*. Table 3.2 shows recent

Table 3.1 Prevalence of protein-energy malnutrition (PEM) in children. Sanitary facilities and educational level.

Country	Under-5 (2003–2008) suffering PEM (%)			% Population using improved drinking-water sources, 2008	% Population using improved sanitation facilities, 2008	% Youth (15–24 years) literacy rate 2003–2007
	Underweight	Wasting	Stunting			Male/female
Afghanistan	33	9	59	48	37	49/18
Angola	16	8	29	50	57	84/63
Australia	–	–	–	100	100	100/100
Bangladesh	41	17	43	80	53	71/73
Burundi	35	7	53	72	46	77/70
Colombia	5	2	15	92	74	98/98
Congo	11	8	30	71	30	99/98
Germany	–	–	–	100	100	100/100
Spain	–	–	–	100	100	100/100
Switzerland	–	–	–	100	100	100/100
Uganda	16	6	38	67	48	88/84
USA	–	–	–	100	100	100/100

Under-five mortality rate – Probability of dying between birth and exactly five years of age expressed per 1000 live births. Main data sources: (i) UNICEF, United Nations Population Division and United Nations Statistics Division (UNICEF Global database). (ii) World Health Statistics, 2008 and 2010. Geneva: World Health Organization (WHO), 2008 and 2010.

Literacy rate – Percentage of persons who can read and write. Data from: United Nations Educational, Scientific and Cultural Organization (UNESCO) and UNESCO/UIS (UNESCO Institute of Statistics), including the Education for All 2000 Assessment.

data from WHO on life expectancy at birth in developed countries versus developing countries, noting the disparity among them [27].

The following factors are involved in protein-energy malnutrition, mainly in childhood:

1. **Poverty, natural disasters, wars and political affairs.** Intake is generally insufficient as a consequence of extreme poverty. The low availability of food is aggravated by ravenous hunger, droughts, and natural disasters (floodings, hurricanes, earthquakes, etc.), political affairs, and wars.

2. **Deficient health-sanitary conditions.** The lack of potable water supply, together with the poor sanitary conditions and the immunological deficit caused by malnutrition favor the occurrence of frequent viral or bacterial infections, or parasite diseases. Table 3.1 outlines data on the percentage of the population using improved drinking-water sources and on the population using improved sanitation facilities [28]. It shows that in those countries with high under-5 mortality rate and malnutrition, there is lack of access to potable water and sanitary facilities in almost half of the population. These sanitary

Table 3.2 Life expectancy and mortality under 5 years.

Country	Life expectancy at birth (years)		Under-5 mortality rate (probability of dying age 5 per 1000 live birth)	
	1990	2008	1990	2008
Afghanistan	43	42	260	257
Angola	42	46	260	220
Australia	77	82	9	5
Bangladesh	54	65	149	54
Burundi	50	50	189	168
Colombia	69	75	35	20
Democratic Republic of the Congo	49	48	199	199
Germany	75	80	9	5
Romania	70	73	31	13
Spain	77	81	9	4
Switzerland	77	82	9	5
Uganda	47	52	186	134
USA	75	78	11	8

Life expectancy at birth – The number of years that newborn children would live if subject to the mortality risks prevailing.
Data from: World Health Statistics, 2010. Geneva: World Health Organization (WHO), 2010.

conditions trigger high levels of environmental contamination and exposure to the risks of worm infestation and microbial infection (such as hepatitis or cholera).

3. **Intrauterine growth retardation.** Women often become pregnant early in their life but with an impaired nutritional status exhibiting multiple deficiencies (iron, zinc, iodine, folate and vitamins A and B12) that often lead to intrauterine growth retardation and low birth weight. As a consequence of these deficiencies, infants exhibit more risk of cognitive and somatic deficits and are prone to infections.

4. **High nutritional requirements of the infant.** A child under two years of age has especially high dietary requirements of calories and specific micronutrients (iron, vitamins A and D), especially when pre-term and with low birth weight.

5. **Lack of breastfeeding or insufficient milk secretion.** Suboptimal breastfeeding is considered responsible for 35% of child deaths due to undernutrition. In most of these cases, breastfeeding is replaced by inadequate or poor nutrition.

6. **Infections.** The immune impairment of PEM, together with associated deficits of micronutrients, such as zinc, and the deficient sanitary facilities promote the development of frequent infections that worsen the nutritional status. Subsequently malnutrition aggravates the risk of infection, which in turn enhances malnutrition in a vicious circle.

When undernutrition is severe and occurs early in life, it may impair the development of the nervous system, which together with the social and educational environment negatively affect the learning and intellectual capabilities of the child. Indeed, the lack of education and training together with malnutrition are associated with less intellectual capabilities. Table 3.1 shows percentage of persons who can read and write (% youth literacy rate) in different developing and developed countries.

3.6
Strategies against Poverty: A General Approach with Context-Specific Solutions

Since the United Nations Millennium Summit in 2000, the Millennium Development Goals are the world's reference for dramatically reducing extreme poverty. The first Goal is to eradicate extreme poverty and hunger, and its first target to halve, between 1990 and 2015, the proportion of people whose income is less than $1 a day. This target is far from being achieved and much more effort is needed to meet the Goal in the near future.

The many dimensions of poverty require diverse and context-specific solutions [13]. Therefore, a general approach to consider all different aspects of poverty is needed together with a case-specific background. States, not-for-profit organizations, microfinance institutions, and others must help poor people to develop their human capital by promoting health, education, and skills [13]. We should take into account that the poor are not a homogeneous group; they exhibit different needs and objectives depending on the place where they live.

3.6.1
Renewable Energy Sources and Sustainable Development

Access to reliable, affordable, modern energy services is a pre-requisite for sustainable development and poverty reduction. Access to energy reduces hunger, improves access to safe drinking water through pumping facilities, and reduces disease and child mortality through refrigerating medicines, sterilizing medical equipment and providing transport to clinics. It also reduces the time spent on basic survival activities and increases affordable transport to look for employment and other economic opportunities, promoting socioeconomic development.

During the period 1970–1990, around 40 million people per year gained access to modern energy services, but the need for services is growing in the developing world. In fact, almost 100 million people should gain connection to modern energy systems each year to achieve universal access by 2020.

The energy issue is interdisciplinary because it involves chemists as well as engineers, physicists, and biologists. The ample and relevant contribution of chemistry to energy supplies covers the provision of fuels from crude oil, natural gas, coal, and biomass; the production and storage of hydrogen; the generation of energy from sunlight; the development of fuel cell technology, of new types of batteries; the provision of thermoelectric devices, of materials for collectors, and for superconductors, of luminescent materials for light-emitting diodes, and nano-porous foams [5].

The poor often suffer from a "poverty penalty" and pay more than rich consumers for essential products and services. People in the slums of Jakarta, Manila and Nairobi pay 5–10 times more for water than people in high-income areas of these cities and more than consumers in London or New York [13]. This "poverty penalty" is similar in electricity, health care, and credit.

Renewable energy supplies are continuously obtained from natural repetitive and persistent flows of energy occurring in the immediate environment [29]. Access to energy is essential for eradication of poverty. There is need for an effective commercial and industrial development policy, a favorable environment for energy end-users and an appropriate framework [16]. Two major challenges in this regard are to increase access of the poor to electricity and to deal with the high usage of traditional biomass fuels by the poor for cooking and heating, especially in Africa. Indeed, a major task in developing countries is the provision of electricity to rural communities far from a network grid.

Introductory stages of electricity supply have relied on government action and ownership. Market economies and liberalization require the involvement of competitive private companies to improve services and reduce energy tariffs [30]. However, the energy plans of a country rarely have a special focus on poverty reduction, and, far from reducing energy poverty, market-oriented power sector reforms in Africa may have even increased energy poverty in many sub-Saharan countries. Many of these reforms have increased the cost of electricity together with an associated reduction or removal of subsidies for the poor. Consequently, liberalization of utility services should be carried out to protect the supply at affordable tariffs to the very poor, which generally requires low-cost electrification options and continued provision of subsidies to the poor.

A successful economy depends on both supply and use of energy being safe and efficient [16]. Electrical power plants in developing countries are considered 20–40% less efficient than those of developed countries, with additional 20% transmission losses. Accordingly, more than half of the energy put to use in industrial countries is lost in the developing world. African oil and gas extraction plant continuously flare gas as waste, and this wasted energy is estimated as equivalent to twelve times the energy that the continent is using [31]. An example of good practice is the sugar cane industry in Africa, which is self-heated and self-electrified by waste bagasse combustion, and gives bio-ethanol as a by-product [16]. Another adequate strategy would be the single wire earth return systems, used in Australia, that may save between 30 and 50% of the network cost when compared with conventional systems. The energy efficiency of an incandescent light bulb is around 4% with a lifetime of 1000 h, whereas a compact fluorescent light with electronic starter exhibits 22% efficiency with a lifetime of 10 000 h [31].

In most low-income countries, most of the energy (up to 90%) is used for cooking and relies on firewood, alone or in combination with coal [16]. The need for a renewable energy supply and increased energy efficiency involves the utilization of hydroelectric, solar, wind, or modern biomass as local resources (see Box 3.2). Hydro, geothermal, fuel cell, and biomass (crops and waste) energy should be used on a large scale whereas solar water-heaters and wind generators would be for a small scale, especially for urban poor in developing countries. Alternatively, photovoltaic systems, mini hydroelectric stations and small-scale wind generators would be preferred to meet the needs of the rural poor in areas far from national electricity grids. A regional approach to energy planning provides joint infrastructure and resource development and sharing to reduce costs and reverse constraint factors, such as sub-critical grid and market capacities, limited institutional infrastructure and technical know-how, and work-force requirements [16].

3.6.2
Infrastructure, Science, and Technological Progress

Practical strategies to reduce poverty and promote development require investments in infrastructure, knowledge, and technological innovation. Policy actions

---

Box 3.2   Renewable energy and environmental sustainability

The need for renewable energy supply and increased energy efficiency involves the utilization of fuel cell technology together with hydroelectric, solar, wind, or modern biomass as local resources with acceptable emissions, avoiding fossil-fuel costs and contamination. Good examples of renewable sources of energy and environmental sustainability are:

- The sugar cane industry in Africa is self-heated and self-electrified by waste bagassse combustion, and gives bio-ethanol as by-product [16].

- Craftskills Enterprises from Nairobi manufactures and installs small-scale wind generators to run domestic lighting, fridges, televisions, and computers.

- In Brazil, the food processing company Sadia provides swine producers with biodigesters which use bacteria to ferment swine waste, producing crop fertilizers and food for fish (UNDP Overview 2008).

- In the Philippines, Cocotech transforms coconut husk waste into cocofiber nets to prevent soil erosion. Most of Cocotech's suppliers and operators come from the rural poor.

- In Mali, Électricité de France in collaboration with local and international partners, established rural energy services to generate energy through photovoltaic facilities and diesel generators [13].

should focus on platform technologies, infrastructure services, higher education in science and engineering, and on promoting business activities in science, technology, and innovation [32].

A major constraint for alleviation of poverty in developing countries is the absence of appropriate infrastructure services, that is, public utilities, public works, transport services, and telecommunication networks [32]. Thus, a challenge for local and regional policymakers is to identify and implement the infrastructure services needed to develop basic capabilities in science and technology in order to achieve the Millenium Development Goals. The creation and diffusion of technology, and hence employment, relies on the availability of infrastructure.

Development is largely an expression of local initiative and international partnership [32]. International partnerships are needed to make available the benefits of new technologies, especially information and communication. The private sector is a great untapped resource for investment and innovation to achieve the MDGs [13].

Science and technology are a critical component of development, and hence in poverty reduction [33]. Developing countries need to acquire global knowledge and to participate more actively in international trade to acquire technological capabilities and innovation, and for an efficient transfer of know-how. Responding to challenges in areas such as economic productivity, agriculture, education, health, water, environment, and global economics require increased use of scientific and technical knowledge. Technological progress makes the difference between fast-growing economies that have made great progress in reducing poverty, and slow-growing economies that have been much less successful [34]. Asia's green revolution is a good example of how modest technological advances may greatly contribute to poverty reduction (see Box 3.3). Technological progress may contribute in this regard, on the one hand by lowering costs, improving quality, creating new products, and helping reach new markets, and on the other hand by providing skills for getting employment [34]. Developing countries need to create and improve science and technology advisory institutions at the national and international levels. The role of higher education and training in science, engineering, and technology is crucial for development [35–37], (see Box 3.4).

Although protecting intellectual property rights is a critical aspect of technological innovation, intellectual property protection systems need to be revised to take into account the special needs of developing countries [32]. In this regard, the

---

Box 3.3 Asia's green revolution: simple technological advances may have a dramatic impact on poverty reduction (The World Bank, 2008)

- Between 1970 and 1995, better agricultural technologies doubled Asia's cereal production while increasing by only 4% the land area used for cultivation. The improvement was related to pesticides, irrigation, fertilizers, and the use of high-yield varieties of maize, wheat, and rice.

- By the late 1990s, the poor benefited from higher income, cheaper food, and more employment.

Box 3.4   The role of science and technology is crucial for development

- In the conference in Mexico in 2002 [35] on Science and Technology for sustainable development, it was proposed that there should be a new contract between science, technology, and development to allow a dialog between scientists and policymakers in order to promote research and produce technology to meet the priorities, needs and real problems of developing countries. The importance of transfer of technology from developed countries to developing countries with the aid of local scientists was also highlighted.

- In 2004, the Prime Minister of Canada [36] achieved the agreement to use 5% of Canada's research and development investment to address developing world challenges (Government of Canada, 2004).

- In 2010, the Spanish Ministry of Science and Innovation together with the European Community organized a Conference entitled "Science Against Poverty" (http://www.scienceagainstpoverty.es/) where they called for a more active role of scientific and technological research towards social integration, inclusion and the fight against poverty [37]. They highlighted the challenges which should be accomplished in this regard, such as a universal diffusion of scientific knowledge, to add the cooperative approach and the social impact assessment into the scientific programs, and to transfer technology to developing regions. They also underlined the need for chemistry for an adequate treatment of water from natural sources and for the reutilization of domestic waste water in cities.

Global Responsibility Licence (GRL) designed to allow licensing of intellectual property for humanitarian uses is very helpful. It unlocks intellectual property solely for humanitarian purposes that would otherwise infringe the patents, thus allowing the use of patented inventions to improve the conditions of people living in poverty and reducing the transaction costs of negotiating licenses. In addition, the Global Responsibility License Initiative aims at identifying existing technologies that can address critical problems among the very poor and provide practical tools in using these technologies to impact poverty.

In addition, international rule-making and standards-setting institutions, such as the World Trade Organization, the International Organization for Standardization, and the Bretton Woods Institutions, should revise these guidelines to better meet the needs of developing countries [32]. Finally, as pointed out by Foladori and Ivernizzi [38], we cannot forget that the efficiency and implications of the application of science and technology depend on the social context. Thus, organizing people committed towards a helpful goal may be as important as the technological progress to reduce poverty.

3.6.3
Microcredits and Inclusive Business Models

Microcredit is probably the first model to show that business with the poor may be successful [13]. Some years ago conventional banks refused the possibility of lending to the poor because they were not creditworthy. Muhammad Yumus founded the Grameen Bank as a socially-minded non-profit business and it has become a commercial business which provides valuable help to more than 7 million borrowers [13]. It is worth noting that one out of five borrowers moved out of poverty in around four years. At present it is well recognized that the major contribution of microfinance to poverty reduction is to promote income-generating employment. In developing countries there is high demand for saving opportunities among poor families and microenterprises [39]. Nevertheless, most poor people still have no access to microcredits and without them they have very limited opportunities for managing their financial resources [13].

The UN Advisors Group on Inclusive Financial Sectors was established in 2006 and engages in high level advocacy to support financial inclusion and microfinance management. The United Nations Capital Development Fund (UNCDF) supports least developed countries with local development programs according to priorities established by the beneficiaries themselves.

Inclusive business models include the poor not only as customers, but also as employees, producers, and business owners [13]. Inclusive business models may deal with the challenges of doing business where markets suffer from lack of information, infrastructure, and institutions. They are very important in the fight against poverty because they provide higher productivity, sustainable earnings, and greater empowerment for the poor, and enable them to escape poverty using their own means. Among the poor people, much business is informal and unregulated giving rise to fragmented markets.

According to the United Nations Development Programme [16], there is a need for scale economics and for a critical mass in institutions and markets. Larger regional infrastructures may ensure profitability providing better quality and lower cost. Moreover, combining small national markets into a regional market may create the critical mass needed to attract international investors.

According to Ted London, Director of the Base of the Pyramid Initiative at the University of Michigan, success of an integrative business development approach would require three tools: the leader method, immersion, and innovation workshops. The leader method was developed by Professor Eric von Hipper at the Massachussetts Institute of Technology, and consists basically in developing ideas about how to meet the need of users using or adapting existing products. Immersion requires integration into poor communities of a company representative or project facilitator, at least for a short period.

3.6.4
Health Promotion and Malnutrition Prevention

Poverty, together with its related diseases and chronic malnutrition, create a dreadful poverty-promoting vicious circle which needs to be stopped. Although the issues related to health and hunger have been extensively treated in other chapters of this book, this item provides some relevant examples and strategies for health promotion and malnutrition prevention which should be considered in the fight against poverty.

Nowadays more than 1 billion people, that is, approximately one sixth of the world's population, have no access to adequate water sources [13]. As an example, a recent study performed in a region of Guinea-Bissau revealed that 79% of the wells showed fecal contamination [25, 40]. Despite this pollution, the water from these wells could be suitable for domestic use after appropriate treatment. In several developing countries, including Haiti, Viet Nam and Pakistan, Procter& Gamble, in collaboration with nonprofit organizations, sell at low price point-of-use purification powder to provide safe and clean water. In South Africa Amanz'abantu provides free access to 25 liters of clean water a day for each person and access to extra water at a low cost to periurban and rural poor in the Eastern Cape, where a quarter of people lack potable water [13]. In India, entrepreneur Bindheshwar Pathak offers clean and cheap sanitation systems to 1.2 million households. In Pakistan, UNDP, in collaboration with food corporations, developed a program to provide hands-on training in livestock health management to rural women [39]. They were given toolkits with medical instruments, medicines, and vaccines. They not only raised significantly household income but also increased health conditions and food security. Moreover, over 2000 of those women became self-employed.

At present, HIV/AIDS and poverty-related diseases are at the center of poverty reduction and national development strategies [39]. In the Philippines, RiteMed, the generics division of pharmaceutical Unilab, sells generic drugs to more than 20 million low-income clients at prices 20–75% lower than leading brands [13]. In Kenya, Advanced Bio Extracts produces low-cost pharmaceutical grade artemisin and artemisin-based derivatives to treat malaria, and to get the drugs the company purchases from local farmers, who earn more from the crop of *Artemisia annua* than they would do from maize [13].

Among the numerous cooperative activities considered in the Conference "Science Against Poverty" that took place in Spain in 2010, the activities of the Innovations in International Health Programme, which has developed vaccines to inhale and disease control systems by radiofrequency, and the Napo Network in the Amazon developed by the EHAS Foundation to provide health aid in isolated rural areas should be underlined [37]. In addition, they highlighted the critical problem concerning the access to water at present and in the future, because it is estimated that in 2025 around 3000 million people will not have access to water. In this regard, chemistry is needed for an adequate treatment of water from natural sources and for the reutilization of domestic waste water in cities.

One of the major advantages of developing countries is their capability of manufacturing low-cost products [41] (see Box 3.5).

The core health indicators from WHO show that many developing countries still exhibit high rates of chronic malnutrition (>30%). Maternal nutrition and health is a prime determinant of health, learning abilities, skills, and future employment opportunities of the child [42]. Food aid from developed countries is only considered in emergency situations, such as wars or tsunamis, but not as

Box 3.5   Low-cost products: a major advantage of developing countries

---

- India is currently the world's leading manufacturer of diphtheria-pertussis-tetanus vaccine, Brazil of yellow fever vaccine, China of penicillin, and Cuba of meningitis B vaccine.

- At present, around 70% of India's and Brazil's drug exports go to other developing countries [41].

---

post-emergency. The aid for development should also include food aid in those cases of children and pregnant women with chronic malnutrition [43]. A significant reduction in the costs of the World Food Programme (WFP) would help in this regard, since its activities are extremely expensive in comparison with NGOs that work on similar goals on a small scale.

On the other hand, to avoid increases in the prices of basic food, the FAO should consider buying low-cost arable lands for agriculture production, to provide food for those regions where children are suffering from hunger and chronic malnutrition [43]. Moreover, the excessive agricultural productivity in Western countries often leads to a surplus of non-perishable food, which could be sent to the World Food Programme if they worked at much lower prices. It is wrong to limit the agricultural productivity when there are millions of people dying of hunger in other countries.

3.6.5
Involvement of the Local Government: The Ijebu-Ode Experiment

The Ijebu-Ode experiment is a project on poverty reduction that began in 1998 in the Nigerian city of Ijebu-Ode, with an estimated population of 163 000 inhabitants in 1999 [24]. At the beginning of the experiment, and excluding the economic help from relatives abroad, more than 90% of the population lived below the extreme poverty line of $1 per person per day. The major aim of the experiment was to test whether poverty can be dramatically reduced by mobilizing the entire community along with its diaspora through a city consultation process. City consultation is a process of civic engagement in which all stakeholders in a city and the official authorities (local, state, and federal government) are brought together to share knowledge and experiences in order to seek consensus in the design and implementation of poverty reduction programs [24]. The city consultation included the creation of working groups that examined the four dimensions of poverty, that is, socioeconomic, natural resources, governance/infrastructural, and human/cultural. These working groups identified enterprises that could provide opportunities for capability development, employment, and income generation. To guarantee proper implementation and sustainability, the Ijebu-Ode Development Board for Poverty Reduction was created.

After seven years, poverty in the city was reduced significantly, mainly through the microfinancing of existing and new productive activities and the creation of more than 8000 jobs [24]. These jobs were generated through the establishment of cooperatives and the development of new enterprises in specialty crops, small animal, and fish production.

Microfinancing was also essential for poverty reduction. For minimizing risks of loans default, stakeholders formed self-selected cooperatives and the loans were cross-guaranteed by the cooperative group. The default of one member of the group in the repayment of the loan was the default of the entire group. The Ijebu-Ode experiment demonstrates that poverty reduction is achieved when the whole community is mobilized, together with civil society organizations and the local and national government, to pursue a well articulated poverty reduction strategy. It also shows that poverty reduction should be based on enhancing the knowledge, skills, and capability of the people. This was achieved through training workshops conducted by scientists as well as practitioners versed in local production experience, technical skills, and accumulated indigenous knowledge [24].

3.6.6
UN, CSOs, and Governments from Developed Countries: a Joint Crucial Effort

United Nations efforts are central in the fight against poverty and are mostly coordinated by the United Nations Development Programme (UNDP). UNDP constitutes the UN's global development network and hence it is mainly engaged in poverty eradication. UNDP's work on the MDGs focuses on coordinating global and local efforts to achieve the MDGs through advocacy, monitoring and reporting progress, and supporting governments. Many of UNDP's reduction programmes are a direct result of needs identified by local communities [39]. Through regional Energy for Poverty Reduction programmes, UNDP provides capacity development support as well as technical and financial assistance for the implementation of policies on energy access to the poor. UN Secretary General Ban Ki-moon outlined that the current financial crisis cannot be allowed to deflect attention from the basic injustices in our world represented by the MDGs.

Many nongovernmental civil society organizations (CVOs) help communities to escape poverty. The CVOs aid was especially prevalent during the last two decades of the last century, when the role of the state in social welfare and development was markedly reduced, leading to the expansion of poverty [44]. NGOs (or CSOs) usually serve as a bridge between UN organizations and local populations. It is only when local civic organizations together with business interests participate in designing and implementing sustainable development programs that they have a chance of success.

CVOs can be divided into three types, north-based nongovernmental organizations (NGOs), Africa-based NGOs, and community-based organizations [45]. North-based NGOs, such as Oxfam, provide aid against extreme poverty and in response to natural disasters, and individual and family tragedies. African-based NGOs usually depend for their funding on partnership through north-based NGOs. Community-based organizations are often rooted in traditional groupings of various social organizations. In India there are several excellent CVOs, such as Missionary Sisters and Brothers of Charity and the Vicente Ferrer Foundation, which may serve as a reference throughout the world. CVOs alleviate poverty through credit provisioning, support for microenterprises, low-income housing and slum upgrading, community development, education, health care, preservation of the environment, family planning, and services for mothers and children [46]. Frequently they provide this aid in circumstances in which the local government fails to give the service expected or does not include these actions among the solidarity programs. Although being efficient and helpful, in some cases their efforts may be sporadic, disperse, or of short duration. Hence, in the long-term it is crucial to join and coordinate the actions of all actors – UN, CSOs, as well as local government and governments from developed countries – in the fight against poverty. In this regard, costs should mainly focus directly on the actions in developing countries rather than on the planning and assessment of the results, which have been already extensively studied during the last decades. Sometimes, the expenses corresponding to UN activities, such as those of the World Food Programme, are very expensive and abrogate the initiation of a project or hinder the cost-efficiency of a project. UN costs should be reduced to achieve the maximum cost-efficiency in the development of strategies to eradicate poverty.

3.6.7
Additional Efforts towards Eradication of Poverty

If we really want to eradicate poverty in the long term it is necessary to have peace all over the world. Eradication of wars is needed because they promote poverty, directly and indirectly. Conflict and persecution force the people to leave their homes and without employment, permanent residence, and access to social safety nets they quickly fall into poverty. In fact, the number of internally displaced persons worldwide has increased from 21 million to 26 million during the last decade and the number of refugees has remained at 15 million [14]. Wars also promote poverty indirectly because they dramatically increase the budget of governments on army acquisition, lowering the investment on development goals. Therefore, we propose to add "peace" to the millennium development goals (MGDs) and to consider it as the 9th MDG, as was first suggested by the Afghanistan's Millennium Development Goals Report 2005.

Another strategy which should be considered in the fight against poverty is birth rate control. This is a question of responsibility taking into account that basic services cannot be provided yet to all children living in developing countries. Birth rates in developing countries are very high (more than 40 per 1000 persons) in comparison with developed countries (less than 12 per 1000 persons) and, as a consequence, the number of people living in extreme poverty may even increase in the following decades. As an example, the per capita consumption of electricity has declined in some of these countries because population growth rates have been higher than the increase in electricity production rates [47].

Finally, all efforts towards eradication of poverty are based on the absence of corruption. Thus, eradication of corruption from any country – developing or developed – is required for safe investment and for achieving the MDGs.

3.7
Chemistry is Essential for Poverty Alleviation

Chemistry plays a key role in the development and applications of new scientific areas, such as those of nanoscience, which will decisively contribute to the eradication of poverty. Chemistry also strongly helps development through the supply of biofuels and the applications of combinatorial chemistry.

3.7.1
Nanotechnology and Nanochemistry

Nanotechnology is the study, design, creation, synthesis, manipulation, and application of functional materials, devices, and systems through control of matter at the nanometer scale (1–100 nanometers), that is, at the atomic and molecular levels [48]. The term nanotechnology was first used by Professor N. Taniguchi in 1974 and it was greatly developed during the last decades of the last century. Nanotechnology is based on nanoscience, which studies the structure, properties and manipulation of materials at the nanometer scale. Although nanoscience is multidisciplinary, chemistry, particularly nanochemistry, plays a central role in this new science. Nanochemistry is a new discipline within chemistry concerned with the unique properties associated with the reactions and assemblies of atoms or molecules on the nanometer scale.

Since the 1980s, nanotechnology has developed from a basic research field to one of the most promising reseach fields with high impact in chemistry, physics, and biology [49]. The global market for nanotechnology is forecasted to lead possibly to up to 2 million jobs globally during the next years [49]. Its applications range from manufacturing over life sciences to traditional industries like electronics or textiles.

Nanotechnology applications may be used for water treatment, energy storage, production and conversion, disease diagnosis, drug delivery systems, food processing and storage, and agricultural productivity, which are critical for developing countries to eradicate poverty. Nanotechnology, together with emerging technologies such as biotechnology and information technology, will contribute decisively to technological progress and economic growth in developing countries. Nanotechnological products require little labor and maintenance, are relatively inexpensive and highly productive.

Several successful nanotechnological projects are already going on in developing countries. In this regard, high-level projects on pharmacological research, clinical trials, and patent applications of nanotechnology bone scaffold, nanoparticle drug delivery, or carbon nanotube field emission display are underway in China, India, and the Republic of Korea. Several centers for research on nanoscience and nanothechnology in Brazil, Chile, the Philippines, South Africa, and Thailand are also contributing significantly to the field [32]. Research groups with interest in the area and great potential are also present in Argentina and Mexico.

More than 50 000 patent references on nanotechnology were registered overall from 1985 to 2007, and among them 12 979 were Chinese and 2901 German [49]. The above-average growth rate in nanotechnology patents for China is a consequence of the great efforts undertaken at the end of the 1970s by the Chinese government – especially by Deng Xiaoping's Open Door Policy – to decentralize research, development, and engineering [49]. At the end of the 1970s, China lagged far behind industrial nations in economic and technological development, but with the Open Door Policy they realized the necessity of technological knowledge and foreign investments [49]. Universities and research institutions became more autonomic in order to achieve international competitive research findings, and market-based resource allocation mechanisms were implemented together with a regulatory framework for private-owned corporations and spin-offs from universitites.

Salamanca-Buentello *et al.* [48] identified the top ten applications of nanotechnology most likely to benefit developing countries (see Box 3.6) Nanotechnology offers great opportunities to develop renewable energy sources through solar cells, fuel cells, and hydrogen storage systems based on nanostructure materials [32]. Nanophotovoltaic devices based on quantum dots or ultrathin films of semiconducting polymers would reduce the costs of conventional solar cells. Carbon nanotubes are versatile because they can be used in composite film coatings for solar cells, in lightweight materials for tanks and liquid hydrogen vessels in hydrogen storage systems, and in flexible conduits for electricity distribution networks.

Photo- and thermo-chemical nanocatalysts can generate hydrogen from water at relatively low cost. Hydrogen fuel cells are expected to provide electricity to rural, remote, and poor urban communities, apart from being essential for transportation in sustainable development [32]. Indeed, future cars will use compressed hydrogen gas as fuel to achieve less than half energy consumption and they will not give on-road emissions with the exception of water vapor.

One of the major contributions of nanotechnology should be through affordable agricultural applications related to increased soil fertility and crop production. Nanosensors may efficiently monitor the presence of pathogen agents. Nanoporous materials, such as zeolites, may contain absorbed or adsorbed compounds used for slow release and efficient dosage of fertilizers, drugs, and nutrients. A special strain of rice with shorter stems and resistant to sunlight has been developed by nanotechnology at the Chiang Mai University in Thailand. This rice strain is more resistant to generally inadequate climate conditions and requires reduced storage cost.

Nanotechnology can also contribute to health promotion by developing accurate and affordable methods for diagnosis, drug and vaccine delivery, surgical devices and prosthetics [32]. Thus, relatively low cost nanosensors, including microfluidic devices, carbon nanotube-based biosensor arrays, fluorescent semiconductor nanoparticles, magnetic nanoparticles, and quantum dots, have been designed for diagnostic kits. Of special relevance may be dendrimers and nanomaterials, such as atomic wires and nanobelts, for early detection of cancer. Nano-encapsulation

Box 3.6   Top ten applications of nanotechnology for developing countries [48]

Applications	Deliverables
Energy storage, production, and conversion	Novel hydrogen storage systems with carbon nanotubes
	Solar cells with carbon nanotubes
	Photovoltaic cells and organic light emitting devices with quantum dots
	Nanocatalysts for hydrogen generation
Agricultural productivity	Nanoporous zeolites for slow release of water and fertilizers
	Nanocapsules for herbicide delivery
	Nanosensors for soil quality and plant health monitoring
	Nanomagnets for removal of soil contaminants
Water purification and desalination	Nanomembranes, nanoporous zeolites, nanoporous polymers, attapulgite clays, magnetic nanoparticles, TiO_2 nanoparticles
	Nanosensors for detection of contaminants and pathogens
Disease diagnosis	Lab-on-a-chip, nanosensor arrays, quantum dots
	Magnetic nanoparticles, nanowire and nanobelt nanosensors
	Antibody-dendrimer conjugates
Drug delivery	Nanocapsules, dendrimers, buckyballs, nanobiomagnets, attapulgite clays
Food processing and storage	Nanocomposites for plastic film coating
	Antimicrobial nanoemulsions for decontamination
	Antigen-detecting biosensors
Air pollution and Remediation	Photocatalytic degradation of air pollutants by TiO_2 nanoparticles
	Nanocatalysts, nanosensors, and gas separation nanodevices
Construction	Nanomolecular structures to improve asphalt and concrete
	Nanomaterials for cheaper and durable housing
Health monitoring	Nanotubes and nanoparticles as sensors
Vector and pest detection and control	Nanosensors for pest detection
	Nanoparticles for pesticides, insecticides, and insect repellents

systems allow slow delivery of drugs. Polymers for long-term delivery of drugs would be especially useful for the treatment of tuberculosis. Nanotechnology can also reduce transportation and storage costs by improving shelf-life, thermo-stability and resistance to changes in humidity. Nanocapsules, liposomes, den-drimers, and buckyballs also provide selective delivery of drugs. On the other hand, nanoceramics can be used to obtain long-lasting medical prosthetics.

Poverty-associated diseases are responsible for the deaths of more than 2 million children each year [32]. Most of these deaths are caused by diarrhea, cholera, or typhoid due to inadequate water supply. Indeed, in some cities of developing countries only 10% of sewage is treated. Nanomembranes and nanoclays are affordable and easily transportable systems to purify, detoxify, and desalinate water. In addition, they are cleaned easily in comparison with conventional bacterial filters, which should be changed often. A collaboration between Bandaras Hindu University in Varanasi, India, and the Rensselaer Polytechnic Institute of the United States, has yielded very useful filters made of carbon nanotubes.

Nanosensors can also be used for quality control of water and food to detect contaminants and pathogens. In addition, nanoelectrocatalysts may break down organic pollutants and also serve to remove salts and heavy metals from polluted water which may then be used for irrigation, sanitation, or even for drinking [32]. In this regard, zeolites and nanoporous polymers may also be used to purify water and absorb metals; attalpugite clays can remove heavy metals, oils, organic mate-rial, and bacteria from water; magnetic nanoparticles adsorb metals and organic compounds; titanium dioxide and iron nanoparticles catalytically degrade pollut-ants. Titanium dioxide nanoparticles have also been used in self-cleaning coatings to photocatalyze air pollutants and diminish fossil fuel emissions. Semiconducting nanospheres with organic light-emitting devices may also provide lighting in rural areas. Engineered matrixes with energy transduction proteins may also produce affordable electricity.

Housing should be affordable to low-income populations, however, the con-struction industry in developing countries faces the problem of material shortage, aggravated by rising prices and by the world economic crisis. As a consequence, housing costs are too expensive and even continue to increase and unplanned urban settlements expand. The use of innovative composite materials from local resources, such as forestry, agriculture, natural fibers, plant materials, and indus-trial wastes may be cost-effective building materials. Although this alternative may be of help, nanochemistry and nanotechnology may reduce the cost also, providing the innovative building materials and construction technology needed to achieve the general access of the poor to appropriate housing.

The supply of low-cost building materials can be of critical importance in build-ing houses and schools for those suffering from extreme poverty. Ceramic com-posites, ceramic coatings, ceramic films, and glass ceramics, are valuable materials for industrial development [32]. Solid-state ionic materials are presently used as batteries, fuel cells and sensors and provide a great potential for economic growth in developing countries.

Box 3.7   Industrial biotechnology and biofuels

Industrial biotechnology, also known as white biotechnology, is able to produce bio-based chemicals (such as vitamins, amino acids, or enzymes for textile finishing and the detergent industry), biomaterials (such as biodegradable plastics) and biofuels from biomass out of agricultural products [50]. Industrial biotechnology will allow the eco-efficient use of renewable resources as industrial raw materials and, consequently, will enable manufacturing of products in an economically and environmentally sustainable way.

- Rural bio-refineries might replace port-based oil refineries wherever it is economical feasible [5].

- Biomass can be converted into liquid and gas fuels and to electricity by using digesters together with improved and more energy-efficient cookstoves.

- Biodiesel for diesel compression engines may be obtained from oil-producing plants, such as palm oil, jatropha, sunflower, canola, or coconut.

3.7.2
Industrial Biotechnology and Biofuels

Industrial biotechnology, also known as white biotechnology, is able to produce bio-based chemicals, biomaterials and biofuels from biomass out of agricultural products [50] (see Box 3.7). Biotechnological syntheses are going to outpace many classical chemical synthetic routes for established chemicals by higher cost-efficiency, saving energy resources and offering benefits towards sustainability by lowering or even avoiding greenhouse gas emissions [50].

At present, oil, coal, and natural gas resources provide around 90% of all world commercial energy requirements, whereas nuclear energy and large-scale hydro-electric power together account for most of the remaining 10%. Unfortunately, alternative technologies combined (small hydro, geothermal, wind, solar, tidal) provide less than 1% of the world's commercial energy. Consequently, currently, renewable sources of energy play only a limited role globally, but soon they may be very useful locally and certainly this distribution has to change in the future if we really want sustainable development and poverty reduction. The World Bank estimates that alternative energy systems will provide 20% of the world's energy by the year 2100.

In Africa, biomass energy accounts for almost 50% of the total primary energy supply, whereas renewable energy sources only account for 0.6% of total electricity production [16]. Presently, the best way to satisfy the energy needs of most of the urban population in developing countries is considered to be the use of modern biomass and domestic fuels [16]. Modern devices range from small-scale domestic boilers to multi-megawatt size power plants. Accordingly, in developing countries, biomass energy is likely to be an important global energy source during this century.

Fossil fuels are not ultimately sustainable, are often expensive, and their emissions must be reduced considering the climate change. The world's oil production is likely to reach its maximum during the following two decades which will lead to volatile uncontrolled prices. Our efforts in the future should be centered on biofuels as an alternative. Thus, liquid biomass (bioethanol and biodiesel) can be used to power cars. The fuel crisis during the 1970s prompted the Brazilian government to reduce its dependence on oil by manufacturing ethanol-only cars, which reached almost 80% of cars produced in Brazil during the 1980s. This initiative promoted sugarcane cultivation and its conversion into bioethanol, which increased agricultural employment and income generation. At present, three quarters of Brazil's new cars can burn either bioethanol or gasoline because flex-fuel engines are relatively inexpensive and avoid the dependence on conventional fuels.

In 2007, the South African Cabinet started a strategy for mandatory biofuel components of petroleum and diesel. These mandatory components were bioethanol (8% in petroleum) and biodiesel (2% in diesel). Bioethanol is obtained from fermentation of molasses, a by-product of the sugar cane industry, and from fermentation of yellow maize [16]. The benefits of this government strategy are the creation of more than 50 000 new jobs, especially in rural areas, the reduction in national unemployment, and the reduction in imported oil. Recently, in Mali was inaugurated a village of 3000 inhabitants electrified with a diesel engine that uses jatropha oil to drive a generator able to provide 10 hours of electricity daily [16].

The exploitation of energy from wastes would reduce by more than 60% the amount of urban wastes and would also markedly diminish the costs of waste disposal, especially in developing countries [16].

Other uses of industrial biotechnology, such as biocatalytic conversions of fine chemicals and the manufacture of high-value products, such as nutraceuticals, and cosmeceuticals offer growth opportunities for chemical industries [50].

3.7.3
Combinatorial Chemistry

Combinatorial chemistry was first developed in the 1980s and has greatly contributed to the discovery and development of new drugs, such as new HIV protease inhibitors, antimicrobial agents, and opiate receptor ligands [51]. Hence, it may be very useful for treatment of AIDS, malaria, and other poverty-related diseases. As mentioned previously, child mortality is largely caused by pneumonia, diarrhea, and malaria, which are acquiring resistance to available treatments. Combinatorial chemistry may provide new or more effective and affordable drugs for these diseases [32]. Combinatorial synthesis is a means of producing thousands and even millions of novel structures in the time that it would take to synthesize a few dozens by conventional means [51]. It offers automated techniques to obtain a great number of different chemical agents at a low cost. Furthermore, robots may be in charge of the preparation and screening of these compounds to identify the most active compounds.

References

1 Little, A.D. (2009) The place of chemistry in business. *J. Bus. Chem.*, **6** (2), 57–63.
2 Aftalion, F. (2001) *A History of the International Chemical Industry: from the Early Days to 2000*, Chemical Heritage Press, Philadelphia, PA.
3 Aruvian's Research (2010) Chemical Industry in China http:// www.researchandmarkets.com/reports/ c62143 and http://www.redorbit.com/ news/science/995226/the_chemical_ industry_in_china_accounts_for_10_of_ the/index.html
4 CYGNUS Business Consulting and Research (2007) Indian Chemical Industry. http:// www.researchandmarkets.com/ reportinfo.asp?report_id=452577
5 Müllen, K. (2009) Chemists of the future. *J. Bus. Chem.*, **6** (2), 65–67.
6 Kates, R.W. and Dasgupta, P. (2007) African poverty: a grand challenge for sustainability science. *Proc. Natl. Acad. Sci. U. S. A.*, **104**, 16747–16750.
7 Agence Française de Développement (2003) Poverty, Inequality and Growth Proceedings of the AFD-EUDN Conference. 1–344.
8 Chen, S. and Ravallion, M. (2007) Absolute poverty measures for the developing world, 1981–2004. *Proc. Natl. Acad. Sci. U. S. A.*, **104**, 16757–16762.
9 Sen, A. (1999) *Development as Freedom*, Oxford University Press, Oxford.
10 Maxwell, S. (1999) *The Meaning and Measurement of Poverty*, ODI Poverty Briefing, Overseas Development Institute, London.
11 Foster, J. (1981) *A Class of Decomposable Poverty Measures*, Cornell University
12 Hansen, B.C. and Bray, G.A. (2008) *The Metabolic Syndrome: Epidemiology, Clinical Treatment, and Underlying Mechanisms*, Humana, Totowa, NJ.
13 UNDP (2008) *Business with the Poor – Creating Value for All*, United Nations Development Programme. New York.
14 UN (2009) *The Millennium Development Goals*, United Nations, New York.
15 United Nations (1990–2005) *Progress Towards the Millennium Development Goals*, Statistic Division, Department of Economic and Social Affairs, New York.
16 UN-Energy/Africa (2007) *Energy for Sustainable Development; Policy Options for Africa*, UN-Energy/Africa.
17 ECOWAS (2006) *White Paper for a Regional Policy*, Economic Community of West African States, Abuja, Nigeria.
18 Jamison, D.T. (2006) *Disease Control Priorities in Developing Countries*, Oxford University Press, New York; World Bank, Washington, DC.
19 Collier, P. (2007) Poverty reduction in Africa. *Proc. Natl Acad. Sci. U. S. A.*, **104**, 16763–16768.
20 Sanchez, P., Palm, C., Sachs, J., Denning, G., Flor, R., Harawa, R., Jama, B., Kiflemariam, T., Konecky, B., Kozar, R., Lelerai, E., Malik, A., Modi, V., Mutuo, P., Niang, A., Okoth, H., Place, F., Sachs, S.E., Said, A., Siriri, D., Teklehaimanot, A., Wang, K., Wangila, J., and Zamba, C. (2007) The African millennium villages. *Proc. Natl. Acad. Sci. U. S. A.*, **104**, 16775–16780.
21 Okwi, P.O., Ndeng'e, G., Kristjanson, P., Arunga, M., Notenbaert, A., Omolo, A., Henninger, N., Benson, T., Kariuki, P., and Owuor, J. (2007) Spatial determinants of poverty in rural Kenya. *Proc. Natl. Acad. Sci. U. S. A.*, **104**, 16769–16774.
22 Hyden, G. (2007) Governance and poverty reduction in Africa. *Proc. Natl. Acad. Sci. U. S. A.*, **104**, 16751–16756.
23 Hyden, G. (1980) Population research potentials in Africa. *Afr. Link.*, 1–3.
24 Mabogunje, A.L. (2007) Tackling the African "poverty trap": the Ijebu-Ode experiment. *Proc. Natl. Acad. Sci. U. S. A.*, **104**, 16781–16786.
25 Sachs, J.D. (2004) Seeking a global solution. *Nature*, **430**, 725–726.
26 Black, R.E., Morris, S.S., and Bryce, J. (2003) Where and why are 10 million children dying every year? *Lancet*, **361**, 2226–2234.
27 WHO (2010) *World Health Statistics 2010*, World Health Organization, Geneva, http://www.who.int/whosis/ whostat/EN_WHS10_Full.pdf (accessed June 2010).

28 WHO (2008) *The Global Burden of Disease: 2004 Update*, World Health Organization, Geneva.

29 Twidell, J. and Weir, A.D. (2006) *Renewable Energy Resources*, Taylor & Francis, London.

30 Hunt, S. and Shuttleworth, G. (1996) *Competition and Choice in Electricity*, John Wiley & Sons, Ltd, Chichester.

31 EIA (2000) Carbon emission data are mostly for year 2000. Energy Information Agency of the US DoE, http://www.eia.doe.gov/emeu/cabs/South_Africa/Background.html (accessed July 2010).

32 UN Millenium Project (2005) Task Force on Science, Technology, and Innovation: Applying Knowledge in Development.

33 Sachs, J. (2002) The essential ingredient. *New Sci.*, **2356**, 175.

34 World Bank (2008) *Technological Progress and Development. Global Economic Prospect (GEP)*, The World Bank, Washington, DC.

35 Mexico City Synthesis Workshop (2002) ICSU Series on Science for Sustainable Development No. 9: Science and Technology for Sustainable Development. 30 pp. ISSN 1683-3686. http://www.icsu.org/Gestion/img/ICSU_DOC_DOWNLOAD/70_DD_FILE_Vol9.pdf (accessed July 2010).

36 Government of Canada, Office of the Prime Minister (2004) http://www.pm.gc.ca/eng/news (accessed 2004).

37 L. G. S. Ministry of Science and Innovation, Spanish Government (2010) Conference "Science Against Poverty". http://www.scienceagainstpoverty.es/ (accessed July 2010).

38 Foladori, G. and Invernizzi, N. (2005) Nanotechnology for the poor? *PLoS Med.*, **2**, e280.

39 UNDP (2009) Poverty reduction: Maintaining the focus on achieving the MDGs. United Nations Development Programme Annual Report.

40 Bordalo, A.A. and Savva-Bordalo, J. (2007) The quest for safe drinking water: an example from Guinea-Bissau (West Africa). *Water Res.*, **41**, 2978–2986.

41 Gardner, C.A., Acharya, T., and Yach, D. (2007) Technological and social innovation: a unifying new paradigm for global health. *Health Aff. (Millwood)*, **26**, 1052–1061.

42 Crawford, M.A. (2008) The elimination of child poverty and the pivotal significance of the mother. *Nutr. Health*, **19**, 175–186.

43 Sastre, J. (2008) The state of global hunger. *Science*, **322**, 1788–1789.

44 Mitlin, D. (2001) *Civil Society and Urban Poverty : Overview of Stage One City Case Studies*, University of Birmingham, International Development Department.

45 Baldwin, G.B. (1990) *The Long-Term Perspective Study of Sub-Saharan Africa: Background Papers*, World Bank, Washington, DC.

46 Arrossi, S. (1994) Funding Community Initiatives: The Role of Ngos and Other Intermediary Institutions in Supporting Low Income Groups and Their Community Organizations in Improving Housing and Living Conditions in the Third World, Published for the United Nations Development Programme by Earthscan Publications, London.

47 International Energy Agency (2006) Key World Energy Statistics. http://www.iea.org (accessed July 2010).

48 Salamanca-Buentello, F., Persad, D.L., Court, E.B., Martin, D.K., Daar, A.S., and Singer, P.A. (2005) Nanotechnology and the developing world. *PLoS Med.*, **2**, e97.

49 Preschitschek, N. and Bresser, D. (2010) Nanotechnology patenting in China and Germany – a comparison of patent landscapes by bibliographic analyses. *J. Bus. Chem.*, **7** (1), 7–13.

50 Schneider, B.W. (2009) Capital market's view on Industrial Biotechnology – proper valuation is the key for picking the right investment opportunities in stormy times. *J. Bus. Chem.*, **6** (3), 108–110.

51 Patrick, G.L. (2005) *An Introduction to Medicinal Chemistry*, Oxford University Press, New York.

4

The Human Element: Chemistry Education's Contribution to Our Global Future

Peter Mahaffy

4.1
The International Year of Chemistry Educational Challenge

"... education in and about chemistry is critical in addressing challenges such as global climate change, in providing sustainable sources of clean water, food and energy and in maintaining a wholesome environment for the well-being of all people ..."

UN Resolution declaring 2011 as an International Year of Chemistry

In declaring 2011 the International Year of Chemistry (IYC), the United Nations recognized that chemistry professionals and educators, working hand in hand with those in many other disciplines, have a crucial role to play in solving the global challenges needed to build a sustainable future for our planet. These global challenges and steps required to meet them are outlined fully in the UN Millennium Development goals [1] and the UN Decade of Education for Sustainable Development [2]. The role for chemistry in contributing to solutions is addressed in the other chapters of this book. Chemistry plays critical roles in providing potable water, sustainable energy, in maintaining human and environmental health; and in contributing to the scientific understanding of and solutions to global climate change, stratospheric ozone depletion, loss of biodiversity, world hunger, and poverty. As the discipline of chemistry has responded to these challenges, its very contours have evolved through creation of new research domains such as green chemistry, environmental chemistry, medicinal chemistry, agricultural chemistry, biochemistry, toxicology, and materials chemistry. In 2011, chemists rarely work in isolation, but they contribute to interdisciplinary research teams that come from many other disciplines. Most members of each interdisciplinary research team addressing these challenges, both those who are now chemistry professionals and those practicing other scientific professions, will have received education in chemistry. They comprise one important part of the human element, as described in the title of this chapter.

The Chemical Element: Chemistry's Contribution to Our Global Future, First Edition.
Edited by Javier Garcia-Martinez, Elena Serrano-Torregrosa
© 2011 Wiley-VCH Verlag GmbH & Co. KGaA. Published 2011 by Wiley-VCH Verlag GmbH & Co. KGaA.

Is education *in* and *through* chemistry, as it is currently practiced, adequately equipping the next generation of scientists and citizens to meet the International Year of Chemistry challenge outlined in the UN resolution? How can it do so better in the decades following IYC-2011? This central question is the primary focus of this chapter.

We also touch on a second question. Each human being who benefits from developments of chemistry in every aspect of daily life, comprises an equally important part of the human element in our chapter title. Do the members of these various publics recognize the fundamental importance of the molecular world? Has education *about* the nature and role of chemistry succeeded in creating the public climate needed to support the fundamental and applied research required to tackle these IYC global challenges?

But the year 2011 is not the first time the profession of chemistry has received a compelling call to play a role in global challenges. Let us picture a scene.

4.2
Scene 1–Chemistry to the Rescue of Threatened Communities

Over time, the Western world has come to depend on extracting and processing large deposits of an international commodity for both civilian and military use. It would not be an overstatement to say that assured supplies are required for sustaining life as we have come to know it. Known deposits of this substance should last for at least several hundred years. Yet obtaining the resource requires extracting material from the ground, transporting it thousands of kilometers across the ocean, and converting it into a form that can readily be used. The world's superpowers have created legal frameworks as well as resorted to military force to ensure access to this resource. As political tensions build, ready access to global markets seems increasingly unsure. Without finding other supplies or more sustainable alternatives, dire predictions of how life will be affected will soon be realized. "No nation can advance in material interests, or even maintain strict independence, without possessing (it) within its boundaries or the sources from which it can be drawn at all times" [3]. Scientific leaders urgently appeal for society to tackle this challenge, stating that the present course of action is unsustainable. "Serious peril," contends the president of the British Association for the Advancement of Science, awaits those "who contentedly pursue the present wasteful system ..." His compelling voice is to make a plea for the tools of chemistry to be used to develop alternatives and help make the transition to a sustainable future. "It is the chemist who must come to the rescue of the threatened communities" [4].

The scene you have just pictured might sound like a modern day call to use the tools of chemistry to ensure sustainable development of alternative energy resources. However the year is not 2011 but 1898. The chemical substance is not black tarry petroleum from the Middle East or Northern Alberta, formed by the action of pressure and geological time on plant material. Rather, it is yellowish-white crystalline sodium nitrate obtained from massive deposits of saltpeter formed from solidified bird excrement from arid tropical islands and from large deposits in the rainless plains of Tamarugal in the northern provinces of Chile. The global challenge is not the limited supply and environmental consequences of the unsustainable use of fossil fuel combustion to produce energy. Rather it is the recognition that without additional or alternative sources of fertilizers containing fixed nitrogen to replace saltpeter, insufficient food would be produced for a rapidly growing world, and famine is predicted within 30 years.

It took a hundred years after the discovery of nitrogen as an element late in the 18th century to understand its importance in plant and animal growth. An important chemistry textbook by Justus von Liebig provided a key part of the understanding leading to the 1898 call for rescue. Von Liebig described the critical importance of agriculture in the production of "digestible nitrogen" [5] and he showed that plant growth is limited by the availability of nitrogen atoms in a form usable for metabolism. While elemental nitrogen in the atmosphere surrounds every plant, comprising 78% of our atmosphere, it needs to be "fixed" or converted into more reactive forms for use by plants such as wheat and other grains, which in turn are harvested and processed to make flour for bread. In 1898, the only available source of fixed nitrogen for use on the required global scale was saltpeter, imported mostly from Chile. And global political tensions were threatening secure supplies of this commodity for Western countries. This led to the assessment by Sir William Crookes in his presidential address to the British Association for the Advancement of Science, that "England and all civilized nations stand in deadly peril of not having enough to eat" unless chemists could provide a solution (Figure 4.1).

The calls of Crookes were echoed by those of military strategists, for at the beginning of the 20th century sodium nitrate was also the limiting reagent in the production of nitrogen-based explosives. Over the next 15 years, leading chemists

Figure 4.1 Vanity Fair portrait of chemist Sir William Crookes, 1903. Courtesy of the Chemical Heritage Foundation. (Please find a color version of this figure in the color plates.)

of the day rose to the challenge, seeking a catalyst that would make possible the industrial scale production of reactive nitrogen. After fierce competition and an intense effort, a way was found to produce reactive or fixed nitrogen from dinitrogen, $N_2(g)$, in the air. Ultimate success came to the German chemist, Fritz Haber, in 1909 for developing a catalyst and finding ways to control the yield and rate of the reaction of nitrogen, $N_2(g)$, with hydrogen, $H_2(g)$, to form ammonia, $NH_3(g)$, thus turning "air into bread". Carl Bosch solved some overwhelming engineering problems created by the very high pressures needed to optimize and fully implement this reaction. Both Haber (1918) and Bosch (1931) were awarded Nobel Prizes for their accomplishments.

What "rescue" of threatened communities through application of chemical technology would William Crookes see, if he could glimpse into the future to the International Year of Chemistry, just over a century after the development of the Haber–Bosch process for the synthesis of ammonia? This process has been described as 'the most important technological invention of the 20th century". Hundreds of millions of tonnes of fertilizer are produced each year based on this synthesis, consuming about 1% of the world's energy supply in the process [6]. Smil estimates that the application of nitrogen fertilizers, such as urea, derived from ammonia, is responsible for the survival of 40% of our planet's human population [7]. Approximately half of the nitrogen atoms in each human body have come at some point through the Haber–Bosch process [8].

"When you travel in Hunan or Jiangsu, through the Nile Delta or the manicured landscapes of Java, remember that the children running around or leading docile water buffalo got their body proteins via the urea their parents spread on the fields, from the Haber–Bosch synthesis of ammonia. Without this, almost two fifths of the world's population would not be here – and our dependence will only increase as the global count moves from six to nine or ten billion people."

Vaclav Smil

But strictly technological solutions almost never provide quick fixes to complex global problems. The cycles of transformation of an element into other species do not occur in isolation from cycles involving other elements. During the period since the Industrial Revolution, as atmospheric concentrations of CO_2 have increased by 30%, the concentration of reactive nitrogen species has gone up by 300%. Human activity is now responsible for about half of the entire global cycle of nitrogen. Other human activities besides ammonia synthesis produce large amounts of reactive nitrogen. The effects on health and the environment of nitrogen oxides (NO, NO_2, HNO_3, and NO_3^-) in the troposphere, produced during combustion of fossil fuel, are well known. Nitrous oxide (N_2O), which has a very long atmospheric lifetime, is released from complex reactions in soils and plays roles in both tropospheric warming and stratospheric ozone depletion. In some parts of our world, the production of food is severely nitrogen-limited; hunger and malnutrition result. Other places experience severe environmental degradation from excess reactive nitrogen produced in fossil fuel combustion and from nitrate run-off into ground and surface water as a result of intensive agriculture. An integrated approach to understanding the beneficial and harmful effects of this escalation of N-containing species is just beginning, and chemistry has a central role to play. A series of international conferences has been established to bring together an inter-disciplinary community to understand reactive nitrogen science [9].

4.3
Sequel to Scene 1 – An Education in Chemistry

It's now 2009, 100 years after the invention of the Haber–Bosch process. You peer through a window into a chemistry classroom full of 18 or 19 year old students half-way through their academic year. They are in their first year of post-secondary education or perhaps their final year of secondary school. They close their textbooks at the end of a unit on chemical equilibrium. Who are these students? What and how are they being taught? What are they learning? How is their learning assessed?

If this is a first-year undergraduate course offered in many places in the world, each student may sit in a lecture session with 300 or more others (Figure 4.2), and care will be taken to ensure that his or her learning experience and assessment is as consistent as possible among the multiple sections of the course. Laboratory sessions, where they meet in smaller groups with laboratory instructors, used to be offered every week, but due to fiscal and staffing constraints, hands-on

Figure 4.2 A scene typical of many first year university chemistry classrooms. Photo reproduced with permission, © by Thomas Wirtz 2011.

experience of chemistry might now be limited to every other week. In any given lecture, about a third of the students are missing. While there is no such thing as a typical secondary school, if this scene depicts a final year secondary school class, there might be 35–50 students in the class, and they may be fortunate enough to receive 2–4 short laboratory experiences each academic term.

This chemistry course and the textbook resource used by the students you observe are organized along the same themes that have been used for almost half a century. This instructor takes a strictly "atoms first" approach, which emphasizes that no chemical reactions should be introduced until students have a strong foundational knowledge of the structure of atoms, elements, and chemical compounds. The first part of the course then focuses on the mastery of atomic theory, the ability to write electron configurations, followed by an introduction to chemical bonding and the structures of chemical compounds. Late in the term, following the introduction of chemical reactions and explanations of principles of reactivity, the first applications of chemistry are discussed. In the textbook, most of the applications of chemistry and human interest items are visually set apart from the core content in shaded boxes. Examinations do not normally include questions based on material in the shaded boxes.

The unit on equilibrium that the class has just finished was introduced by the reaction:

$$N_2(g) + 3H_2(g) \rightleftharpoons 2NH_3(g)$$

The textbook mentions briefly that Fritz Haber optimized this reaction, and that it has been used to make ammonia, an important agricultural commodity. A small box in the margin shows a picture of Fritz Haber and states that he received the Nobel Prize in 1918 for his synthesis of ammonia, despite some controversy. The textbook and classroom treatment of chemical equilibrium focuses on numerical problem solving, making use of algorithms described as "ICE" tables. Extensive coverage is given to the direction a reaction shifts when a chemical equilibrium is disturbed. Student learning of this topic is assessed through a multiple choice examination that focuses almost entirely on mathematical calculations relating concentrations and pressures of substances to equilibrium constants. The experienced teacher is frustrated by the evidence of incomprehension of basic concepts, despite the two weeks spent with the class on the topic. Many of the misconceptions this group of students demonstrate are the same ones he has seen for the past 20 years. At least part of the conceptual confusion comes from students having used concepts of equilibrium in physics class, with the term having fundamentally different connotations [10].

Learning objectives for the unit did not include connecting this simple chemical reaction to the survival of 40% of our planet's human beings. Nor did students learn about the challenges posed by too much reactive nitrogen in our environment, as outlined in a recent assessment of how human activity is beginning to transgress acceptable planetary boundaries in *A Safe Operating Space for Humanity* [11].

The classroom we have peered into above is clearly a caricature of secondary and post-secondary chemistry classrooms. But the stereotype portrays substantial realities about flaws in the present state of chemistry education that serve as barriers to providing relevant chemistry education to young learners. Alex Johnstone, acclaimed Professor Emeritus of chemistry and science education at the University of Glasgow, assesses the state of chemistry education as follows: Concepts are introduced that are inappropriate for the stage of learning of students. Ideas are clustered in indigestible bundles, and theoretical ideas are not linked to the reality of students' lives. Chemistry has become so irrelevant, uninteresting and indigestible, says Johnstone, that many students progress in their attitudes toward chemistry from "I can't understand" to "I shall never understand," and finally "I don't care if I understand." [10] He sounds an urgent call for chemistry educators to catalyze change in chemistry education, making it relevant, interesting, and digestible – by applying research about how students best learn chemistry.

"Many schools are teaching their chemistry courses with the same content and in the same manner as 30 years ago."

Harry Lewis, Journal of Chemical Education, 1948 [12]

In the next section we examine steps that can be taken by chemistry educators to begin to address these concerns.

4.4
Equipping the Human Element with Relevant Education *in*, *about*, and *through* Chemistry

The other chapters in this book describe the fundamental importance of chemistry in tackling the global challenges articulated in the United Nations IYC 2011 resolution and the Millennium Development Goals. The humans who must shoulder most of the responsibility for finding solutions to those challenges in the decades following IYC are presently in school and university classrooms. How well will their chemistry education serve them in preparing them to contribute solutions? Many are in classrooms similar to that described in Section 4.3.

Just as technological solutions to the global challenge in *Scene 1* are complex, the solutions to our global challenge of making chemistry education more relevant to our global future requires more than tinkering with curricula. We begin thinking about how to address this global educational challenge by understanding and deliberately designing the learning environments, and the curriculum, pedagogy, and physical spaces that shape and enrich the experiences of learners (Figure 4.3) [13].

The chemistry education system is, of course, very different in different countries, cultures, school and university systems, and levels, and no solutions for educational reform will be simple or universal. The right questions to ask about

Figure 4.3 The learning environments, curriculum, pedagogy, and physical spaces for effective science learning need to be appropriate to local cultures and education systems. Reproduced with permission © UNICEF/KENA2010-00321/Noorani. (Please find a color version of this figure in the color plates.)

enriching the learning environment will differ from course to course, and will depend on the nature of the course, the level of the learners, their overall learning objectives, and the experience, strengths, and teaching style of their instructor. There is no single "best way to teach" [14]. Nevertheless, we can learn a great deal from a pot pourri of practices that are based on evidence for what works to support student learning. Reflecting on and implementing the practices outlined below as they apply to a particular context may help catalyze changes to better support student learning and make education in chemistry more relevant for our global future. Following on our glimpse into a classroom of 18–19 year old students, our focus here is on helpful practices for upper level secondary and particularly first year post-secondary chemistry education.

4.4.1
Identify the Learners, Understand Their Overall Learning Objectives and Career Goals, and Ensure Education in Chemistry Meets Their Needs

Did the instructor of science majors we saw through the window know who the students in the classroom are, and their learning needs in chemistry? Which students in this school or university do not have access to instruction in chemistry that is relevant to their needs, perhaps because they have no secondary school prerequisites?

A very small percentage of university students in the arts, humanities, and social sciences take elective courses in chemistry. Often this is because there are no, or few, courses designed to meet their needs for basic chemistry literacy. For most university students, a primary objective for a chemistry course should be to introduce them to the pervasive molecular world, empower them as citizens to learn problem solving through the study of chemistry and to use basic tools of chemistry in decision making.

The approach taken to first-year university chemistry courses for science majors is often based on meeting the needs of students pursuing chemistry as a vocation, and many final year secondary school courses have, as a primary goal, preparing students for their first post-secondary course in chemistry. Yet, in many countries, first year university courses are populated by a diverse group of students pursuing careers in health and life sciences, engineering, materials science, and education. Chemistry majors often form a small minority of the learners. The starting point for better equipping learners in chemistry courses is to know *who they are*. To what use will they put their chemistry education and will their learning have been thoughtfully designed to be relevant to their needs and vocational choices? The answer to this question has significant implications for the way chemistry is contextualized and the applications of chemistry that are highlighted. Curriculum and pedagogy in chemistry needs to reflect the learning needs of the students who take chemistry courses and those who would benefit from chemistry courses, but do not find them accessible.

"... learner-centered environments include teachers who are aware that learners construct their own meaning, beginning with the beliefs, understandings, and cultural practices they bring into the classroom. If teaching is conceived as constructing a bridge between the subject matter and the student, learner-centered teachers keep a constant eye on both ends of the bridge. The teachers attempt to get a sense of what students know and can do as well as their interests and passions – what each student knows, cares about, is able to do, and wants to do. Accomplished teachers 'give learners reason,' by respecting and understanding learner's prior experiences and understandings, assuming that these can serve as a foundation on which to build bridges to new understandings."
How People Learn. US National Research Council [15]

4.4.2
Build and Support Active Learning Communities

As learners are placed in fewer and much larger chemistry classrooms, and content is "covered" through lecture formats to hundreds of students, meaningful mentoring relationships between faculty and students are frequently lost, and learning can become a solitary and isolating experience [16]. When it is necessary to work within this constraint, strategies to facilitate group work during and outside of lecture periods can bring learners together and support their active learning, even in large sections of courses [17]. Laboratory instruction, where class sizes make the construction of learning communities easier, can include open-ended team projects that tackle real-life problems and model professional science practice [18].

Helpful Practices: Active Laboratory Learning Communities

Even in large university classrooms, laboratories provide rich opportunities for hands-on experience of chemistry, for more individualized instruction, and for the formation of smaller learning communities within a course.

One example of an effort to construct a laboratory learning environment that incorporates collaborative inquiry activities, cooperative negotiation of conceptual understanding, and individual writing and reflection is the Science Writing Heuristics initiative [19].

ACELL, Advancing Chemistry by Enhancing Learning in the Laboratory, is an international model for providing professional and personal development and facilitating improved student laboratory learning outcomes [20].

4.4.3
Engage Students with Curriculum and Pedagogy that Takes Account of Research about How They Best Learn and How They Best Learn Chemistry

---

"An examination of learning outcomes from disciplinary and professional communities suggest there are many 'verbs' on the road to engaged learning: Knowledge is *constructed* in the mind of the learner by the learner; students *build* for themselves a workable understanding of sophisticated concepts from the field; students are involved with *solving* problems, *designing* and *constructing* experiments, *dissecting* problems, *collaborating* and *communicating* effectively, and more..."

Project Kaleidoscope, What Does Engaged Learning Look Like? [13]

It would be impossible to summarize adequately in this short section the now massive research literature reporting evidence on pedagogical strategies that foster student engagement in learning more effectively than the teacher-centered, lecture-based approach in the classroom we peered in at in Section 4.3. Rather than make any attempt to be comprehensive, I single out a few strategies that have been important in my own development over the past decade as a Professor of Chemistry and student of learning:

- Nurture active learning [21], where students are empowered to take responsibility for making sense of chemical structure and reactivity through solving open-ended and investigative problems, collaborating in small groups, and communicating what they have learned to others [22].

- Understand and accommodate the diverse learning styles of students, and help them understand how they learn best through the use of learning activities and inventories.

- Provide motivation by seamlessly drawing out the intriguing connections between the molecular world and the everyday world a student experiences, as well as global challenges.

- Prepare for a new topic by consulting research literature on student beliefs and preconceptions, and make use of inventories of misconceptions about that topic [23].

- Be guided in approach and pace by models of learning that identify the demands placed on working memory by intellectual performance such as chemistry problem solving [24].

- *Employ* and *adapt* to fit ones own teaching style and the learning environment, pedagogies of engagement such as Process Oriented Guided Inquiry Learning (POGIL) [25], Problem-Based Learning (PBL) [26], Just-in Time Teaching [27], Peer-led Team Learning (PLTL) [28] and creating and using context-rich problems [29].

- In addition to algorithmic, "chug and plug" problems that usually dominate formative and summative assessment, include problems at higher orders in the cognitive domain that require learners to apply, analyze, evaluate, and create [30].

- Make use of the overlap between subject matter expertise and pedagogical knowledge in teaching chemistry. *Pedagogical content knowledge* is the repertoire of conceptual and pedagogical knowledge grounded in the beliefs and practices of the teacher that is requisite for equipping chemistry educators to teach specific aspects of the subject matter to a particular group of students [31, 32]. Subject matter for effective teaching of chemistry includes not only knowledge of chemistry, but also its many interfaces: with the life sciences, materials, engineering, and professional ethics.

- Pay careful attention to the three levels of learning (observable, symbolic, and the molecular) at which students need to operate to make sense of chemistry. Be constantly aware that expert learners can flit seamlessly from one level to another, but novice learners need to be reminded at each step along the way which level they are currently operating at.

- Ensure that to whatever extent possible, assessment is resonant with pedagogy.

Helpful Practices: Sharing Ideas about "What Works"

Project Kaleidoscope (PKAL) is an informal alliance of the leading advocates in the United States for what works in building and sustaining strong undergraduate programs in the fields of science, technology, engineering and mathematics (STEM). Over its 21 year history, PKAL has provided resources that are adaptable to many different post-secondary contexts to help strengthen student learning. The PKAL network shares ideas about "what works" under seven umbrella themes:

- Institutional transformation (exploring what works in engaging the people, policies and practices that make it happen)

Continued

- The human infrastructure (exploring what works in nurturing STEM leaders, at all career stages)

- The physical infrastructure (exploring what works in shaping spaces that support 21st century STEM learning)

- The academic program (exploring what works in undergraduate STEM courses, from the very first courses for all students through capstone courses for majors)

- The pedagogical tools (exploring what works in designing, implementing and assessing teaching approaches that have an impact on student learning)

- The national (US) context (exploring the societal context for attending to the quality of undergraduate learning in STEM fields)

- The 21st century student (exploring the nature of current and emerging generations of students) [33].

4.4.4
Provide Education about Chemistry, and through Chemistry, as well as in Chemistry

While the focus of secondary and post-secondary courses, such as in the classroom we looked in at, is often on providing detailed content *in* chemistry, the skills in problem-solving and application of knowledge that can be obtained *through* chemistry instruction have been undervalued. A recent pilot study of university chemistry graduates in the UK was carried out to find out what was demanded of them in terms of skills, techniques and chemical knowledge, and to assess how well their university experience prepared them [10 (pp. 27–28)]. Analysis of the survey results highlighted the need for greater emphasis on process skills such as independent learning, team work and management, problem solving, information retrieval, report writing, oral presentation, experimental design, and data interpretation. A challenge for chemistry educators is to find ways to introduce these process skills *through* instruction in chemistry. This requires intentional design of both curriculum and pedagogical approach (Figure 4.4).

Education *about* the nature and role of chemistry has also been identified as a priority for the International Year of Chemistry and beyond. Attitudes toward chemistry and science, and learner conceptions are developed early, and primary schools have a vital role to play in laying the conceptual framework for understanding by learners at higher levels. Students deserve authentic portrayals of the scientific process at all levels of chemistry instruction [34], rather than the "sanitized" development of historical ideas [35] and linear textbook descriptions of "the scientific method" [36] sometimes introduced early in chemistry courses. In reality, chemistry is carried out by human beings engaged in a wide range of scientific methods of inquiry and application. A contextual understanding of the practice of science shows that chemists, like all scientists, rarely work in objective isolation. Many forces shape research and development, and chemical discoveries often happen at the human interfaces between between objectivity, logic, intuition, and passion [37].

Figure 4.4 Interactive digital learning objects, such as those developed at the King's Centre for Visualization in Science, can nurture active learning and help students understand concepts that are difficult to visualize.

Facilitating the understanding and appreciation of the role of chemistry in meeting world needs by adult citizens is another crucial dimension to gaining the knowledge base needed to make informed decisions about health and medicine, water resources, food supply, climate change, and alternative energy. Increasing public understanding of chemistry is particularly challenging, and requires first the careful identification and understanding of the "publics" that chemists and chemistry educators are best equipped to communicate with, so as to meet their needs for information about chemical substances and reactions that are such an important part of everyday life [38]. The International Year of Chemistry provides a golden opportunity to move the bar with respect to public understanding of our molecular world.

Helpful Practices:

Chemistry for Whom?
   What Chemistry?
      How to be Taught?
         How to be Assessed?

The Centre for Science Education and the University of Glasgow has used evidence from empirical research to define a set of guidelines to inform curriculum planning in chemistry. The executive summary of "Factors Influencing Curriculum Development in Chemistry" suggests that a chemistry curriculum should:

1. **Meet needs of all learners.** Meet the needs of the majority of school pupils (who will never become chemists or even scientists), and most students who will undertake chemistry degrees but never become bench chemists. Thus, the curriculum must seek to educate through chemistry as well as in chemistry.

2. **Relate to life.** At school level, be strongly "applications-led" in construction, while university courses should relate tightly to applications.

Continued

3. **Reveal chemistry's role in society.** Reflect attempts to answer questions like: what are the questions that chemistry asks? How does chemistry obtain its answers? How does this chemistry relate to life?

4. **Have a low content base.** Not be too "content-laden", so that there is adequate time to pursue misconceptions, to aim at deep understanding of ideas rather than content coverage, and to develop the appreciation of chemistry as a major influence on lifestyle and social progress.

5. **Be within information processing capacity.** Not introduce sub-micro and symbolic ideas too soon or too rapidly; avoid developing topics with high information demand before the underpinning ideas are adequately established to avoid overload and confusion.

6. **Take account of language and communication.** Be set in language which is accessible (especially at school level) and offer learners opportunities to express chemical ideas verbally and in writing (especially at university).

7. **Aim at conceptual understanding.** Be couched in terms of aims which seek to develop conceptual understanding rather than recall of information, being aware of likely alternative conceptions and misconceptions.

8. **Offer genuine problem solving experience.** Offer experiences of more open-ended problems (along with algorithmic exercises), with emphasis on the use of group work to solve "real-life" problems in chemistry.

9. **Use lab work appropriately.** Involve laboratory work with very clear aims: these should emphasize the role of lab work in making chemistry real as well as developing (or challenging) ideas rather than any focus on practical hands-on skills; lab work should offer opportunities for genuine problem solving.

10. **Involve appropriate assessment.** Involve assessment which is integrated into the curriculum and reflects curriculum purpose, is formative as well as summative and aims to give credit for understanding rather than recall, for thinking rather than memorization [39].

4.4.5
Move beyond the Fractionation of Knowledge

"Complexity requires specialization in the pursuit of discovery as we deepen our understanding of the modern world and create the knowledge needed to resolve current dilemmas and improve the quality of life. In this process, we continually fractionate knowledge, analyzing the pieces in greater and greater depth. We have trained our 20th century professional quite well in this task—it's a global strength we must sustain—but what additional skill will be demanded of 21st century leaders?

The ability to make connections among seemingly disparate discoveries, events, and trends and to integrate them in ways that benefit the world community will be the hallmark of modern leaders. They must be skilled at synthesis as well as analysis, and they must be technologically astute."

Joseph Bordogna [40]

An ironic metaphor (given that we are discussing *chemistry* education) to describe specialization in disciplines is the reference above to the *fractionation* of knowledge, that involves breaking down what we know into smaller and smaller pieces that are analyzed at greater and greater depth as a learner moves through the educational system and into research. While narrow specialization has led to enormous depth of understanding in the various sub-domains of chemistry, learners also need other skills. These include experience in working at interfaces with a wide range of other disciplines, seeing connections, and putting pieces together to tackle global challenges.

4.4.6
Show the Integral Connection between Chemical Reactivity and Human Activity

The global challenges of the 21st century are about people – their health, energy and food requirements, and interactions with their environment. Chemistry, too, is fundamentally about people – creating and investigating substances and benefiting from them. Our curriculum and pedagogy needs to stress in an integral way that chemistry is a current, creative, fascinating, and worthwhile human activity, seamlessly connected to the lives of students and to the global challenges our planet faces. Understanding chemistry is needed to understand our world and to inform decision making.

A visual metaphor that emphasizes this point for chemistry educators extends the three levels at which we know learners need to engage chemistry (visually represented by a triangle showing the macroscopic, symbolic, and molecular) [41] into a tetrahedron – with the addition of a third dimension that stresses the human activity involved in doing and learning chemistry. *Tetrahedral chemistry education* emphasizes our need to situate chemical concepts, symbolic representations and processes in the authentic contexts of the human beings who create substances, the culture that uses them, and the students that try to understand them (Figure 4.5) [42].

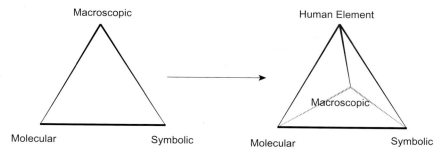

Figure 4.5 Chemical concepts, symbolic representations and processes need to be situated in human contexts.

Helpful Practices: Thinking Outside the Box

Unfortunately, in many learning resources for science majors taking general chemistry, the human element is lost or marginalized. Meaningful human contexts are often relegated to boxes and margin comments, and seemingly unconnected to core learning objectives. How do students perceive what's inside and outside the boxes? Aren't the boxes a signal that what is inside, interesting though it may be, is not part of the core curriculum, and will not be assessed through examinations?

An innovative and successful example of seamlessly integrating context and content is the issues-based university chemistry textbook for *non-science majors*, developed by the American Chemical Society. A primary goal for *Chemistry in Context* is to "enable students to learn chemistry in the context of their own lives and significant issues facing science and the world. The integrated activities help students develop an informed understanding of topics such as air quality, global warming, alternative fuels, acid rain, polymers, and genetic engineering." [43]

Can we restructure appropriately our general chemistry curriculum to place content in rich and motivating learning contexts for our *science majors*? This means learning how to think outside the boxes that separate content and context, without sacrificing the rigor that is required.

One way to situate learning in the rich context of global challenges and the everyday lives of both chemistry and other science majors is through the use of case studies. Only recently have case studies, which have been used extensively in medicine and law, seen application to teaching and learning science. Extensive resources on how to write and use different kinds of case studies in science teaching are available at the US National Center for Case Study Teaching in Science. C. Herreid from the Center describes the approach: "The case method involves learning by doing, the development of analytical and decision-making skills, the internalization of learning, learning how to grapple with messy real-life problems, the development of skills in oral communications, and often team work. It's a rehearsal for life." [44].

A special interest group in the Physical Sciences Centre at the University of Hull, UK has been set up by T. Overton and K. Moss to share ideas and resources for improving student learning through context and problem based approaches and case studies [45].

In the general chemistry textbook *Chemistry: Human Activity, Chemical Reactivity* [9], we use "trigger" [46] cases, integrated into the content coverage, to motivate students and to introduce each new chapter topic. Topics for trigger cases particularly relevant to global challenges include: methane clathrate hydrates as a way to introduce both supramolecular chemistry and the planetary carbon cycle; ocean acidification by CO_2 as a way to introduce solubility, precipitation, and complexation; photochemical smog to introduce spontaneous direction of change and equilibrium; and the chemistry of mountain pine beetle pheromones to introduce carbonyl compounds and organic functional groups.

4.4.7

Integrate Sustainability Themes into Chemistry Education

Sustainability has been defined as "the ability to provide a healthy, satisfying, and just life for all people on earth, now and for generations to come, while enhancing the health of ecosystems and the ability of other species to survive in their natural environments." [47] Education for sustainability has received a global emphasis since the 1992 Earth Summit, and the United Nations Decade of Education for Sustainable Development (2005–2014) has been established to "integrate the principles, values, and practices of sustainable development into all aspects of education and learning, in order to address the social, economic, cultural and environmental problems we face in the 21st century." [2] Chemistry education is well positioned to make a key contribution, by introducing learners to the structure, composition, energetics, properties, and reactivity of materials–all key components in the multidisciplinary understanding that will move us toward a more sustainable future.

One study has identified several "grand challenges" for sustainability, to which chemists are uniquely qualified to provide a molecular level approach. These include new approaches to energy and energy conversion to replace the combustion of carbon-based fuels; materials processing to use alternative and renewable feedstocks to replace petroleum-derived substances; and minimizing waste and environmental impact of chemical processing. Key research areas that will enable the addressing of these challenges include nanomaterials with applications in sustainability; bioprocesses that mimic or use environmental materials to increase efficiency and decrease hazardous waste; catalysts to harvest, store, and use energy and clean water, air and soil; and measurement science to improve the efficiency of chemical reactions and avoid production of waste products [48].

Other "grand challenges" discussed in this book – where education in the molecular sciences will contribute to solutions – include health, food security, and the provision of potable water.

Despite the contributions already being made by chemistry research in these areas, at present, sustainability is not found in the glossary of most chemistry textbooks, nor are the links between chemistry and sustainability overt themes, integrated into the content of most chemistry courses. Many topics in a chemistry course could be introduced and developed so as to emphasize these connections. Examples include introducing atom economy and atom efficiency as alternative ways of "counting atoms" when introducing stoichiometry; discussing a hydrogen economy in the context of energy and thermodynamics; introducing acidification of the oceans by CO_2 when studying acid–base chemistry and equilibria; using atmospheric gases important in preserving earth's radiation balance when studying the properties of gases; introducing concepts of equilibria and reaction rates in the context of nitrogen fixation and reactive nitrogen required to sustain human life; using molecular errors of disease and interventions to exemplify structure–reactivity relationships; and covering modern materials and photovoltaic applications.

Helpful Practices: Green Chemistry and New Ways of "Counting Atoms"

Students in the classroom we observed in Section 4.3 will have practiced, perhaps endlessly, reaction stoichiometry–the calculation of measurable relationships between the number of moles of reactants and products in the balanced equations for chemical reactions. But there are also other important ways of "counting atoms" that they will have missed in the curriculum. Green chemistry (described in Chapter 6) is a new emphasis by chemical industry on using and producing safe materials that do not damage the environment or human health. If mentioned in general chemistry textbooks, it is usually relegated to a box that lists green chemistry principles, rather than integrated into the core content. One central feature of green chemistry is to account for all of the atoms that go into

industrial processes and try to ensure that, as far as possible, all of the atoms in the starting materials should be in the desired products, to minimize waste.

These new ways of accounting for atoms include the concepts of *atom economy* or *atom efficiency*, and can be integrated into chemistry curricula, in the context of principles of green chemistry, alongside conventional discussion of reaction stoichiometry. The concept of atom economy can also be usefully applied to global challenges [9 p. 457]. It is estimated that of every 100 nitrogen atoms produced in the synthesis of NH_3 fertilizer through the Haber–Bosch process, only 14 end up in the mouth of a vegetarian and 4 in the mouth of a meat-eater [49].

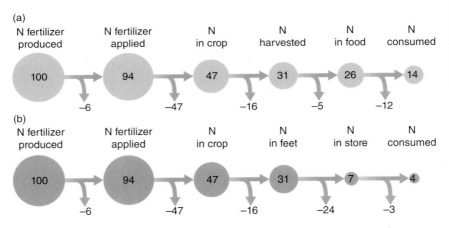

(a)

| N fertilizer produced | N fertilizer applied | N in crop | N harvested | N in food | N consumed |

(b)

| N fertilizer produced | N fertilizer applied | N in crop | N in feet | N in store | N consumed |

The number of nitrogen atoms that end up in the mouth of a vegetarian (a) and meat eater (b) starting with 100 nitrogen atoms produced in the synthesis of NH_3

fertilizer through the Haber Bosch process. [49] Reproduced with permission, © Royal Swedish Academy of Sciences.

The remainder are returned to the environment as many different compounds during fertilizer production, storage, transport, an application; uptake of fertilizer by crops; crop harvesting; uptake by livestock; and the formulation of

food products from plants and animals for humans. And, as seen in Section 4.2, many of these "waste" nitrogen compounds cause environmental problems.

4.5

An Example of Integrating Sustainability and Chemistry Education Curriculum: Visualizing the Chemistry Underlying Climate Change

"Re-stabilization of earth's climate is the defining challenge of the 21st century. The unprecedented scale and speed of global warming and its potential for large-scale, adverse health, social, economic and ecological effects threatens the viability of civilization. The scientific consensus is that society must reduce the global emission of greenhouse gases by at least 80% by mid-century at the latest, in order to avert the worst impacts of global warming and to re-establish the more stable climatic conditions that have made human progress over the last 10 000 years possible. Without preventing the worst aspects of climate disruption, we cannot hope to deal with the other social, health and economic challenges that society is facing and will face in the future."

US College and University President's Climate Commitment, 2007 [50]

Understanding, addressing, and mitigating climate change is one of the critical, systemic challenges faced by the world in the 21st century. Does climate science currently have a place in secondary or post-secondary chemistry curriculum? Should it?

The US interagency Climate Change *Climate Literacy* [51] initiative suggests that over the next several decades encompassing the professional careers of the students who are currently entering university classrooms, climate change is expected to have an increasing impact on human and natural systems – affecting human health, biodiversity, economic stability, national security, and accessibility to food, water, raw materials, and energy. To prepare graduates to adapt to these new conditions (and benefit from the new economic opportunities they create) will require both the ability to understand climate science and the implications of climate change, as well as the capacity to integrate and use that knowledge effectively.

Yet there is a pervasive and global disconnect between actual climate science knowledge and perceived knowledge [52]. Recent research on public attitudes and misconceptions about climate change shows that most adults believe that climate change is happening [53]. However, without a solid public understanding of the causes of anthropogenic climate change and potential solutions, individuals are left with "overwhelming, frightening images of potentially disastrous impacts, no clear sense of how to avert this potentially dark future, and therefore no way to direct urgency toward remedial action." [54]

A review of five decades of science education relating to climate and climate change in particular demonstrates that basic climate science has not been well addressed at either school or post secondary levels. Key misconceptions and misinformation about basic climate science; the role of human activities and reliance on fossil fuels on the climate system; and the consensus among the climate research community about the issues, are commonly held by students, teachers, politicians, and members of the public [55].

What might be a meaningful role for chemistry education in facilitating climate literacy? An analysis of the *seven essential principles of climate science*, suggested as necessary for achieving climate literacy [51], suggests compelling and urgent roles for secondary school and university chemistry education in addressing some of the key cognitive gaps and misconceptions documented by research. Modern chemistry is defined by its interdisciplinary boundaries, some of the most important of which (energy, green chemistry, and the environment) play pivotal roles in the quest for sustainability. Yet the evolution of the chemistry curriculum to support context-rich learning about complex systems such as our planet's changing climate has lagged far behind the new frontiers in research.

To fill this gap, an IUPAC task group [56] has been formed to produce a set of interactive lessons for global dissemination to help 16–19 year old students visualize and understand the science underlying climate change. We started with research literature reports of common misconceptions that students have about climate change, and the fundamental principles of climate science needed to create climate literate students and teachers. Digital learning objects have been developed at the King's Centre for Visualization in Science (Canada), and integrated with written materials prepared by The Royal Society of Chemistry (UK) and the American Chemical Society (USA) into a set of critically reviewed interactive lessons [57]. Task group members from IUPAC's Committee on Chemistry Education, UNESCO, and the Federation of African Societies of Chemistry have participated in the review of materials and dissemination through national and international networks.

Screen captures from two of the interactive visualizations that could readily be integrated into the chemistry curriculum are shown below [58]:

The interactive flash learning object in Figure 4.6 helps to answer the question: Why are chlorofluorocarbons such as CF_2Cl_2 (CFC-12) such potent greenhouse gases? Absorption of infrared radiation in the region 700–1200 cm^{-1} causes excitation of the C—F stretching vibrational modes of CFC molecules. This occurs in a region of the IR spectrum where water and carbon dioxide, the two best known greenhouse gases, are transparent (a spectral window). Thinking of the earth as a giant IR source, CFCs absorb energies of IR radiation which have historically escaped into space, thus cooling our planet. This is also in a region of the spectrum close to the peak of the earth's emission band (shown in Figure 4.6 as a blue overlay).

A second interactive learning object (Figure 4.7) helps learners visualize the molecular level mechanism by which increased amounts of greenhouse gases such as CO_2 cause tropospheric warming. Most (99%) of the earth's atmosphere consists of only two gases: nitrogen, $N_2(g)$, and oxygen, $O_2(g)$, neither of which are able to directly absorb IR radiation and are not greenhouse gases. Carbon dioxide molecules, however, absorb photons of infrared radiation if they are of exactly the right energy match to cause stretching or bending vibrations. In itself, that absorption of energy does not cause warming. Atmospheric warming takes place when that vibrational energy is transferred to other molecules in the atmosphere (such as nitrogen and oxygen) that are unable themselves to absorb

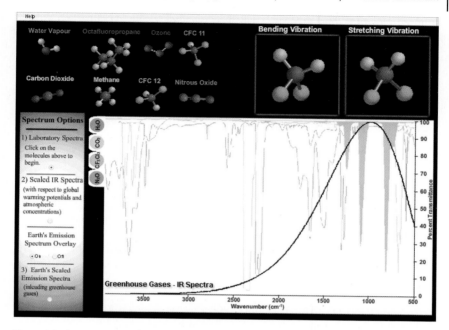

Figure 4.6 Interactive visualization showing the connection between the infrared spectra of gases and their global warming potential. (Please find a color version of this figure in the color plates.)

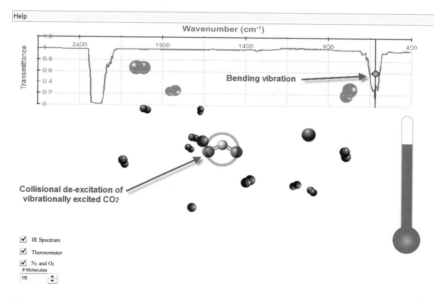

Figure 4.7 Interactive visualization showing the molecular-level mechanism for tropospheric warming by greenhouse gases. (Please find a color version of this figure in the color plates.)

infrared radiation. This process is called *collisional de-excitation* of carbon dioxide [9 p. 93].

The complete set of interactive lessons from this IUPAC project will provide tools for chemistry educators to make important connections in their classrooms to help students understand climate change, one of the defining challenges of the 21st century. A creative educational challenge is to do so in a manner that integrates deep coverage of core content for parts of the secondary and post-secondary chemistry curriculum and supports pedagogical engagement.

4.6
Scene 2 – Chemistry Education and Our Global Future

As we began this chapter (Section 4.2), we conclude it with a new scene

> "Over time, the Western world has come to depend on extracting and processing large deposits of an international commodity for both civilian and military use ..."

Does this sound familiar? This time our scene is set in the 21st century and describes the interrelated web of global challenges associated with our insatiable appetite for energy; the dwindling resources of easily extractable fossil fuels for combustion to provide that energy; the rapidly escalating emission of carbon compounds and other greenhouse gases into our atmosphere; the dwindling resources of water for drinking, livestock, agriculture, and industrial processes; and the compelling need to develop sustainable alternatives.

It would not be an overstatement to say that the quality of life as we know it now depends on assured supplies of fossil fuels. At present, we carry out combustion reactions directly on most of the fossil fuels we extract from the ground to obtain energy to fuel modern life. But we know, too, that these fossil fuels are the feedstocks for the polymers that we have come to depend on, and that future generations will need to make strong, lightweight materials.

Known deposits of fossil fuels should last for hundreds of years, and longer in the case of coal, although enormous environmental challenges need to be overcome to continue using them for fuels. Obtaining the resource requires extracting material from the ground, transporting it thousands of kilometers across the ocean, and converting it into a form that can readily be used. The world's superpowers have created legal frameworks and resorted to military force to ensure access to this resource. As political tensions build, ready access to global markets seems increasingly unsure. Without finding other supplies or more sustainable alternatives, dire predictions of how life will be affected will soon be realized.

In addition, present global practices of large scale combustion of fossil fuels to produce energy sends carbon dioxide gas into our atmosphere, and atmospheric concentrations of CO_2 have increased from 295 ppm at the time Sir William Crookes made his address to the British Academy of Sciences in 1898 to 393 ppm at the beginning of the 2011 International Year of Chemistry. Human activity has also placed a large fingerprint on the atmosphere through alteration of global cycles of other greenhouse gases such as CH_4, N_2O and introduction in the 20th century of trace quantities of new "super" greenhouse gases such as CFCs.

Some of the greatest impacts of planetary-scale changes in the earth's climate will be felt far from the largest sources of greenhouse gases. For example, while its climate is diverse, on average Africa is hotter and drier than most other regions of the world, and livelihoods from the individual to regional levels depend heavily on climate. Models suggest that large areas of Africa will become even drier and have more variable precipitation over the next century. These changes make Africa particularly vulnerable to even small changes in climate [59].

As a result of energy production and use, population growth, industrialization, urbanization, agricultural practices, and climate change, management of supplies of another international commodity, water, pose unprecedented global challenges. By 2025, 1.8 billion people will live where water is scarce. H_2O is another chemical substance for which global cooperation is urgently needed to ensure sustainable supplies to maintain life as we know it.

What is the role for chemistry and chemistry education amid these interlinked 21st century challenges related to energy supply and efficiency, climate change, and fresh water resources? A modern day Sir William Crookes (Section 4.2) appeal for a rescue of our threatened global community with the aid of chemistry comes from Daniel Nocera, Professor of Chemistry and Henry Dreyfus Professor of Energy at Massachusetts Institute of Technology. Nocera summarizes analysis of global energy requirements as doubling by 2050 and tripling by 2100 [60]. After examining alternatives to meet the energy needs of all people and the environmental consequences of those alternatives, he gives a compelling call to find innovative ways to make use of energy from the sun. All the energy stored in Earth's reserves of coal, oil, and natural gas is matched by the energy from just 20 days of sunshine [61]. Chemists, working in concert with others, he suggests, need to carry out research and development to (i) uncover the secrets of bioenergy conversion and biocatalysis, (ii) design novel solar capture and conversion materials and (iii) create catalysts that can store energy in hierarchical materials or in the form of the chemical bonds of fuels [60]. Nocera's research group tackles his vision for "personalized energy for 1×6 billion" by trying to artificially recreate the functional elements of photosynthesis. Their vision is to use solar energy powered photovoltaic assemblies to produce electricity during the day. This powers a home while the sun shines, and also splits water to produce H_2 gas and O_2 gas, which are stored during the day and recombined at night in a fuel cell to supply energy when the sun does not shine [62]. Just as was the case a century ago, where the development of catalysts was the key to optimizing the fixation of nitrogen, water-splitting catalysts are required to improve the efficiency of the electrolysis reaction.

Nocera has put his brush to canvas to begin painting a new scene. And like Sir William Crookes, he draws chemists in the foreground.

"The Earth, who has sustained us so well until now, needs us. Of all the sciences that can respond, chemistry will most succinctly answer our planet's distress calls. The chemist will develop new ways to recover elements in forms to which they were lost and re-enter them to our ecosystem and to the life cycle that sustains our humanity. The chemist will develop new reactions that preserve and use our elements efficiently... The chemist will develop the new materials for energy conversion and discover the new reactions needed for the bond rearrangements for fuel-forming reactions from carbon-neutral sources. Only a modest investment of time will convince anybody that the greatest challenge confronting our planet in the coming century is the need for new energy sources that do not suffocate our planet in carbon dioxide. With societal success in increasing the life expectancy of those living in the developed world (combined with an always vigilant commitment not to leave behind our sisters and brothers in the developing world), we are on a nexus of change. Energy efficiency and new carbon-neutral energy sources will become the most pressing issues confronting this planet in the coming century. These are concerns for the chemist. The chemical bond is the currency of energy. The answer to the energy problem confronting this planet deals in the chemist's currency."

Daniel Nocera, MIT [63]

Will the tools of chemistry be used, this time in concert with those of the other sciences, social sciences, and humanities, to rescue our threatened 21st century-and-beyond global communities? Will chemistry educators see their role and catalyze the changes needed to equip the human element to address these challenges – both the next generation of chemists and scientists and the general public who votes and sets policy?

During the International Year of Chemistry and the years to follow, it is you and I, the human element, who will fill in the rest of this scene and its sequels.

References

1 United Nations (2010) Millennium Development Goals. http://www.un.org/millenniumgoals/ (accessed 9 January 2011).

2 UNESCO (2011) UN Decade of Education for Sustainable Development, 2005–2014. http://www.unesco.org/en/esd/ (accessed 9 January 2011).

3 Thaxter, C., Jewett, S.O., and Dickinson, E. (1862) Saltpetre as a source of power. Atlantic Monthly, p. 593.

4 Crookes, W., Sir (1917) *The Wheat Problem*, 3rd edn, Longmans, Green & Co, London, p. 3.

5 Shenstone, W.A. (1901) *Justus Von Liebig: His Life and Work*, Cassell & Co, London, pp. 80–125.

6 Smith, B.E. (2002) Nitrogenase reveals its inner secrets. *Science*, **297** (5587), 1654.

7 Smil, V. (1999) Detonator of the population explosion. *Nature*, **400**, 415.

8 Fryzuk, M. (2004) Ammonia transformed. *Nature*, **427**, 498.

9 Mahaffy, P., Bucat, B., Tasker, R., *et al.* (2011) *Chemistry: Human Activity, Chemical Reactivity*, Nelson Education, Toronto, p. 458.

10 Johnstone, A. (2010) You can't get there from here. *J. Chem. Educ.*, **87** (1), 22–29.

11 Rockstom, J. *et al.* (2009) A safe operating space for humanity. *Nature*, **461**, 472–475.

12 Lewis, H. (1948) Chemical education today. *J. Chem. Educ.*, **25** (10), 576.

13 Project Kaleidoscope. What does engaged learning look like?, in *Then, Now & In the Next Decade*, http://www.pkal.org/documents/WhatDoesEngaged LearningLookLike.cfm (accessed 9 January 2011).

14 Heller, K. and Heller, P. (2004) Using the learning knowledge base: The connection between problem solving and cooperative techniques, in *What Works, What Matters, What Lasts*, http://www.pkal.org/documents/Vol4Using TheLearningKnowledgeBase.cfm (accessed 9 January 2011).

15 Narum, J. (2004) Realizing a learner-centred environment, in *What Works, What Matters, What Lasts*, http://www.pkal.org/documents/Vol4ALearnerCenteredEnvironment.cfm (accessed 9 January 2011).

16 Armour, M.-A. (2010) Reflections on teaching and learning, in *Academic Callings: The University We Have Had, Now Have, and Could Have* (eds J. Newson and C. Polster), Canadian Scholars Press, Toronto, pp. 228–233.

17 MacGregor, J., Cooper, J.L., Smith, K.A., and Robinson, P. (eds) (2000) *Strategies for Energizing Large Classes: From Small Groups to Learning Communities*, Jossey-Bass Publishers, San Francisco.

18 Mahaffy, P., Newman, K., and Bestman, H. (1993) From lead solder to kiwi fruit: reshaping introductory chemistry labs with investigative projects. *J. Chem. Educ*, **70** (1), 76–79.

19 Science Writing Heuristics (n.d.) The Process of the Science Writing Heuristic Home Page. http://avogadro.chem.iastate.edu/SWH/homepage.htm (accessed 9 January 2011).

20 Buntine, M.A., Read, J.R., Barrie, S.C., Bucat, R.B., Crisp, G.T., George, A.V., Jamie, I.M., and Kable, S.H. (2007) Advancing Chemistry by Enhancing Learning in the Laboratory (ACELL): a model for providing professional and personal development and facilitating improved student laboratory learning outcomes. *Chem. Educ. Res. Pract.*, **8** (2), 232–254.

21 Meyers, C. and Jones, T.B. (1993) *Promoting Active Learning: Strategies for the College Classroom*, Jossey-Bass Publishers, San Francisco.

22 Barkley, E.F. (2010) *Student Engagement Techniques: A Handbook for College Faculty*, Jossey-Bass Publishers, San Francisco.

23 (a) Horton, C. (2007) Student Alternative Conceptions of Chemistry, http://assessment-ws.wikispaces.com/file/view/chemistry-misconceptions.pdf (accessed 9 January 2011); (b) Garnett, P., Garnett, P., and Hackling, M. (1995) Students' alternative conceptions of chemistry: a review of research and implications for teaching. *Stud. Sci. Educ.*, **25** (1), 69–96; (c) Mulford, D. and Robinson, W. (2002) An inventory for alternate conceptions among first-semester general chemistry students. *J. Chem. Educ.*, **79** (6), 739.

24 Herron, D. (1990) Research in chemical education: results and directions, in *Toward A Scientific Practice of Science Education* (eds M. Gardner, *et al.*), Lawrence Erlbaum, Associates, Inc., Hillsdale, NJ.

25 Process Oriented Guided Inquiry Learning (2010) http://www.pogil.org/ (accessed 9 January 2011).

26 Rhem, J. (1998) Problem based learning: an introduction. *Natl Teach. Learn. Forum*, **8** (1). http://www.ntlf.com/html/pi/9812/pbl_1.htm (accessed 9 January 2011).

27 Novak, G.M., Patterson, E.T., Gavrin, A.D., and Christian, W. (1999) *Just-in-Time Teaching: Blending Active Learning with Web Technology*, Prentice Hall, Upper Saddle River, NJ.

28 Peer Led Team Learning (2011) http://www.pltl.org/ (accessed 9 January 2011).

29 University of Minnesota, Department of Physics (2010) Creating Context-Rich Problems, http://groups.physics.umn.edu/physed/Research/CRP/crcreate.html (accessed July 7, 2010).

30 Anderson, L.W. and Krathwohl, D.R. (eds) (2010) *A Taxonomy for Learning, Teaching, and Assessing: A Revision of Bloom's Taxonomy of Educational Objectives*, Longman, New York.

31 Bucat, R.W. (2004) Pedagogical content knowledge as a way forward: applied research in chemistry education. *Chem. Educ. Res.*, **5**, 215–228.

32 Kind, V. (2009) Pedagogical content knowledge in science education:

perspectives and potential for progress. *Stud. Sci. Educ.*, **45** (2), 169–204.

33 Project Kaleidoscope (2011) http://www.pkal.org (accessed 9 January 2011).

34 Martin, B., Kass, H., and Brouwer, W. (1990) Authentic science: a diversity of meanings. *Sci. Educ.*, **74** (5), 541–554.

35 Bent, H. (1977) The use of history in teaching chemistry. *J. Chem. Educ.*, **54** (8), 462–466.

36 Mahaffy, P. (1992) Chemistry in context. *J. Chem. Educ.*, **69** (1), 52–56.

37 Grinnell, F. (2009) *Everyday Practice of Science*, Oxford University Press, Oxford.

38 (a) Mahaffy, P. (2006) Chemist's understanding of the public. *Chem. Int.*, **28** (4), 12–16; (b) Mahaffy, P., Ashmore, A., Bucat, R., Do, C., and Rosborough, M. (2008) Chemists and the public: IUPAC's role in achieving mutual understanding. *Pure Appl. Chem.*, **80** (1), 161–174.

39 Mbajiorgu, N. and Reid, N. (2006) Factors influencing curriculum development in chemistry. Physical Sciences Centre, University of Hull, Hull, UK. http://www.heacademy.ac.uk/assets/ps/documents/practice_guides/practice_guides/ps0074_factors_influencing_curriculum_development_in_chemistry_Nov_2006.pdf (accessed January 9, 2011)

40 Bordogna, J., Fromm, E., and Ernst, E. (1995) An integrative and holistic engineering education. *J. Sci. Educ. Technol.*, **4** (3), 192.

41 (a) Johnstone, A.H. (2000) Chemistry teaching: logical or psychological. *Chem. Educ. Res. Pract. Eur.*, **1**, 9–15; (b) Gabel, D. (1999) Improving teaching and learning through chemistry education research: a look to the future. *J. Chem. Educ.*, **76**, 548.

42 Mahaffy, P. (2006) Moving chemistry education into 3D: a tetrahedral metaphor for understanding chemistry. *J. Chem. Educ.*, **83** (1), 49–55.

43 Eubanks, L., Middlecamp, C., Heltzel, C., and Keller, S. (2009) *Chemistry in Context*, 6th edn, McGraw-Hill.

44 Herreid, C. (1994) Case studies in science: a novel method of science education. *J. Coll. Sci. Teach.*, February, 221–229.

45 UK Physical Sciences Centre (2011) Context and Problem Based Learning SIG, http://www.heacademy.ac.uk/physsci/home/networking/sig/CPBL (accessed 9 January 2011).

46 National Center for Case Study Teaching in Science (2011) http://sciencecases.lib.buffalo.edu/cs/ (accessed 9 January 2011).

47 From http://www.earthethics.com, cited in Grassian, V.H., Meyer, G., *et al.* (2007) Chemistry for a sustainable future. *Environ. Sci. Technol.*, **42**, 4840–4846.

48 Grassian, V.H., Meyer, G., *et al.* (2007) Chemistry for a sustainable future. *Environ. Sci. Technol.*, **42**, 4840–4846.

49 Galloway, J.N. and Cowling, E.B. (2002) Reactive nitrogen and the world: 200 years of change. *Ambio*, **31** (2), 64–71.

50 American College and University Presidents' Climate Commitment (2011) http://www.presidentsclimatecommitment.org/ (accessed 9 January 2011).

51 US Global Change Research Program (2011) Climate Literacy: The Essential Principles of Climate Science, http://www.globalchange.gov/resources/educators/climate-literacy (accessed 9 January 2011).

52 Dupigny-Giroux, L.-A.L. (2008) Introduction-climate science literacy: a state of the knowledge overview. Physical geography. Special Issue, *Clim. Lit.*, **29** (6), 483–486.

53 (a) Krosnick, J. (2006) The origins and consequences of democratic citizens' policy agendas: a study of popular concern about global warming. *Clim. Change*, **77**, 7–43; (b) Leiserowitz, A., Maibach, E., and Roser-Renouf, C. (2008) Global Warming's Six Americas: An Audience Segmentation. http://www.climatechangecommunication.org/resources_center.cfm (accessed 9 January 2011).

54 (a) Moser, S.C. and Dilling, L. (2004) Making climate hot: communicating the urgency and challenge of global climate change. *Environment*, **46** (10), 32–46; (b) Niepold, F., Herring, D., and McConville, D. (2007) The case for climate literacy in the 21st century. Proceedings of 5th International Symposium on Digital

Earth, June 5, http://www.isde5.org (accessed 9 January 2011).

55 (a) McCaffrey, M.S. and Buhr, S.M. (2008) Clarifying climate confusion: addressing systemic holes, cognitive gaps, and misconceptions through climate literacy. Physical Geography Special Issue, *Clim. Lit.*, **29** (6), 512–528; (b) Dikmenli, M. (2010) Biology students' conceptual structures regarding global warming. *Energy Educ. Sci. Technol., Part B Soc. Educ. Stud.*, **2** (1-2), 21–38; (c) Sundblad, E.-L., Biel, A., and Gärling, T. (2009) Knowledge and confidence in knowledge about climate change among experts, journalists, politicians, and laypersons. *Environ. Behav.*, **41** (2), 281–302.

56 Mahaffy, P., Martin, B., Kirchhoff, M., *et al.* (2010) Visualizing the Science of Climate Change (IUPAC Project), http://www.iupac.org/web/ins/2008-043-1-050 (accessed 9 January 2011).

57 Mahaffy, P., Martin, B., *et al.* (2011) Visualizing and Understanding the Science of Climate Change. http://www.ExplainingClimateChange.com (accessed January 26, 2011).

58 (2011) Infrared Spectral Windows and Collisional Heating by CO_2 in the Atmosphere. http://www.kcvs.ca (accessed 9 January 2011).

59 (a) ICSU Regional Office for Africa (2008) Global Environmental Change (Including Climate Change and Adaptation) in Sub-Saharan Africa, International Council on Science, http://www.icsu-africa.org/docs/sp_gec.pdf (accessed 9 January 2011); (b) Boko, M., Niang, I., Nyong, A., Vogel, C., Githeko, A., Medany, M., Osman-Elasha, B., Tabo, R., and Yanda, P. (2007) Africa. Climate change 2007: impacts, adaptation and vulnerability, in *Fourth Assessment Report of the Intergovernmental Panel on Climate Change* (eds M.L. Parry, O.F. Canziani, J.P. Palutikof, P.J. van der Linden, and C.E. Hanson), Cambridge University Press, Cambridge. http://www.ipcc.ch/pdf/assessment-report/ar4/wg2/ar4-wg2-chapter9.pdf (accessed 9 January 2011).

60 Nocera, D. (2009) Living healthy on a dying planet. *Chem. Soc. Rev.*, **38**, 13–15.

61 Union of Concerned Scientists (2009) How Solar Energy Works, http://www.ucsusa.org/clean_energy/technology_and_impacts/energy_technologies/how-solar-energy-works.html (accessed 9 January 2011).

62 Nocera, D. (2009) Personalized energy: the home as a solar power station and solar gas station. *ChemSusChem*, **2**, 387–390.

63 Nocera, D. (2008) Great challenges ahead. *ChemSusChem*, **1**, 8.

5
The Impacts of Synthetic Chemistry on Human Health

René Roy

5.1
The Molecules at the Origin of Drug Discoveries

The roles of chemistry in human health are numerous. In this regard, among the multitude of subdisciplines in which chemistry has key responsibilities, organic chemistry, in particular, is taking a leading position. Of course, other disciplines such as analytical chemistry, material sciences, and immunochemistry, to name but a few, also contribute greatly to making the daily lives of the human and animal kinds more comfortable. Unfortunately, they cannot all be covered within the limited scope of this single chapter. Several of these parallel purposes are some-what overlapping with the subject of other chapters covered in this book dedicated to the International Year 2011 devoted to chemistry. Hence, this chapter will concentrate primarily on various aspects involved in drug discovery and development. It will end with a personal success story from the author's own research activities that led to the first commercial human synthetic vaccine against bacterial infections causing the death of more than half a million infants each year.

The use of plants to treat diseases has been a common practice in all great civilizations and is often referred to as *Folk Medicine* (Figures 5.1 and 5.2) [3, 4]. Nowadays, a certain practice among pharmaceutical companies is to seek information from "Shaman" in different parts of the world to identify plants in common use for illness treatment. This strategy complements other modern approaches to widen the scope of existing chemical entities such as: the synthesis of chemical libraries, the use of "in-house" chemical libraries, solid-phase syntheses, all of which are assisted by high throughput screening (HTS), virtual (computarized) drug screenings ("*in silico*") and related technologies.

For instance, the history of the popular Aspirin™ (ASA—acetyl salicylic acid) and its medical use can be traced back to antiquity even though pure ASA (Figure 5.3) has only been manufactured and marketed since 1899 [5–7]. Medicines extracted from the willow tree date back to at least 3000 BC. Willow bark extract became recognized for its specific effects on fever, pain and inflammation in the mid-18th century. Chemists at the drug and dye firm Bayer investigated ASA in 1897.

The Chemical Element: Chemistry's Contribution to Our Global Future, First Edition.
Edited by Javier Garcia-Martinez, Elena Serrano-Torregrosa.
© 2011 Wiley-VCH Verlag GmbH & Co. KGaA. Published 2011 by Wiley-VCH Verlag GmbH & Co. KGaA.

Figure 5.1 Medicine man in American Indian healing [1]. (Please find a color version of this figure in the color plates.)

Figure 5.2 A moment in the life of an Egyptian physician of the 18th dynasty (1500–1400 BC) [2]. (Please find a color version of this figure in the color plates.)

By 1899, Bayer had named this drug *Aspirin* and started selling it around the world. Aspirin's popularity grew even further during the first half of the 20th century because it was shown to be efficacious in the wake of the Spanish flu pandemic of 1918. Aspirin's profitability led to fierce competition and the proliferation of aspirin brands and products. It is estimated that over 100 million people use the drug regularly. The annual production of aspirin is still in excess of 40 000 tons worldwide. However, the popularity of aspirin declined after the development of acetaminophen in 1956 and ibuprofen in 1962. Today, aspirin is present in the cocktail of medications used as an anticlotting agent for the prevention of heart attack.

The branch of chemistry dealing with the discovery of new medication lies in organic synthesis and is often referred to as medicinal chemistry. Chemists are thus responsible for creating, designing or discovering new molecules having the

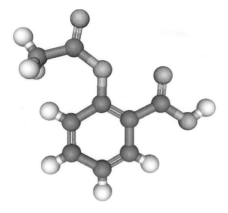

Figure 5.3 Structure of Aspirin™ shown as ball and stick representation. (Please find a color version of this figure in the color plates.)

properties to cure, stop or prevent certain diseases or symptoms. With the assistance of biologists and biochemists, the team has to identify the most active drugs (low dosage), most efficient (target specific), and most innocuous (nontoxic) molecular entity that can target the disease causative sources. The arsenal of modern techniques assisting the different steps involved in the complex process of drug discoveries has expanded greatly over time. However, even now the role of natural products or plant extracts in new drug development is still preeminent. Curiously, almost 60% of modern cancer curing drugs originate from natural sources or are derived from them. Even the most potent drug Paclitaxel (Taxol™) introduced in 1992 by Bristol-Myers Squibb and used to treat ovarian and breast cancer was first extracted from the Pacific yew tree (Figure 5.4) [8, 9].

A refreshing and creative strategy in organic synthesis that has led to more effectively arrived at novel chemical entities has been described by the group of S. L. Schreiber from Harvard University in the US [10–12]. They proposed that, as opposed to target-oriented synthesis (TOS) and medicinal or combinatorial chemistry, seeking limited regions of the chemical space, diversity-oriented synthesis (DOS) would allow one to occupy chemical space more broadly with small molecules having diverse structures.

Organic chemistry has recognized the strength of Nature in designing potent antibiotics and drugs. As stated, natural products, their derivatives, and their analogues constitute the most important sources for new drug candidates and tools for chemical biology and medicinal chemistry research. Therefore, there is a need for the development of efficient synthesis methods which give access to natural-inspired compound libraries [13–15].

Visualization of the chemical space available for lead optimization is made possible with the aid of a wide range of software that can even operate on simple PCs. In the example illustrated in Figure 5.5, the antihypertensive drug Captopril™ (Bristol-Myers Squibb) is shown with atoms circled by their hydrodynamic volumes

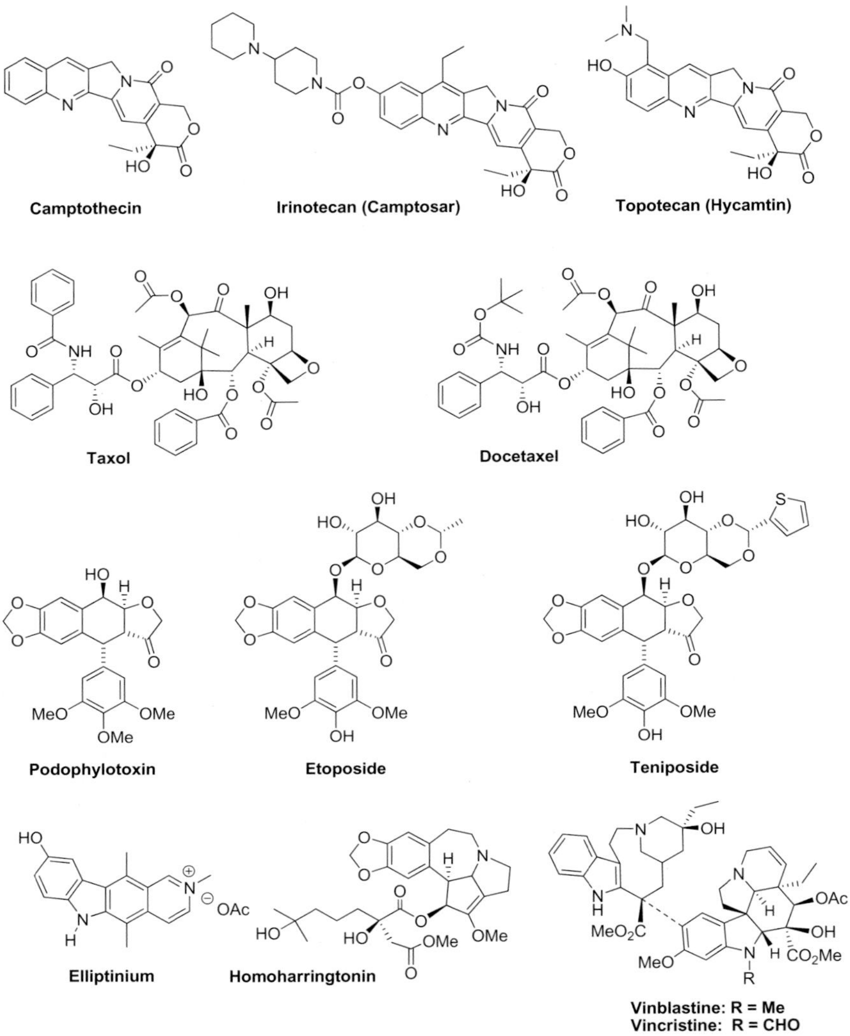

Figure 5.4 Plant-derived anti-cancer agents in clinical use [8].

wherein binding interactions can occur through electrostatic, hydrogen bonding or hydrophobic interactions.

5.2
From Bench to Market Place

Enormous efforts and money have been put in place for each new drug that has reached the market place [16, 17]. It is estimated that 12 to 15 years and more than

Figure 5.5 The hypertension drug Capoten™ as a ball and stick 3D model showing chemical space available for binding interactions. (Please find a color version of this figure in the color plates.)

half a billion dollars are necessary to get one drug approved by the Food and Drug Administration (FDA) (Figure 5.6). Surprisingly, and in spite of all the modern techniques and knowledge involved in drug discovery, the number of successful drug candidates has steadily decreased over time. Admittedly, however, the rewards for every winning candidate are huge since each of the "blockbuster drugs" can bring in several billions of retail dollars per year (see Table 5.1). Hence, the market for the drug Lipitor™ (Pfizer), used to lower Low Density Lipoprotein (LDL) cholesterol levels and the number one drug in 2008 was valued at $5.88 billion sales per year. As of May 2010, the FDA has registered 1495 drugs with a molecular weight < 2000. The number of new drugs that have been approved by the FDA per year (also called NME for "New Molecular Entity") averages 20 NME per year (Figure 5.7) [19].

Statisticians estimate that nearly half of the life expectancy increase attained over the last 15 years in industrialized countries can be attributed to new drugs. Private industry tends to fund virtually all discoveries and development of new medicines, often building on basic medical hypotheses developed through universities and publicly funded research institutions. Industry scientists searching for a new drug typically must screen (test) 5000–10000 new promising chemical structures in order to identify a pool of approximately 250 compounds that then enter into preclinical testing. Of those 250 compounds, fewer than 10, on average, will show enough potential to qualify for Phase I human testing to establish basic safety.

Phase I trials include a very small group of healthy volunteers who are tested to determine whether the drug candidate is both safe and effective. A compound that makes it through Phase I then enters small-scale Phase II trials in patients with a specific condition to test whether the compound has the targeted effect on the disease. If shown promising, it undergoes Phase III trials, which are wide-scale

Bringing a new drug to market can take 15 years

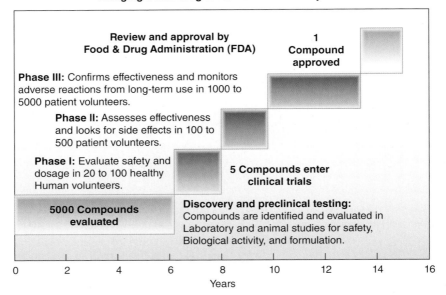

Figure 5.6 Timescale and steps involved in new drug development. Adapted from [17].

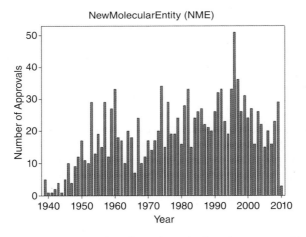

Figure 5.7 The number of new drugs that have been approved by the FDA per year [19].

tests involving thousands of patients in carefully controlled clinical settings. A few drug candidates undergo several different types of Phase III trials in order to test for various effects. On average, for every five compounds that make it into human trials (of the original 5000–10 000 studied), US government authorities will grant the pharmaceutical company approval to market just one.

Continued investor confidence has enabled large-scale research and development to continue in the pharmaceutical industry, in the United States and in other countries. Pfizer alone is investing over $8 billion this year and employing more than 12 000 scientists in the search for new cures, with significant investments in cardiovascular disease, cancer, HIV/AIDS and other infectious diseases, central nervous system (CNS) afflictions, and a wide range of other chronic and acute diseases [20].

Table 5.1 below, extracted from a list of the top 200 brand-name drugs classified by their retail dollars in 2008, provides a useful overview of the top 20 chemical entities and their targeted mode of action that have gone through the entire drug development process [18]. As can be seen by the total amount of retails (almost $60 billion) for only those 20 "blockbuster drugs" sold worldwide, the drug industry may appear to be in a pretty good financial situation. However, given the high cost spent for development, the late attribution of residual patent life expectancy, the international competition on generic, and the fierce competition between "big pharma" for a given medical target, the increasing cost for late clinical trials, these amounts of money may not necessarily secure survival.

Table 5.1 Top 20 brand-name drugs (2008) [18].

Rank	Name	Sales ($ billion)	Company	Mode of action	Structure
1	Lipitor	5.88	Pfizer	HMG-CoA reductase inhibitor used to lower LDL cholesterol levels	
2	Nexium	4.79	AstraZeneca	Proton pump inhibitor used to treat heartburn and esophagitis	
3	Plavix	3.80	Bristol-Myers Squibb	Platelet aggregation inhibitor used to reduce stroke and heart attack risk	

Table 5.1 *Continued*

Rank	Name	Sales ($ billion)	Company	Mode of action	Structure
4	Advair Diskus	3.57	GlaxoSmith Kline	Corticosteroid and bronchodilator used to treat and prevent asthma	
5	Prevacid	3.30	Abbott	Proton pump inhibitor used to treat gastric reflux disease	
6	Seroquel	2.91	AstraZeneca	Antipsychotic used to treat schizophrenia and bipolar mania	
7	Singulair	2.90	Merck	Leukotriene receptor antagonist used to treat asthma and allergies	
8	Effexor XR	2.66	Wyeth	Serotonin and norepinephrine reuptake inhibitor used to treat depression	

Table 5.1 *Continued*

Rank	Name	Sales ($ billion)	Company	Mode of action	Structure
9	OxyContin	2.50	Purdue	Opioid analgesic used to treat moderate to severe pain	
10	Actos	2.45	Lilly	Used to treat type 2 diabetes	
11	Lexapro	2.41	Forest Laboratories Inc.	Selective serotonin reuptake inhibitor used to treat depression and anxiety	
12	Abilify	2.37	Ostuka	Antipsychotic used to treat schizophrenia and bipolar mania	
13	Topamax	2.18	Johnson-Johnson	Carbonic anhydrase inhibitor used to treat seizures and prevent migraines	
14	Cymbalta	2.17	Lilly	Serotonin and norepinephrine reuptake inhibitor used to treat depression	

Table 5.1 *Continued*

Rank	Name	Sales ($ billion)	Company	Mode of action	Structure
15	Zyprexa	1.75	Lilly	Used to treat schizophrenia and bipolar mania	
16	Valtrex	1.68	GlaxoSmith Kline	Antiviral agent for treating shingles, cold sores, and genital herpes	
17	Crestor	1.68	AstraZeneca	HMG-CoA reductase inhibitor used to lower LDL cholesterol levels	
18	Vytorin	1.55	Merck	Statin and a cholesterol absorption blocker used to treat high cholesterol	
19	Lamictal	1.54	GlaxoSmith Kline	Anticonvulsant used to treat seizures and bipolar disorder	
20	Celebrex	1.53	Pfizer	COX-2 inhibitor NSAID used to treat arthritis pain	

Interestingly, only number 16 (Valtrex™) is used to fight viral infections, while most of the others are essentially directed at "modern" human diseases such as depression, gastric ulcers, and cardiovascular diseases caused by high blood cholesterol. From these top 20, AstraZeneca ranked first with 3 drugs ($9.38 billion), followed by Pfizer (2 drugs–$7.41 billion), GlaxoSmithKline (3 drugs–$6.79 billion), and Lilly (3 drugs–$6.37 billion).

5.3
General Concepts of Drug Design

The following aspects concern the roles of medicinal chemists and chemistry in general in new drug development. There are actually two main stream approaches in the drug discovery process [21]. In the classical one, the synthetic chemists were initially confronted with already identified protein or enzyme targets, several of which having their 3D structures determined by either X-ray crystallography or sophisticated nuclear magnetic resonance (NMR) techniques. The strategy was then to screen thousands of chemical entities originating from the three main sources identified above ((i) in-house library, (ii) synthetic–often commercial library, (iii) natural product collections from plants or marine products). The entire process is now supported by robotized HTS technologies that can screen thousands of compounds daily, hence putting a great pressure on traditional organic chemistry. Once a "hit" is identified, it needs to be optimized into a "lead" compound which is not yet, or sometimes not even close to, what the final drug may look like. This step is facilitated by well-trained medicinal chemists who can usually recognize the "hot spots" (functionality) of the molecules that can undergo bioisosteric modifications. That is to say that some functional groups can be replaced by others having similar physico-chemical properties or others that have to be ignored because they can be readily predicted as rendering the molecules metabolically unstable. In this situation, the medicinal chemists are assisted by the so-called Lipinski's rule of five [22, 23]. This rule helped by making suggestions on chemical modifications that will be needed in order for the compounds to go through the ADMET (absorption, distribution, metabolism, excretion, toxicity) assessment (Figure 5.8). The goal of this particular stage is to target orally active drugs that need to go through the intestinal tract, pass through the lipid bilayers of the epithelium, and eventually get into the blood stream were the molecules can reach their intended targets.

Through the optimization phase, the drug candidate will undergo several chemical modification cycles that will make the molecules more active (low IC_{50}s), less toxic (low LD_{50}s) and more effective. They are likely to obey most of the Linpinski's rule, although several exceptions to it exist [22, 23]. The chemists are also assisted in this task by structural biologists or, nowadays, by computer chemists, who can make the candidates "fit" into the protein's active site physically or virtually on computer screens.

(a)

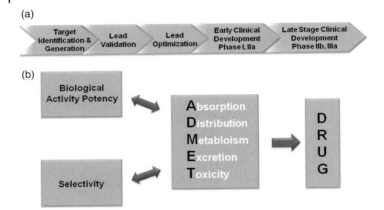

(b)

Figure 5.8 (a) The sequence of events involved in drug discovery and selected key technologies/activities. There is an overlap of their activities along the value chain, some might be needed along the entire value chain, for example, genomics/proteomics or biomarker (eADMET = early absorption/distribution/metabolism/excretion/toxicity; QSAR = quantitative structure activity relationship. (b) Schematic representation of the reiterative steps involved in the drug development cycle. Adapted from [21].

Lipinski's rule of five

According to the rule, poor absorption is likely if:

- the active substance contains more than five hydrogen bond donors,
- the molecular weight is higher than 500,
- the distribution coefficient LogP is more than 5, and

- there are more than ten hydrogen-bond acceptors.

This rule explicitly does not apply to active transporters. The rule of five is so named because all classification parameters are multiples of five.

In the second, more recent strategy, the process undergoes chemical biology sophisticated approaches that were initially thought to accelerate the lead identification timescale by skipping the target protein/enzymes/receptors isolation and characterization [24–26]. This planning concentrates on cell phenotype modifications that then needs system biology in order to identify possible biosynthetic pathways that the drug candidates down- or up-regulate.

5.3.1
Tasks and Bottlenecks in Medicinal Chemistry

Consequently, the main contributions and tasks of chemistry in pharmaceutical research now consist of: (i) identification of new leads, (ii) their optimization to clinical candidates, and (iii) the provision of sufficient amounts of these substances for further studies and for development [21]. The fundamental issues in lead optimization are:

- What biological testing systems are required to select the clinical candidate?
- What question do we want to answer by using a particular compound and what can we learn from the answer(s)?
- How many and which compounds need to be synthesized to recognize the potential of a lead?
- Which structural elements (substructures) determine the profile of biological characteristics?
- Can the profiles of a class of substances be predicted or extrapolated to other leads?
- Which synthetic methods yield the highest throughput?

As a result, the demand to be made on chemists and the tasks that lie ahead are obvious:

- The lead must be rapidly optimized,
- The improvement of several parameters must be worked on simultaneously,
- Ecologically and economically acceptable syntheses must be developed, and
- The ADMET parameters must be optimized and their efficacy increased as early as possible in the development.

In summary, the following few notions can be presented. The bottleneck in lead generation lies in the provision of new, biologically relevant substances and, hence, largely in the field of chemistry. The high-throughput technologies have not led to the desired innovation and productivity, and it is only by means of new, knowledge-based approaches in joint efforts undertaken by chemistry and biology that we will be able to "synthesize into" the complexity of biological structure spaces (Table 5.2) [21].

5.3.2
Lead Validation

The lead validation involves demonstrating the relevance of a target in a disease process, that is, whether the desired effect on the disease can be achieved by influencing a biological system. Definitive proof of a causal relationship such as this can only be provided by clinical studies on patients. Clearly, it is vital that target validation should be performed in the early phases and repeated in the further course of drug research development. Target validation in the early phases is based on several criteria, for example, different gene expression in healthy and diseased tissue, functional studies at the protein level, and studies in animal models on the disease-related phenotypes. Gene chips, antisense technologies, ribosomes, neutralizing antibodies, and knockout of transgenic mice are some of the methods and instruments used in this modern context [21].

Table 5.2 Expertise and technologies needed at different phases of drug development.

• Genomics/Proteomics including Bioinformatics	• High Throughput Screening	• eADMET	• Formulation
• Transgenics	• Combinatorial Chemistry	• Parallel Synthesis	• Kinetics, Metabolism • Toxicology
• Assay Development	• Virtual Screening	• QSAR-Models including Chemoinformatics	• Clinical Pharmacology
• Structural Biology	• Drug Design	• Chemical Development	• Biomarker
		• Early Drug Supply and Large-Scale Synthesis	

5.4
Patent Protection Issues

After having committed so much time, financial resources and work, it should be no surprise that pharmaceutical companies wish to protect their investment for the longest possible period of time. This is why new drugs are patented in as many countries as possible. This implies that no other competitor in those countries can sell the same drug. Additionally, it is also very expensive to maintain patents as maintenance fees need to be paid annually to enforce protection. Once the duration of a patent protection is terminated, the competitors can then freely enter into the market and start to sell copies of the medication. The drugs coming from competitors are the very same molecules and are called "generics". Most people are generally afraid to buy generics because they ignore this fact. You can always ask your pharmacist if a generic exists, if so, the cost is usually much less [27].

A patent's term is the maximum period of time during which protection occurs. It is normally expressed in number of years starting from the filing date of the patent application or from the date the patent is issued (granted). In most patent laws, renewal annuities or maintenance fees have to be paid regularly in order to keep the patent in force, otherwise the patent lapses before its term.

Significant international harmonization of patent term across national laws was provided in the 1990s by the implementation of the World Trade Organization's (WTO) Agreement on Trade-Related Aspects of Intellectual Property Rights (TRIPs Agreement). The agreement insures that the: "The term of protection available for patents shall not end before the expiration of a period of 20 years counted from the filing date".

Consequently, in most patent laws nowadays, the term of patent is 20 years from the filing date of the application. In the United States, under current patent law, for patents filed on or after 1995, the term of the patent is 20 years from the earliest claimed filing date. For patents filed prior to 1995, the term of patent is either 20 years from the earliest claimed filing date or 17 years from the issue date, whichever is longer. The exact date of termination is an important issue, particularly where daily profits from a pharmaceuticals patent amount to billions of dollars.

It usually takes at least a few years between the granting of patents and marketing approval. This means that, despite the standard 20-year patent life, the average effective patent life for a new drug – the amount of time where the product is sold under patent protection – is roughly 10 to 12 years. In addition to the direct costs of development, firms must pay returns on the capital they invest on behalf of shareholders over the course of a decade or more, a cost that increases as development time increase.

At current levels of return, economists estimate that only 30% of new drugs actually earn enough revenue during their patented product life cycle to cover the average development cost. If a firm incurred the average cost of drug development and only invented "average" drugs, it would quickly go out of business.

5.5
Drug Metabolism and Drug Resistance or Why Make Big Pills?

Drug dosages have to be established early on in the marketing goals and this is another criterion to establish during the initial clinical trials. Indeed, a drug may appear to be very active in a test tube or in a cell assay (low IC_{50}) but after all, it has to be active in an average size human body. These trials are, therefore, keys for accurate dosage determination. In this regards, several factors need to be accounted for, among which, the capacity of the body to more or less rapidly eliminate foreign substances. After all, this is what a drug is – a non-natural chemical entity that the body sees for the first time. Several mechanisms are in action for this process and metabolism/excretion is the major one. Consequently, the amount of active ingredient in a given pill must be carefully determined to provoke the medical illness healing before adverse effects can be observed. This dosage does not take into consideration the pills' excipients (an excipient is generally a pharmacologically inactive substance used as a carrier for the active ingredients of a medication), color, preservatives, coatings, and so on. Our understanding of drug metabolisms has steadily improved over time and, for some of them, prediction is possible. When this is the case, the medicinal chemists provide the necessary efforts to chemically modify the "lead" compounds in order to take their possible site of metabolic transformations into consideration. This is often achieved, for example, by adding a non-reactive atom in the appropriate position of the molecules, such as replacing an aromatic hydrogen by an inert fluorine atom [28–30].

5.5.1
Drug Metabolism

Once a drug enters into the blood stream, in order to reach its target receptors, several factors may contribute to its deactivation or excretion in the urine, thus accounting for certain loss in the real quantity absorbed. This important factor is always taken into consideration when dosages are prescribed. The consumers always prefer small and attractive pills (for example, Viagra as a blue pill–blue for men) as well as the minimum daily uptake because they tend to forget. Parts of these issues are taken care of at the formulation level. Drug metabolisms constitute pathways for their deactivation and drug elimination. This part and the entire ADMET process constitute one of the greatest obstacles in new drug discoveries and medicinal chemists are taking great care of them. There are two kinds of drug metabolism: Phase I transformations and Phase II transformations [28].

5.5.1.1   Phase I Transformations

Drug metabolisms are biochemical processes that your body utilizes to "destroy" these foreign and unnatural chemical entities. The liver is the most important organ involved in this action. The liver contains several different kinds of enzymes that are present to clean these unwanted molecular species from your body and the family of enzymes called cytochrome P-450 is particularly responsible for transformation of drugs into novel species that are eventually rapidly excreted in urine, all this in order to protect your body from noxious substances. As you can imagine, when new drugs are under development, these potential, most of the time unpredictable, transformations have to be taken into account because they may considerably reduce the real amount of the drugs necessary to counteract the deleterious action of the diseases or infections. Moreover, these transforming enzymes may change the chemical structures of the drugs into inactive species or, sometimes, into active metabolites for which we cannot predict whether they would be toxic or not and, hence, produce undesirable side effects. Chemical Phase I transformations most often involve oxidation, reduction, and hydrolysis by appropriate enzymes.

5.5.1.2   Phase II Transformations

Once transformed into these new chemical entities, your body will further carry out additional chemical reactions that normally assist the drugs to be excreted rapidly. This is done by conjugation to polar natural products of your body (amino acids, sugars, etc) and these reactions are called Phase II transformations. Additionally, your intestinal bacterial flora will also contribute to drug conversions.

5.5.2
Drug Resistance

Another important obstacle encountered during the life time of a drug is the high propensity of cancer cells, bacteria and viruses to develop drug resistance. Drug

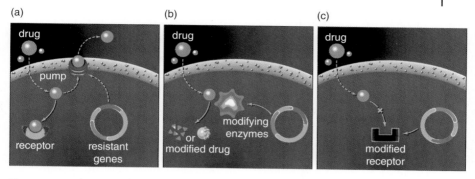

Figure 5.9 Mechanisms of bacterial drug resistance: (a) active efflux; (b) enzymatic modification of the drug; (c) modification of the target receptor or enzyme. Taken from [31]. (Please find a color version of this figure in the color plates.)

resistances are acquired because the pathogenic species have also recognized the drugs as foreign and unwanted molecules that they wish to neutralize. This phenomenon is usually exacerbated by the patients' bad habit of stopping their medication as soon as they feel better, usually after less than 10 days. The consequence is that, in the preliminary stage of the treatment, only the weakest microbes are destroyed, leaving behind the strongest ones. These pathogens then initiate mutations or acquire them (via plasmids) through which novel enzyme machineries develop. The roles of these enzymes are exactly to neutralize the drug by chemical modifications that make the drugs no longer fit into the initially targeted receptors [31, 32]. Even human cancer cells have developed their own aptitude to modify anticancer drugs. There are also other modes of drug resistance, some of which are discussed briefly below (Figure 5.9) [31].

For instance, three life-threatening bacterial species have developed resistance to all available antibiotics. They are: *Enterococcus faecalis, Mycobacterium tuberculosis, and Pseudomonas aeruginosa. E. faecalis* is known to transfer resistance genes readily and a recent outbreak of the virulent *Staphylococcus aureus* suggests that it has also acquired a resistance gene plasmid. The major causes of bacterial drug resistance are twofold: the overuse of antibiotics in humans and animals, and the noncompliance to the prescribed treatment duration. Drugs' bacterial resistance can be mediated by one or more of the following mechanisms, the first three being dominant factors:

1) Prevention of the drug from reaching its target, either by active efflux from or reduced uptake into the cells, as well as by sequestration of the antibiotic by receptor/protein binding. The latter aspect often occurs with human serum albumin normally present in high concentration in the plasma.

2) Deactivation of the antibiotic by enzymatic modification.

3) Modification of the drug's target, thereby eliminating or reducing the binding of the antibiotic.

4) Metabolic bypass of the inhibited reaction.

5) Overproduction of the antibiotic target.

5.6
Antibacterial Agents

In general, antibiotics are classified by target and mechanism of action. Thus, antibiotics that interfere with the biosynthesis and function of the bacterial cell wall provide excellent strategies toward drug development. Bacteria are categorized as Gram-negative or Gram-positive, depending on the structures of their respective cell wall. Figure 5.10 illustrates the fundamental differences that exist between the two categories. The major difference between the two is the second thick outer membrane above a common, but different, peptidoglycan (polysaccharide/polypeptide or murein) coat present in Gram-negative bacteria. The surface of Gram-positive bacteria ends with teichoic acids (phosphorylated polysaccharides) [31].

The peptidoglycan layers in both bacteria provide the required strength and rigidity of the cell envelope, thus allowing them to survive in a hypotonic environment and giving them their characteristic shapes. Therefore, antibiotics that can induce structural defects in the cell wall provoke cell lysis as a consequence of the high internal osmotic pressure. There is no equivalent structural unit in mammalian cells; therefore, medicinal chemists have a unique opportunity to destroy bacteria in a selective manner.

Glycopeptide antibiotics. such as vancomycin and teicoplanin, are other drugs that can inhibit proteoglycan biosynthesis by binding to the D-Ala-D-Ala portion

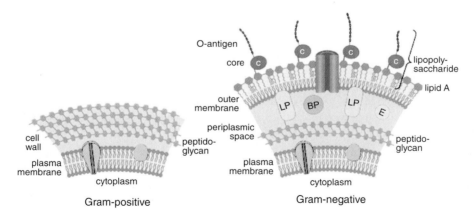

Figure 5.10 The cell envelope of Gram-positive and Gram-negative bacteria. Gram-positive bacteria have a thicker peptidoglycan layer than Gram-negative bacteria, but lack the outer-membrane at the cell surface. Taken from [31]. (Please find a color version of this figure in the color plates.)

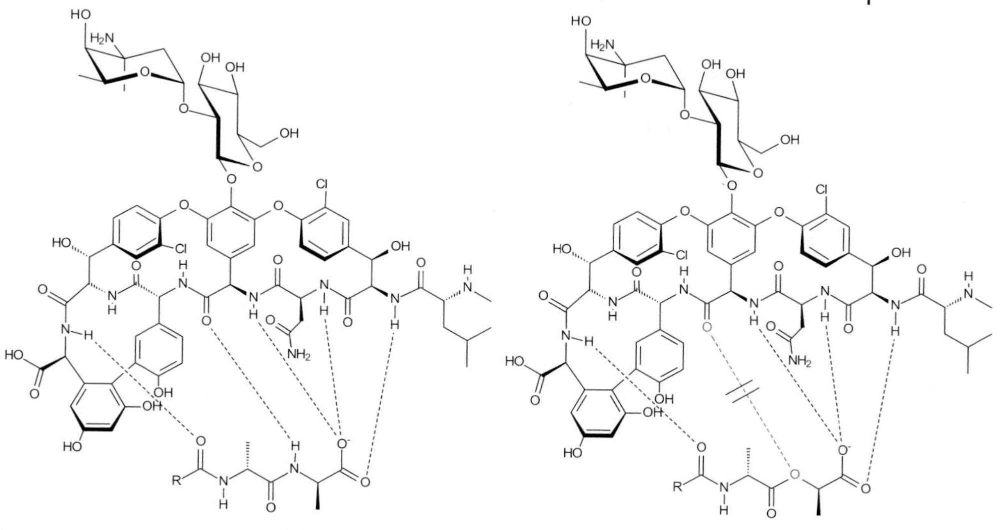

Figure 5.11 An example of bacterial mutation leading to drug resistance. Vancomycin binding to the D-Ala-D-Ala portion of cell wall peptidoglycan precursors is mediated by 5 hydrogen bonds. Mutation of the last D-Ala for D-lactate reduces the number of H-bonds to 4 and a decrease of affinity by 1000-fold.

of the cell wall, thereby preventing *trans*-glycosylation and cross-linking of the peptide side-chains. They are considered as last resource antibiotics since they are useful against bacterial infections that are resistant to other classes of antibiotics, in particular methicillin-resistant *S. aureus* (MRSA). In spite of this, resistance against vancomycin has been observed and occurs when bacteria mutate the composition of their peptidoglycan by using a D-lactate or D-serine instead of the second D-Ala residue (Figures 5.11 and 5.12).

Although intensive structure–activity relationship (SAR) studies have been done on the aglycon portion, the role of the carbohydrate moiety is still unclear. *In vitro* activity of vancomycin lacking the sugar portion has barely any effect, but *in vivo* activity is lost or reduced, thus supporting their implication in the pharmacokinetic properties of the drugs.

5.7
Antiviral Agents: The Flu Virus Story: The Naissance of a Sugar-based Flu Drug

In spite of its apparently innocuous effects, influenza infection continues to be the most serious respiratory disease in both morbidity and mortality. Vaccines are only prophylactic remedies with problems associated with the highly variable mutation of the viral coat proteins. Current available treatments, such as amantidine and its analog rimantidine, were, until recently, the only approved

(a)   (b)

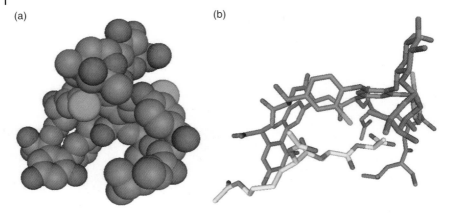

Figure 5.12 CPK model of (a) Vancomycin and (b) Vancomycin bound to the *N*-acetyl-D-Ala-D-Ala residue (PDB No. 1FVM). (Please find a color version of this figure in the color plates.)

compounds against influenza type A. These drugs are ineffective against influenza B and have side effects. They, together with other "cage hydrocarbons" such as homotwistane, are believed to inhibit viral penetration or other early steps of the replicative cycle by increasing the surrounding pH.

Influenza virus particles are spheroidal and approximately 100 nm in diameter. The outer-membrane envelope contains ~500 copies of rod-shaped spikes composed of hemagglutinin (HA) trimers and ~100 copies of mushroom-shaped spikes composed of neuraminidase tetramers (Figure 5.13). The hemagglutinin constitutes the receptors for α-sialoside ligands. Its X-ray structure has been refined at 3 Å resolution including one report in which the HA has been co-crystallized with sialyllactose, part of a ligand present on the natural glycoprotein receptor in our respiratory tracts. The X-ray analysis showed the sialic acid binding pockets to be ~40 Å apart. The neuraminidase catalyzes the hydrolysis of α-sialoside linkages from the host sialyloligosaccharides. It facilitates the transport of the virus to and from the infection site by permitting its passage through the sialic acid-rich mucins and presumably from the infected cells where it helps the release of new-born virion progeny. The X-ray structure of this glycoprotein has also been refined at 2.9 Å resolution.

The initial step in the cascade of events leading to influenza virus infections is mediated by the tight attachment of virus particles to host cell receptors. These receptors are constituted of cell surface sialylated glycoproteins and glycolipids (gangliosides). The following steps involve receptor-mediated endocytosis, where the acidic pH is postulated to trigger viral membrane conformational change that allows its fusion to that of the host with ensuing release of the viral nucleoplasmid (Figure 5.13). The first event responsible for the receptor–virus interaction is, therefore, an attractive target for antiviral intervention.

The discovery of the first anti-flu drug is an interesting one as it was truly developed by rational drug design [33, 34]. Indeed, it was known for a long time that

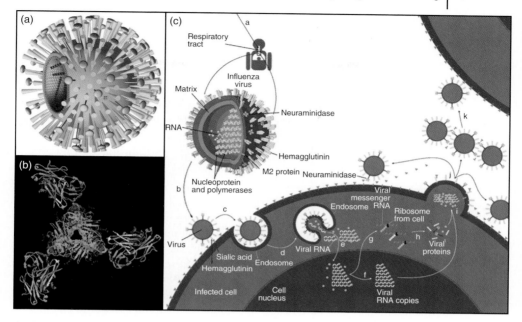

Figure 5.13 Typical representation of a flu virus virion 100 nm wide (a) and covered with hundreds of copies of two surface proteins: a neuraminidase enzyme responsible for infected cell detachment and a hemagglutinin trimers (b) responsible for host receptor adhesion. (c) Influenza flu virus life cycle [48]. (Please find a color version of this figure in the color plates.)

the naturally occurring inhibitor of the viral neuraminidase, the *"original lead"* shown in Figure 5.14 and which constituted a biosynthetic degradation product of *sialic acid* released from its glycoprotein receptor upon the action of the virus enzyme, had low binding affinity against the enzyme. The scientists involved had the opportunity to work in close proximity with an X-ray crystallographer who obtained co-crystals of the enzyme with its bound "lead". The resulting X-Ray data permitted localization of the active functionalities (called pharmacophores) implicated in the intricate enzyme's active site [35]. The first prototype analog resulting from these findings was *Relenza*™ (Zanamivir) from GlaxoSmithKline. Unfortunately, the drug was not orally active because it was too polar and was excreted in the urine very rapidly and has, therefore, to be inhaled. Obviously its sales on the market were quickly overridden by a second generation drug *Tamiflu*™ (Oseltamivir, Roche) that rapidly became a blockbuster drug belonging to the top 200. In fact it is at position 122 with an annual retail value of $0.28 billion.

As can be readily seen from the progression of the chemical structures illustrated in Figure 5.14, the optimization process from the *"original lead"* through *Relenza* and finally into *Tamiflu* is quite impressive. In fact, *Tamiflu* is 4000 times more potent than the original drug candidate. It was the role of the medicinal chemists to recognize the pharmacokinetic weaknesses of *Relenza* and to capitalize

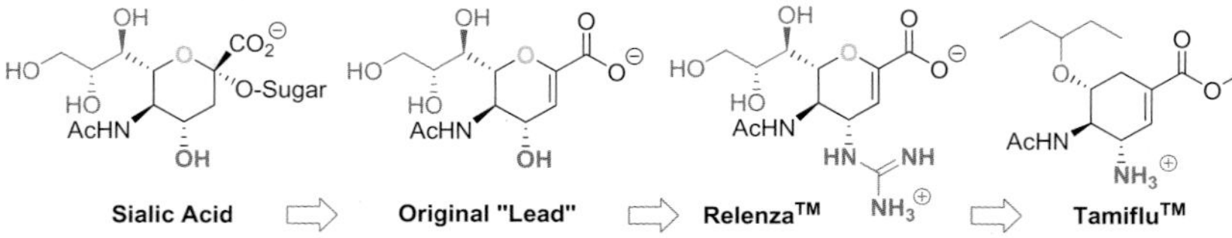

Figure 5.14 A good example of a lead optimization that led to the first anti-flu drug Tamiflu™. (Please find a color version of this figure in the color plates.)

on the fact that the enzyme had a hydrophobic pocket in the glycerol side chain, shown in blue in the structures, an important feature that was not used in *Relenza*. The inventor of *Relenza* had, however, previously noticed the string of anionic amino acids of the enzyme near the hydroxyl group shown in red. This was why they converted the neutral hydroxy group into a counter-acting cationic guanidine function, which later simply became an ammonium group in the final winning *Tamiflu* version. Notice that the anionic carboxylic acid function, known to be a key binding element for the enzyme, was not preserved in the final drug as it is now a neutral ester. Nevertheless, this is simply a "faked" situation since the drug is orally active as it is rapidly hydrolyzed by the body's enzymes. Thus, the *Tamiflu* is considered as a "prodrug".

5.8
The Viagra Story – Serendipity Leading to a Blockbuster Drug

Sildenafil® citrate (Viagra – Reva tio) was discovered by coincidence in 1989. It was introduced onto the market by Pfizer in 1998. Researchers from Pfizer were initially working to identify enzyme inhibitors for the treatment of high blood pressure and angina. During initial clinical trials, it was rapidly observed that the medication was not active against the intended treatment but had interesting side effects as it markedly increased erectile function. It quickly became a blockbuster drug and ranked 38 in the list of top 200 drugs with sales reaching $0.92 billion each year (2008). The exclusive patent will end in June 2011 and it is thus expected that generics will arise at much reduced prices [6].

Viagra was the first selective inhibitor against the phosphodiesterase 5 (PDE-5). It constitutes an orally active treatment against erectile dysfunction, a condition affecting more than 30 million American males. It is also active against pulmonary arterial hypertension caused by narrowed arteries of the lungs, a condition resulting in reduced oxygen supply in the body (ischemia) and other damage, including heart failure. During research on the identification of PDE-5 inhibitors, the lead candidate, later called Sildenafil™, was found potent against the enzyme (Figure 5.15).

Among other factors, erection is initiated by the release of nitric oxide (NO), a short-lived chemical in the body. NO was later found to be the active species in endothelial cells responsible for the relaxation of blood vessels. Nitric oxide was

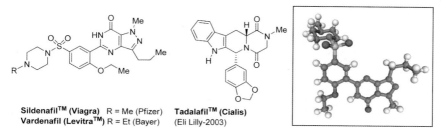

Sildenafil™ (Viagra) R = Me (Pfizer)
Vardenafil (Levitra™) R = Et (Bayer)
Tadalafil™ (Cialis)
(Eli Lilly-2003)

Figure 5.15 Chemical structures of Viagra and related drugs together with its 3D depiction. (Please find a color version of this figure in the color plates.)

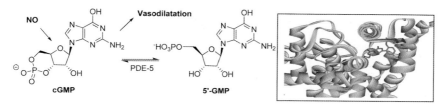

Figure 5.16 Viagra works by blocking the action of the PDE-5 enzyme leading to restoration of nitric oxide (NO) in vasodilatation. Viagra in the active site of PDE-5, the zinc and magnesium ions are indicated by blue and green spheres, respectively (PDB-3JWQ). (Please find a color version of this figure in the color plates.)

later identified as an endothelium-derived relaxing factor (EDRF) that plays many roles in cardiovascular physiology. The physiological process essential for erection is thus caused by the release of NO that activates an enzyme called guanylate cyclase leading to an increased production of cyclic guanosine monophosphate (cGMP) (Figure 5.16). Then, cGMP initiates a cascade of biochemical processes resulting in vasodilatation in the *corpus cavernosum* that causes penile blood flow and hence erection. PDE-5 tends to reverse the process by catalyzing the conversion of cGMP into guanosine monophosphate (5′-GMP). Viagra increases cGMP levels and improves erectile ability. Two other PDE-5 inhibitors Vardenafil (Levitra™ drom Bayer) and Tadalafil (Cialis™ from Eli Lilly) entered the market in 2003 as competitive drugs (Figure 5.15).

Therefore, after hit and lead compounds are identified through early screenings, the role of the medicinal chemists is to provide synthetic schemes that are amenable to large scale synthesis, that is, 1 kg for initial clinical trials. The synthesis has to be as short as possible, take environmental issues into consideration, use low cost reagents, and, whenever possible, the least possible purifications steps through chromatography. The original synthetic schemes initially devised by academic researchers are often not valid under the above limitations. Consequently, the final synthesis is usually modified by the expert team of process chemists that are accustomed to working under these stringent constraints. The synthesis of Sildenafil™, described in Figure 5.17, represents such an accomplishment [36]. As can be noticed, very simple and inexpensive reagents are used throughout.

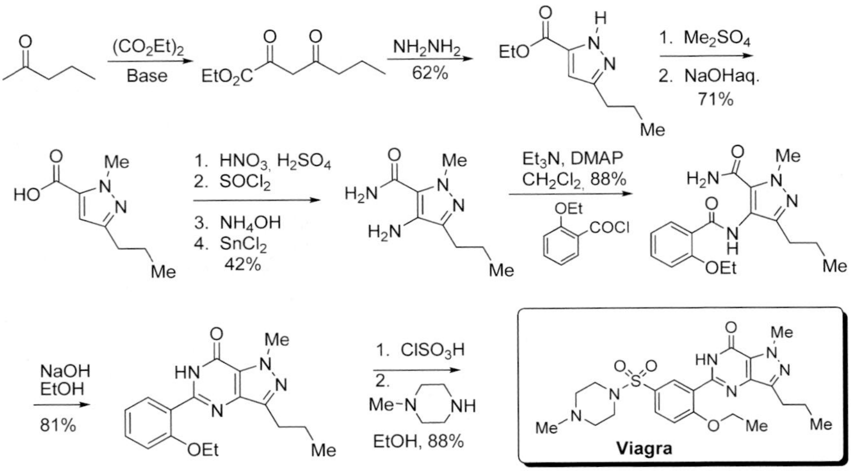

Figure 5.17 Typical chemical synthesis design en route to Sildenafil™ (Viagra) [36].

5.9
Human Vaccines as a Prophylactic Health Remedy

5.9.1
Carbohydrate-based Vaccines

Bacterial, parasitic, and viral infections constitute one of the major health problems in developing countries. As we have seen from the list of top 20 drugs (Table 5.1), in developed countries we are more likely to suffer from cardiovascular diseases, depression and stress. Drug resistance has rapidly increased and new drug development has become prohibitive. Therefore, the development of new prophylactic and potent therapeutic vaccines to fight these infections is of the utmost importance and organic chemistry has made fundamental progress in this direction. The cell-surface of most pathogens and cancer cells exposes several structurally variable and sometimes cryptic carbohydrate molecules that are directly presented to the immune system (Figure 5.18). Hence, by triggering a specific and high affinity immune response against particular pathogenic cell surface carbohydrate antigens, protective and long-lasting immunity can thus be reached. This strategy has been successfully applied against bacterial capsular polysaccharides and several such glycoconjugate vaccines are now commercially available: *Neisseria meningitidis*, *Streptococcus pneumoniae* (23 serotypes), *Haemophilus influenza* type b (Hib), and *Salmonella typhi* [37–42]. For instance, with the advent in the 1990s of protein conjugate vaccines against the Gram-negative organisms *Haemophilus influenzae* type b, the leading cause of meningitis and pneumonia, particularly in infants for which non-conjugated vaccines failed, the disease has been essentially eradicated where national vaccination programs have been implemented. Each

(a)

(b)

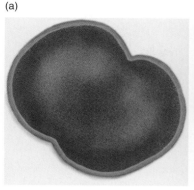
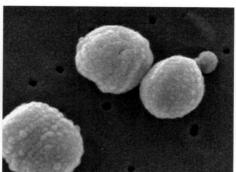

Figure 5.18 (a) The bacteria *Neisseria meningitidis* is surrounded by a capsular polysaccharide against which antibodies can kill the bacteria. (b) *Streptococcus pneumonia* is presently the number one killing bacteria worldwide. (Please find a color version of this figure in the color plates.)

year this bacteria kills more than half a million infants in these countries. This section will briefly highlight recent trends in carbohydrate-based synthetic vaccines using antigens prepared in the author's own laboratory.

5.9.2
The Role of Chemistry in Synthetic Vaccines

Historically, carbohydrate-based vaccines, like others (e.g., flu, small pox, etc.), were prepared from killed or attenuated bacteria or viruses. Nowadays, purified antigens (the chemicals that stimulate an immune response) are required, but for these, generally T-cell independent antigens, conjugation to T-cell dependent protein carriers, capable of immune cell activation through the immune cell machinery, is still necessary. This is because the carbohydrate antigens alone do not stimulate memory effects and protection only lasts for less than two years. Activating immune cell memories by attachment to proteins, acting as T-cell epitopes to produce high affinity antibodies, can provide protection for more than ten years. With increasing and more rigorous regulatory issues for new vaccines approval, it is clear that well-defined synthetic carbohydrate antigens are appealing.

5.9.3
Bacterial Capsular Polysaccharide Vaccines

Despite the inefficacy of bacterial capsular polysaccharides (CPS) alone to mount long-lasting protective immune responses in infants, they are still considered to be the most attractive prophylactic targets. This is because they have highly conserved structures shown to provide reliable sources of anti-bacterial antibodies

(a)　　　　　　　　　　　　　　　　　　　(b)

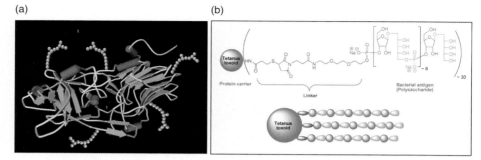

Figure 5.19 (a) Schematic representation of a bacterial polysaccharide antigen (yellow balls) linked in several copies to an immunogenic protein carrier (here Tetanus toxoid). (b) Chemical structure and linkage of the synthetic CPS repeating unit of *Haemophilus influenza* type b bound to T. Toxoid. (Please find a color version of this figure in the color plates.)

when properly exposed to the human immune system. This is why their conjugation to protein carriers have led to successful vaccination strategies and candidate vaccines [37–42]. The many victories encountered using several vaccines composed of natural polysaccharide from *Haemophilus influenza* type b and others, when properly conjugated to suitable protein carriers (Vaxem-Hib® (Chiron), Hiberix® (Glaxo Smith Kline Biologicals)) stimulated chemists to consider bringing vaccine technologies to a higher level of sophistication. A team of Canadian and Cuban scientists has recently succeeded in preparing the first fully synthetic polysaccharide antigen on a large scale which, after covalent attachment to tetanus toxoid, provided a commercial vaccine (Quimi Hib®, Heber Biotec, Cuba) that is equally effective to the existing ones originating from bacterial components [43–47]. The synthetic strategy that was selected differs entirely from the previous attempts using stepwise and lengthy synthesis. Rather, synthetic fragments (oligosaccharides) mimicking the capsular polysaccharide were oligomerized in a single step reaction to provide, under controlled conditions, a synthetic polysaccharide (Figure 5.19). The unique linking arm was then covalently bound to a suitably derivatized and highly immunogenic protein (T. toxoid). Thus, once in hand, the repeating units were condensed in a single step process rather than the 16 or so steps required by other methods, including solid-phase chemistry. The structure and the linking arm of this first semi-synthetic vaccine are illustrated in Figure 5.19.

The successful role of synthetic carbohydrate chemistry in this vaccine accomplishment constituted an important step in health care and has paved the way for several other research groups worldwide to pursue their own activities in this field. Nowadays, considerable efforts are directed at other carbohydrate-based vaccines, including research against HIV and parasite infections together with numerous other antibacterial targets. This first commercial synthetic vaccine bearing a carbohydrate antigen has been approved by the World Health Organization (WHO) and received several awards including the Gold Medal from the World Intellectual

(a)

(b)

(c)

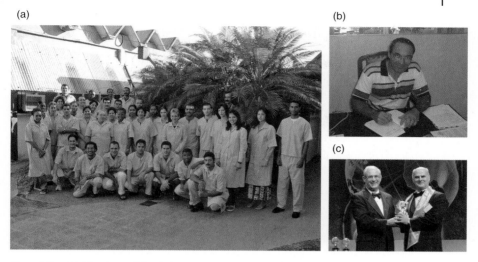

Figure 5.20 (a) Members of the Cuban team that have contributed to the first worldwide semi-synthetic antibacterial vaccines. (b) Dr. Vicente Verez Bencomo heading the Cuban team in his office at the University of Havana, Cuba. (c) Professor René Roy receiving a Tech Museum Award on behalf of the team in 2005. (Please find a color version of this figure in the color plates.)

Property Organization (WIPO) together with an award (health category) from the Tech Museum: Technology Benefiting Humanity in 2005 (Figure 5.20).

5.10
Conclusion

The numerous abilities required of medicinal chemists and the critical role they play in new drug development have been smoothly described throughout this chapter using different case studies. On the outset, it is clear that a deep knowledge of organic chemistry is a basic premise to enter into this demanding, arduous, and somewhat frustrating discipline. The medicinal chemist should also possess a good understanding of biochemical concepts and enzymology together with a fair grasp of structural biology and physical chemistry.

For the general public, this brief introduction to the several obstacles the pharmaceutical companies have to face when dealing with the many steps and ambushes ahead of each new drug to put on the market, will make them aware of the reasons for the high costs of medications. Given that the patent protection of several of the existing drugs will end in 2011, the public will benefit greatly from the flood of generic drugs that will enter the market.

Organic chemistry in general has also benefited greatly from the intellectual pressure put in place by the rapid demand for more chemical entities as well as by the increase environmental regulations and safety issues. Thus, new and creative methodologies have been initiated to generate more compounds more rapidly. In this respect, combinatorial and solid-phase organic chemistry, parallel synthesis, microwave-assisted reaction, and microfluidic reactors have accordingly emerged as useful techniques to ease drug discoveries. We are even talking about green chemistry these days, wherein synthetic "organic" chemistry is increasingly performed under aqueous conditions in order to avoid spreading too much noxious organic solvent into the environment. Microreactors can similarly help perform chemical transformations at the microliter levels.

References

1 Medicine Man of the Cheyenne Painting by Howard Terpning. http://www.firstpeople.us/pictures/Howard Terpning/ls/Howard-Terpning-Medicine-Man-Of-The-Cheyenne.html.

2 Ancient Egypt medicine, in *Days of the Pharaohs. Fresh Look at the Civilization of the Pharaohs*, http://pharaohsdays.blogspot.com/2010/05/ancient-egypt-medicine.html.

3 Steiner, R.P. (ed.) (1986) *Folk Medicine: The Art and the Science*. American Chemical Society, Washington, DC.

4 Lagunin, A., Filimonov, D., and Povoikov, V. (2010) Multi-targeted natural products evaluation based on biological activity prediction with PASS. *Curr. Pharm. Design*, **16**, 1703–1717.

5 Colwell, J.A. (2006) Aspirin for the primary prevention of cardiovascular events. *Drugs Today*, **42**, 467–479.

6 Corey, E.J., Czakó, B., and Kürti, L. (2007) *Molecules and Medicine*, John Wiley & Sons, Hoboken, NJ, p. 38.

7 http://en.wikipedia.org/wiki/Aspirin.

8 Cragg, G.M. and Newman, D.J. (2004/Rev.2006) Plants as a source of anti-cancer agents, in *Ethnopharmacology* (eds E. Elisabetsky and N.L. Etkin), Encyclopedia of Life Support Systems (EOLSS), Developed under the Auspices of the UNESCO, Eolss Publishers, Oxford, UK, http://www.eolss.net.

9 Cancerquest http://www.cancerquest.org/chemotherapy-drug-table.

10 Burke, M.D. and Schreiber, S.L. (2004) A planning strategy for diversity-oriented synthesis. *Angew. Chem. Int. Ed.*, **43**, 46–58.

11 Nielsen, T.E. and Schreiber, S.L. (2008) Towards the optimal screening collection: a synthesis strategy. *Angew. Chem. Int. Ed.*, **47**, 48–56.

12 Altmann, K.H., Buchner, J., Kessler, H., Diederich, F., Krautler, B., Lippard, S., Liskamp, R., Muller, K., Nolan, E.M., Samori, B., Schneider, G., Schreiber, S.L., Schwalbe, H., Toniolo, C., van Boeckel, C.A.A., Waldmann, H., and Walsh, C.T. (2009) The state of the art of chemical biology. *ChemBioChem*, **10**, 16–29.

13 Kumar, K. and Waldmann, H. (2009) Synthesis of natural product inspired compound collections. *Angew. Chem. Int. Ed.*, **48**, 2–21.

14 Arya, P., Joseph, R., Gan, Z., and Rakic, B. (2005) Exploring new chemical space by stereocontrol diversity-oriented synthesis. *Chem. Biol.*, **12**, 163–180.

15 Nandy, J.P., Prakesch, M., Khadem, S., Reddy, T.T., Sharma, U., and Arya, P. (2009) Advances in solution- and solid-phase synthesis toward the generation of natural product-like libraries. *Chem. Rev.*, **109**, 1999–2060.

16 DiMasi, J.A., Hansen, R.W., and Grabowski, H.G. (2003) The price of innovation: new estimates of drug development costs. *J. Health Econom.*, **22**, 151–185.

17 Tuffs Center for the Study of Drug Development (2000) http://csdd.tufts.edu/reports.

18 McGrath, N.A., Brichacek, M., and Njardarson, J.T.A. (2010). Graphical journey of innovative organic architectures that have improved our lives. *J. Chem. Educ.*, **87**, 1348–1349.

19 IPMC. http://bioinfo.ipmc.cnrs.fr/ MOLDB/index.html.

20 Masia, N. (2008) The cost of developing a new drug. New wonder medicines come from years of research, high costs. Focus on Intellectual Property Rights. April 23rd. http://www.america.gov/st/ business-english/2008/April/ 20080429230904myleen0.5233981.html.

21 Wess, G., Urmann, M., and Sickenberger, B. (2001) *Angew. Chem. Int. Ed.*, **40**, 3341–3350.

22 Lipinski, C.A., Lombardo, F., Dominy, B.W., and Feeney, P.J. (1997) Experimental and computational approaches to estimate solubility and permeability in drug discovery and development settings. *Adv. Drug Deliv. Rev.*, **23**, 3–25.

23 Lipinski, C.A. (2004) Lead- and drug-like compounds: the rule-of-five revolution. *Drug Discov. Today Technol.*, **1**, 337–341.

24 Schreiber, S.L., Kapoor, T., and Wess, G. (2007) *Chemical Biology: From Small Molecules to System Biology and Drug Design*, vol. 1, Wiley-VCH Verlag GmbH, Weinheim.

25 Waldmann, H. (2009) *Chemical Biology*, Wiley-VCH Verlag GmbH, Weinheim.

26 Spring, D.R. (2005) Chemical genetics to chemical genomics: small molecules offer big insights. *Chem. Soc. Rev.*, **34**, 472–482.

27 http://en.wikipedia.org/wiki/Patent

28 Silverman, R.B. (1992) *The Organic Chemistry of Drug Design and Drug Action*, Academic Press, San Diego, CA.

29 Patrick, G.L. (2001) *An Introduction to Medicinal Chemistry*, Oxford University Pres, New York.

30 Gringauz, A. (1997) *Introduction to Medicinal Chemistry: How Drugs Act and Why*, John Wiley & Sons, Inc., New York.

31 Ritter, T.K. and Wong, C.-H. (2001) Carbohydrate-based antibiotics: a new approach to tackling the problem of resistance. *Angew. Chem. Int. Ed.*, **40**, 3508–3533.

32 Walsh, C.T. and Wright, G. (2005) Introduction: antibiotic resistance. *Chem. Rev.*, **105**, 391–393. The readers are invited to read the entire issue as it describes drug resistance to several drugs.

33 Wilson, J.C. and von Itzstein, M. (2003) Recent strategies in the search for new anti-influenza therapies. *Curr. Drug Targets*, **4**, 389–408.

34 von Itzstein, M. (2007) The war against influenza discovery and development of sialidase inhibitors. *Drug Discov.*, **6**, 967–974.

35 Russel, R.J., Haire, L.F., Stevens, D.J., Collins, P.J., Lin, Y.P., Blacburn, G.M., Hay, A.J., Gamblin, S.J., and Skehel, J.J. (2006) The structure of H5N1 avian influenza neuraminidase suggests new opportunities for drug design. *Nature*, **443**, 45–49.

36 Terret, N.K., Bell, A.S., Brown, D., and Ellis, P. (1996) Sildenafil (Viagra™), a potent and selective inhibitor of type 5 CGMP phosphodiesterase with utility for the treatment of male erectile dysfunction. *Bioorg. Med. Chem. Lett.*, **6**, 1819–1824.

37 Dick, W.E., Jr and Beurret, M. (1989) Glycoconjugates of bacterial carbohydrate antigens, in *Conjugate Vaccines. Contrib. Microbiol. Immunnol.* (eds J.M. Cruse and R.E. Lewis., Jr), S. Karger, pp. 48–114.

38 Jennings, H.J. and Pon, R.A. (1996) Polysaccharides and glycoconjugates as human vaccines, in *Polysaccharides in Medicinal Applications* (ed. S. Dumitriu), Marcel Dekker, pp. 443–479.

39 Robbins, J.B., Schneerson, R., Szu, S.C., and Pozsgay, V. (1999) Bacterial polysaccharide-protein conjugate vaccines. *Pure Appl. Chem.*, **71**, 745–754.

40 Moreau, M. and Schulz, D. (2000) Polysaccharide based vaccines for the prevention of pneumococcal infections. *J. Carbohydr. Chem.*, **19**, 419–434.

41 Pozsgay, V. (2000) Oligosaccharide-protein conjugates as vaccine candidates against bacteria. *Adv. Carbohydr. Chem. Biochem.*, **56**, 153–200.

42 Ada, G. and Isaacs, D. (2003) Carbohydrate-protein conjugate vaccines. *Clin. Microbiol. Infect.*, **9**, 79–85.

43 Roy, R. (2004) New trends in carbohydrate-based vaccines. *Drug Discov. Today Technol.*, **1**, 327–336.

44 Verez Bencomo, V., Roy, R., Rodriguez, M.C., Villar, A., Fernandez-Santana, V.,

Garcia, E., Valdes, Y., Heynngnezz, L., Sosa, I., and Medina, E. (2007) *Carbohydrate-Based Vaccines*, vol. 989 (ed. R. Roy) ACS Symp. Ser., pp. 71–84.

45 Verez-Bencomo, V., Fernández-Santana, V., Hardy, E., Toledo, M.E., Rodríguez, M.C., Heynngnezz, L., Rodriguez, A., Baly, A., Herrera, L., Izquierdo, M., Villar, A., Valdés, Y., Cosme, K., Deler, M.L., Montane, M., Garcia, E., Ramos, A., Aguilar, A., Medina, E., Toraño, G., Sosa, I., Carbonell, Y., Hernandez, I., Martínez, R., Muzachio, A., Carmenates, A., Costa, L., Cardoso, F., Campa, C., Diaz, M., and Roy, R. (2004) A synthetic conjugate polysaccharide vaccine against *Haemophilus influenzae* type b. *Science*, **305**, 522–525.

46 Icart, L.P., Fernandez-Santana, V., Veloso, R.C., Carmenate, T., Sirois, R., Roy, R., and Verez Bencomo, V. (2007)

T-cell immunity of carbohydrates, in *Carbohydrate-based Vaccines*, vol. 989 (ed. R. Roy), ACS Symp. Ser., pp. 1–20.

47 Fernández-Santana, V., Cardoso, F., Rodriguez, A., Carmenate, T., Peña, L., Valdés, Y., Hardy, E., Mawas, F., Heynngnezz, L., Rodríguez, M.C., Figueroa, I., Chang, J., Toledo, M.E., Musacchio, A., Hernández, I., Izquierdo, M., Cosme, K., Roy, R., and Verez-Bencomo, V. (2004) Antigenicity and immunogenicity of a synthetic oligosaccharideprotein conjugates against *Haemophilus influenzae* type b. *Infect. Immun.*, **72**, 7115–7123.

48 AUF. http://www.lb.refer.org/deriane/cycle.jpg; http://www.drugdevelopment-technology.com/projects/t705/images/1-influenza-virus.jpg.

6
The Greening of Chemistry

Pietro Tundo, Fabio Aricò, and Con Robert McElroy

6.1
Introduction

6.1.1
The History of Green Chemistry

Several Green Chemistry concepts and manufacturing processes, as well as green research fields were already investigated and utilized before the term Green Chemistry came into use. This is because chemists have always pursued the aim to be at the service of humanity with the idea of improving the lifestyle of humankind. An emblematic example is the Solvay process for the production of Na_2CO_3.

The Leblanc process came to dominate alkali production in the early 1800s and consisted of the reaction of salt, limestone, sulfuric acid, and coal to produce soda ash (sodium carbonate).

However, the expense of its reagents and its polluting by-products (HCl) called for the development of new processes. Thus, in the 1860s, Ernest and Albert Solvay developed a new process. The ingredients used for this process were readily available, inexpensive and *green*: salt brine (NaCl) and limestone ($CaCO_3$ from mines) to produce soda ash and, as a by-product, calcium chloride ($CaCl_2$) in aqueous solution. An important key point to be mentioned is that the reaction between NaCl and $CaCO_3$ does not occur directly as it needs a series of steps involving at least five reactions and six chemical intermediates: NH_3, NH_4Cl, CaO, $Ca(OH)_2$, CO_2, and $NaHCO_3$. The Solvay process is very efficient, mainly because the intermediates are all reclaimed and reused.

More recently, but still prior to the exploitation of Green Chemistry, there were other important examples of sustainability in chemistry:

- *Alkyl polyglycosides for the synthesis of fatty alcohols.* Fatty alcohols can be obtained either from petrochemical sources (synthetic fatty alcohols) or from natural renewable resources such as fats and oils. They have several applications as surfactants in hard surface cleaners and laundry detergents. Since the 1980s, the production of alkyl polyglycosides has relied completely

The Chemical Element: Chemistry's Contribution to Our Global Future, First Edition.
Edited by Javier Garcia-Martinez, Elena Serrano-Torregrosa
© 2011 Wiley-VCH Verlag GmbH & Co. KGaA. Published 2011 by Wiley-VCH Verlag GmbH & Co. KGaA.

on sugars: renewable raw materials. The research and development work in this field led to the solving of chemical, performance, and technical problems related to the use of renewable materials. As a result, alkyl polyglycosides became available in industrial quantities (Cognis industries) and a multitude of patents, scientific papers and articles appeared in specialist scientific journals.

- *Phase-transfer catalysis* as a green approach to waste minimization in the chemical industry [1]. This key synthetic methodology was developed in the 1970s and makes use of heterogeneous two-phase systems, that is, water and organic solvent. Importantly it is applied and applicable to a great variety of reactions (for more details see the section on the use of alternative solvents – water) and it is a good example of the transfer of know-how from academia to industry.

- *Chlorofluorocarbon (CFC) substitution.* Paul J. Crutzen, Mario J. Molina and F. Sherwood Rowland, conducted investigations on the ozone layer depletion for which they were jointly award the Nobel Prize for Chemistry in 1995. Their work ultimately led to the substitution of chlorofluorocarbons. A CFC is an organic compound that contains only carbon, chlorine, and fluorine, commonly known by the DuPont trade name Freon. Many CFCs have been widely used as refrigerants, propellants (in aerosol applications), and solvents, the most common being dichlorodifluoromethane (Freon-12). However, since the late 1970s, the use of CFCs has been heavily regulated after Crutzen, Molina and Sherwood Rowland reported on their destructive effects on the ozone layer (due to the presence of chlorine in these molecules). The manufacture of such compounds has been phased out by the Montreal Protocol (1987). Work on alternatives for chlorofluorocarbons led to the use of hydrochlorofluorocarbons (HCFCs) which are less stable in the lower atmosphere, enabling them to break down before reaching the ozone layer. More recently, alternatives without the chlorine, the hydrofluorocarbons (HFCs) have demonstrated even shorter lifetimes in the lower atmosphere. Along with ammonia and carbon dioxide, hydrocarbons have negligible environmental impact and are also used worldwide in domestic and commercial refrigeration applications, and are becoming available in new split-system air conditioners.

The first seeds of Green Chemistry were sown in the early 1960s when environmental statutes and regulations began to proliferate at an exponential rate, especially after publication of the book *Silent Spring* by Rachel Carson, which stimulated widespread public concern with respect to pesticides and pollution of the environment [2]. These regulations established restrictions on the use of chemicals, imposed toxicity tests for chemical substances and, finally, provided incentives for industry to eventually find replacements, substitutes, or alternatives for polluting reagents. So, the public's demand for more information about chemicals had grown rapidly [2]. In the United States, this culminated with the establishment of the Emergency Planning and Community Right-to-Know Act (EPCRA), which made public relevant data on chemicals being released to the air, water, and land by industry (1980).[1] As a consequence, industry was confronted by tremendous pressure, not only to reduce the release of toxic chemicals into the environment, but also to reduce the use of hazardous chemicals overall. Each of these incentives combined to make the 1990s the decade during which green chemistry was introduced and it has found implementation and commercialization on a wide industrial scale. In particular, since 1990 in the USA, sustainable chemistry has been a focus area of the Environmental Protection Agency (EPA),[2] involving a great deal of activity in research, symposia, and education. At the same time, the scientific community was also strongly involved in exploiting sustainable chemistry. In 1993 Anastas and Farris published the first book on the subject in the ACS Symposium series: *Benign by Design, Alternative Synthetic Design for Pollution Prevention* [3]. The book was based on the symposium "Physical Chemistry and the Environment" sponsored by the Division of Environmental Chemistry at the 206th National Meeting of the American Chemical Society in Chicago (22–27 August 1993, Illinois). The book provided a great opportunity for several chemists who were pioneers in the field of "benign by design" chemistry to present their basic research in addition to encouraging many scientists to become involved in environmentally responsible chemistry [4]. In the same year (1993) in Italy, the Consorzio Interuniversitario Nazionale "La Chimica per l'Ambiente", INCA was established, with the aim to unite the academic groups concerned with chemistry and the environment.[3] One of its focus areas being pollution prevention through research for cleaner reactions, products and processes with both academic and industrial applications (Figure 6.1).

Moreover, INCA remains constantly involved in the dissemination of Green Chemistry as demonstrated by the numerous books published (Green Chemistry Series),[4] school awards for undergraduate students,[5] publication of a magazine for young green chemists[6] and the organization of ten editions of the Summer School in Green Chemistry (Figure 6.2).[7]

Despite the continued involvement of the industrial and scientific community in the field of sustainable chemistry, it was only between 1996 and 1997 that the term Green Chemistry was first used. Other terms have been proposed, such as "chemistry for the environment" but this combination of words does not capture the economic and social implications of the concept of sustainability. Herein, the term Green Chemistry will be used according to the IUPAC (International Union of Pure and Applied Chemistry) definition that states: "Green Chemistry includes the invention, design and application of chemical products and processes to reduce or to eliminate the use and generation of hazardous substances" [5]. This

1) For further information see http://www.ecy.wa.gov/epcra/.
2) For further information see http://www.epa.gov/.
3) See Interuniversity Consortium Chemistry for the Environment, http://www.incaweb.org/.
4) For the full collection of Green Chemistry Series see http://www.incaweb.org/publications/gcseries.php.
5) Further details on http://incaweb.org/green/pgsVed/index.htm (in Italian).
6) Further details on http://incaweb.org/green/index.php.
7) For further details see http://www.incaweb.org/education/ssgc.php.

Figure 6.1 Picture representing chemistry in the service of the environment (picture taken from the cover of "Introduzione alla Chimica Verde (Green Chemistry)", eds P. Tundo, S. Paganelli–Lara Clemenza. In Italian).[8] (Please find a color version of this figure in the color plates.)

Figure 6.2 Tenth edition of the summer school on green chemistry organized by INCA (Cà Dolfin, Cà Foscari University, Venezia, 2008). (Please find a color version of this figure in the color plates.)

definition well represents the importance of fundamental research as a base for Green Chemistry development.

In this context, Paul Anastas (EPA) and John C. Warner developed the 12 principles of Green Chemistry [6], which illustrate the definition of this new field in a practical and easy to understand way. The principles cover the main concepts of

8) See footnote 4.

Box 6.1   An Outstanding Father Figure of Green Chemistry

Joseph Breen – born in Waterbury, Connecticut in 1942, dedicated his entire life to public service: he served in the Marine Corps during the Vietnam War, he then moved into the Peace Corps, and finally he spent 20 years at the US Environmental Protection Agency.

Joe Breen played a major role in creating the "Design for the Environment" and "Green Chemistry" programs. Both these programs were funded with the intent to reduce risk and protect human health and the environment. After 20 years at the EPA, he retired in 1997 and co-founded (and served as Executive Director of) the Green Chemistry Institute (GCI), a nonprofit organization in Rockville dedicated to research and education on environmentally friendly chemical synthesis and processing. The GCI promoted Green Chemistry through information and dissemination; research and fellowships; participation in conferences, workshops and symposia; international outreach; awards and recognition; and education. The Green Chemistry Institute now operates as an independent institute within the American Chemical Society. The GCI also provides national recognition for outstanding student contributions to furthering the goals of Green Chemistry. The most famous awards, in fact, are those established in memory of Kenneth G. Hancock and Joseph Breen (died in 1999).[9]

Green Chemistry that are still valid today: (i) prevention, (ii) atom economy, (iii) hazard-free chemical syntheses, (iv) safer chemicals and (v) solvents, (vi) energy efficiency, (vii) renewable feedstocks, (viii) reduce derivatives, (ix) catalysis, (x) degradation, (xi) pollution prevention and (xii) inherently safer chemistry [7].

In August 1996, IUPAC[10] began its involvement in the Green Chemistry field by the foundation of the Working Party on Synthetic Pathways and Processes in Green Chemistry.[11] In September 1997 the First International Conference on "Challenging Perspectives on Green Chemistry" was held in Venice (Figure 6.3) [8]. Since then IUPAC has been actively involved in several projects related to Green Chemistry.[12]

---

9) For the full biography check the following website: http://portal.acs.org/portal/acs/corg/content?_nfpb=true&_pageLabel=PP_SUPERARTICLE&node_id=1343&use_sec=false&sec_url_var=region1&__uuid=15d77960-6a64-4e49-b7dc-ab60577cc875.

10) For further details see http://www.iupac.org/.

11) In August 1996, IUPAC approved the formation of the Working Party on Green Chemistry under Commission III.2, which provided the beginnings of this work, Seoul, Korea 1996.

12) Some examples are: ICOS 13, Mini Symposium on Green Organic Synthesis, Warsaw, Poland, 2000; Special Topic Issue and Symposium-in Print on Green Chemistry, PAC, 2000, the CHEMRAWN (Chemistry Research Applied to World Needs) conference organized by IUPAC in Boulder, Colorado, USA, June 2001, and entitled "Toward Environmentally Benign Process and Products", IUPAC 38th Congress (Environmental Chemistry and the Greening of Industry), Brisbane, Australia, 2001; Workshop on Education in Green/Sustainable Chemistry, Venice, Italy, 2001; ICOS 14 (Symposium on Green Chemistry) Christchurch, New Zealand, 2002, CHEMRAWN-XVII and ICCDU-IX "Conference on Greenhouse Gases" Kingston (Ontario, Canada) 2007.

*Consorzio Interuniversitario Nazionale
"La Chimica per l'Ambiente"
Interuniversity Consortium
"Chemistry for the Environment"*

*European Commission Directorate General XII
Science Research and Development
Directorate D-RTD Action: Environment*

INTERNATIONAL CONFERENCE ON

GREEN CHEMISTRY
CHALLENGING PERSPECTIVES

Venezia (Italy), September 28 - October 1, 1997

Sponsored by the
International Union of Pure and Applied Chemistry

Co-sponsored by ACS, EPA and UNESCO

CHAIRMAN
Pietro Tundo, University of Venezia

CO-CHAIRMAN
Paul Anastas, EPA, USA
Albert E. Fischli, President IUPAC
Switzerland

INTERNATIONAL SCIENTIFIC
COMMITTEE

Joseph J. Breen, ACS, USA
Sergio Facchetti, JRC (Ispra, Italy)
Dieter Lenoir, GSF, Germany
Tim Lester, Deft Technology and Design, UK
Ugo Romano, EniChem, Italy
Martin Scholten, TNO, The Netherlands
Tracy C. Williamson, EPA, USA

NATIONAL COMMITTEE

Ivano Bertini, University of Firenze
Vittorio Carassiti, University of Ferrara,
Salvatore Coluccia, University of Torino
Francesco Fringuelli, University of Perugia
Giovanni Giovannozzi, University of Viterbo
Fernando Montanari, University of Milano
Gennaro Russo, University of Napoli
Ferruccio Trifirò, University of Bologna

LOCAL ORGANIZING
COMMITTEE

Chairman: Maurizio Selva
Mauro Nicolai
Stefano Paganelli
Alvise Perosa

SECRETARIAT ADDRESS

Consorzio Interuniversitario Nazionale
"La Chimica per l'Ambiente"
Calle Larga S. Marta, 2137
30123 Venezia, Italy
Tel. +39-41-5298642
Fax. +39-41-5298620
E-mail. inca@unive.it

- **CHEMICAL PROCESSES: STATE OF THE ART FOR THE ENVIRONMENT**
- **INHERENTLY SAFE PROCESSES**
- **INHERENTLY SAFE PRODUCTS**
- **CATALYSIS AND BIOCATALYSIS**
- **SOLVENTLESS PROCESSES AND ALTERNATIVE SOLVENTS**
- **RESEARCH POLICIES**

Information is available on the Consortium Home Page:
http://www.unive.it/inca.html

Figure 6.3 Poster of the First International Conference on "Challenging Perspectives on Green Chemistry" (Venezia 1997).

In July 2001, IUPAC approved the establishment of the sub-Committee on Green Chemistry (Division III). The committee's primary focus is to establish and carry out educational Green Chemistry programs. Since its conception, the sub-committee has actively organized international workshops, symposia and conferences, in addition to the preparation and dissemination of numerous books on global topics related to green/sustainable chemistry specifically aimed at university students.[13]

In 1997, after more than a year of planning by individuals from industry, government and academia, the Green Chemistry Institute (GCI) was incorporated as a not-for-profit corporation devoted to promoting and advancing Green Chemistry. In January 2001, GCI joined the American Chemical Society (ACS) in

13) Projects and activities include: "Global Climate Change" - Translation and dissemination of a monograph for secondary schools, a IUPAC-INCA joint project on the creation of a web page on Green/Sustainable Chemistry (http://www.incaweb.org/transit/iupacgcdir/INDEX.htm), Green Chemistry in the Arab region.

an increased effort to address global issues at the crossroads of chemistry and the environment.

In the late 1990s, with the new millennium looming, interest in Green Chemistry became widespread. In 1998 upon an EPA proposal, the Organization for Economic Co-operation and Development (OECD), instituted a Directive Committee for the development of sustainable chemistry and finalized a program called "Sustainable Chemistry" that included chemistry aimed at pollution prevention and better industrial performance. The activity commenced with a survey of the Steering Group (USA, Italy, Japan, Germany, Belgium, Canada, Mexico, Sweden, UK), and Business and Industry Advisory Committee to the OECD (BIAC)) on programs and initiatives on Green Chemistry launched worldwide by governments, industries and academia. The USA and Japan were nominated co-leaders in the field of research and development while Italy was appointed leader of the Educational Act. In consideration of the survey results, the policy and programmatic aspects of the sustainable chemistry activity were discussed at the Venice Workshop (October 1998) in the presence of representatives from government, industry and academia from 22 countries, and subsequently approved at the OECD meeting in Paris (June 6, 1999).

As a result of this meeting the following seven research areas in green/sustainable chemistry were identified:

- *Use of Alternative Feedstocks:* the use of feedstocks, which are renewable rather than depleting and less toxic to human health and the environment.

- *Use of Innocuous Reagents:* the use of reagents that are inherently less hazardous and are catalytic whenever feasible.

- *Employing Natural Processes:* use of biosynthesis, biocatalysis, and biotech-based chemical transformations for efficiency and selectivity.

- *Use of Alternative Solvents:* the design and utilization of solvents which have reduced potential for detriment to the environment and serve as alternatives to currently used volatile organic solvents, chlorinated solvents, and solvents which damage the natural environment.

- *Design of Safer Chemicals:* use of molecular structure design – and consideration of the principles of toxicity and mechanism of action – to minimize the intrinsic toxicity of the product while maintaining its efficiency of function.

- *Developing Alternative Reaction Conditions:* the design of reaction conditions that increase the selectivity of the product and allow dematerialization of the product separation process.

- *Minimizing Energy Consumption:* the design of chemical transformations that reduce the required energy input in terms of both mechanical and thermal inputs and the associated environmental impacts of excessive energy usage.

Despite being issued in 1998, these areas of research are still current as they vividly represent the main lines of development of Green Chemistry.

Figure 6.4 1st, 2nd and 3rd International IUPAC conferences on Green Chemistry [6]. (Please find a color version of this figure in the color plates.)

In 1999, the Royal Society of Chemistry introduced a new journal entirely dedicated to sustainable chemistry: *Green Chemistry*. This journal, as it is stated on the RSC webpage, *provides a unique forum for the publication of original and significant cutting-edge research that reduces the environmental impact of the chemical enterprise by developing alternative sustainable technologies.* Since then several new journals dedicated to Green Chemistry have appeared, such as: *ChemSusChem,*[14] *Energy and Environmental Science,*[15] *Environmental Chemistry,*[16] *Journal of Environmental Monitoring*[17], to mention but a few.

In 2006, following the launch of the IUPAC Green Chemistry Subcommittee within the III Division, the Consorzio INCA in collaboration with the German Chemical Society (GDCh) organized the first International IUPAC Conference dedicated to Green-Sustainable Chemistry (ICGC-1), currently in its 3rd edition (Figure 6.4).[18]

It is also important to mention the foundation of the Cost Action D29 (Sustainable/Green Chemistry and Chemical Technology), which is a network

14) For the journal website: http://
www3.interscience.wiley.com/
journal/114278546/home.
15) For the journal website: http://
www.rsc.org/Publishing/Journals/EE/
index.asp.
16) For the journal website: http://
www.publish.csiro.au/nid/
188.htm.

17) For the journal website: http://www.
rsc.org/Publishing/Journals/em/index.asp.
18) The 2nd International IUPAC Conference
on Green-Sustainable Chemistry (ICGC-2)
was held in Moscow-S. Petersburg in
September 2008; the 3rd International
IUPAC Conference on Green-Sustainable
Chemistry was held in Ottawa in August
2010.

comprising 26 COST (European Cooperation in Science and Technology) countries. This network aims to develop sustainable industrial chemicals and chemical-based consumer products utilizing sustainable and environmentally friendly processes and to establish a common understanding of the current status and the future research, development, and educational needs of Sustainable/Green Chemistry and Chemical Technology for Europe. The achievements of this action have been disseminated through Action workshops and Working Group meetings, presentations at international conferences and the publication of many research articles in peer-reviewed journals.

Finally, it should be mentioned that another important step in the history of Green Chemistry has been realized by the introduction of Registration, Evaluation and Authorization of Chemicals Regulation (REACH), which was formally adopted on the 18th of December 2006 by the European Council of Environment Ministers. This new regulation aims to improve the protection of human health and the environment through improved assessment of chemical substances.

This Regulation gives greater responsibility to industry as manufacturers and importers and ultimately calls for progressive substitution of the most dangerous chemicals by greener alternatives.[19]

6.1.2
Green Chemistry in the Economy: the Chinese Circular Economy (CE)

The rapid growth of China's material consumption poses profound challenges to sustainable development in the country and the rest of the world. China is now consuming about half of the world's cement, over 30% of its steel and more than 20% of its aluminum. It is also the leading consumer of fertilizers and the second largest importer of forest products in the world. Because of such rapid growth, the

---

Box 6.2 NOP: a handbook for green reactions

Chemists are well aware that beside theoretical knowledge, it is of the utmost importance to be equipped with a sound manual ability in the laboratory. In Germany, for instance, the Organisches Prakticum (OP)—a handbook for laboratory experimentation—is a must-have for organic synthesis. Now, a group of researchers have published on the net a new version of this handbook called NOP (N for Nachhaltigkeit: Sustainability). Naturally, the content revolves around the concepts of energy consumption, atom economy and (eco)toxicity of organic reactions. It contains an introduction to Green Chemistry, a number of "green" experiments and exhaustive toxicity data of the compounds employed. The website is available in German, English, Arabic, Greek, Indonesian, Russian, Turkey and shortly also in Italian.[20] NOP is a very innovative and powerful tool for the acquisition of eco-friendly laboratory techniques for students and teachers.

---

19) For more details see the website: http://echa.europa.eu/reach_en.asp.
20) NOP website http://www.oc-praktikum.de/en-entry.

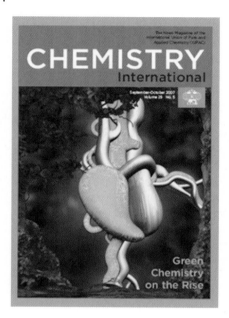

Figure 6.5 A heart representing the concept of Circular Economy: a heart that recycles the waste into new useful green products. This image taken from the cover of Chemistry International [10] the newsmagazine of IUPAC. (Please find a color version of this figure in the color plates.)

natural resources of China are being depleted quickly. To solve this problem, China's leadership, inspired by Japanese and German Recycling Economy Laws, formed a Circular Economy (CE) initiative that started 10 years ago and that has major strategic importance worldwide [9]. The Circular Economy approach to resource-use efficiency integrates cleaner production and industrial ecology in a broader system encompassing industrial firms, networks, eco-industrial parks, and regional infrastructure to support resource optimization (Figure 6.5).

The three basic levels of the Circular Economy action are:

- At the *individual firm level*, managers must seek much higher efficiency through the following three *Rs* of CE: *Reduce* consumption of resources and emission of pollutants and waste, *reuse* resources, and *recycle* by-products.

- The second level is to reuse and recycle resources within industrial parks and clustered or chained industries, so that resources will circulate fully in the local production system (the Chinese use the term "eco-chains" for by-product exchanges).

- The third level is to integrate different production and consumption systems in a region so the resources circulate among industries and urban systems.

According to the principles of CE, state-owned and private enterprises, government and private infrastructure, and even consumers, all have a role to play in achieving CE. CE in many ways resembles the concept of industrial metabolism. It focuses on the input–output analysis of material flows transformed by production and consumption. In fact, the Circular Economy concept brings together cleaner production and industrial ecology, with its application as eco-industrial development.

The essence of the CE concept is the exchange of materials where one's waste, including energy, water, materials – as well as information – is another facility's input.

Comparable to the Chinese concept of CE, is the idea of Green Economy, pursued in Europe and America. Green Economy is a new economic development model born in contrast to the existing black economic model based on fossil fuels. The Green Economy is based on ecological economics, which consider the impact of human economic activities on climate change and global warming. In the midst of the global economic crisis, the United Nations Environment Program (UNEP) called for a global Green New Deal according to which governments were encouraged to support their economic transformation into a greener economy. The green economy supports green and renewable energy as a replacement for fossil fuels and promotes energy conservation for efficient energy use. The green economy aims to create green jobs, ensure real, sustainable economic growth, while preventing environmental pollution, global warming, resource depletion, and environmental degradation [11].

6.1.3
Award for Green Chemistry Research

Green Chemistry research is supported by several awards that incentivize innovation and excellence in the Green Chemistry field. Some examples are:[21]

6.1.3.1   The Presidential Green Chemistry Challenge
In the USA, The Presidential Green Chemistry Challenge was established by President Clinton in 1995 to recognize and promote fundamental and innovative chemical methods that accomplish pollution prevention through reduction at source and that have broad applicability in industry. The Presidential Green Chemistry Challenge Awards Program was established to recognize technologies that incorporate the principles of Green Chemistry into chemical design, manufacture and use. The evaluation of the new technology's impact includes consideration of the health and environmental effects throughout the technology's lifecycle with recognition that incremental improvements are necessary.[22]

21) For more details visit http://www.greenchemistrynetwork.org/awards.htm.
22) See footnote 2.

6.1.3.2   Award for Green Products and Processes [12]

This award was presented by INCA to Italian Companies that excelled in developing green processes and products. Examples of companies that have received the award are: Enichem, Polimeri Europa, Ausimont, Ilva Polimeri, Lamberti, Lonza Group, Mapei and Valagro.

6.1.3.3   The European Sustainable Chemistry Award[23]

In 2010, EuCheMS (the European Association for Chemical and Molecular Sciences), with the backing of the European Environment Agency (EEA) and the support of SusChem (European Platform for Sustainable Chemistry) and CEFIC (European Chemical Industry Association), launched the European Sustainable Chemistry Award. This new award intends to raise the profile of sustainable chemistry and to stimulate innovation and competitiveness. The first Award, a prize of €10 000, was presented during the 3rd EuCheMS Chemistry Congress, on 29 August–2 September 2010 in Nürnberg, Germany.

6.1.3.4   The Institution of Chemical Engineers Award[24]

The IChemE Award is given for Innovation and Excellence in the Green chemical technology and sustainability area.

6.1.3.5   Green and Sustainable Chemistry Network Award (Japan)

The Green & Sustainable Chemistry Network[25] was established in March 2000 to promote research and development for the Environment and Human Health and Safety through the innovation of Chemistry. One of the activities, GSCN established in 2001 was the "GSC Awards". GSC Awards are to be granted to individuals, groups or companies who greatly contributed to promote GSC through their research, development and their industrialization in the fields of development of industrial technologies, reduction of environmental burden (such as carbon dioxide, waste, landfill, harmful by-products etc.) and of establishing new philosophies/methodologies in research. The achievements are awarded either by the Minister of Economy, Trade and Industry, or by the Minister of the Environment, or by the Minister of Education, Sports, Culture, Science and technology, depending on their application.

6.1.3.6   RACI Green Chemistry Challenge Award

The Royal Australian Chemical Institute Green Chemistry Challenge Awards recognize and promote fundamental and innovative chemical methods in Australia that accomplish pollution prevention through reduction at source and that have broad applicability in industry. They also recognize contributions to education in Green Chemistry. The Green Chemistry Challenge Awards are open to all individuals, groups and organizations, both non-profit and for profit, including academia, and industry.

23) For more details see: http://
www.euchems.org/ESCA/index.asp.
24) IChemE is a hub for chemical, biochemical and process engineering professionals

worldwide promoting competence and a commitment to sustainable development.
25) Organization webpage: http://
www.gscn.net/.

Box 6.3   Green Chemistry Research Institutions and Associations

Since 1996 when the term Green Chemistry was coined, several research centers and associations supporting this new research area have been formed.

Some examples of national and international organizations in Green Chemistry are:

National/International Organizations

IUPAC–Subcommittee on Green Chemistry Organic and Biomolecular Chemistry Division (III)[26]

Interuniversity National Consortium "Chemistry for the Environment" (Italy)[27]

Green Chemistry Network (UK)[28]

Green & Sustainable Chemistry Network (Japan)[29]

Environment Protection Agency (USA)[30]
The US EPA's Green Chemistry Program

Green Chemistry Institute (USA)[31]

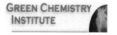

Canadian Green Chemistry Network[32]

European Association for Chemical and Molecular Science (EuCheMS) WP on Green and Sustainable Chemistry[33]

European Association for Chemical and Molecular Sciences

---

26) See footnote 10.
27) See Interuniversity Consortium Chemistry for the Environment, http://www.incaweb.org/.
28) Organization webpage: http://www.greenchemistrynetwork.org/.
29) See footnote 25.
30) See footnote 2.

31) Organization webpage: http://portal.acs.org/portal/acs/corg/content?_nfpb=true&_pageLabel=PP_TRANSITIONMAIN&node_id=830&use_sec=false&sec_url_var=region1&__uuid=d6152ed6-fb6b-488c-b256-a72021645d43.
32) Organization webpage: http://www.greenchemistry.ca/.
33) Organization webpage: http://euchems.org/.

Continued

(Please find a color version in the color plates.)

Some Research Institutes[34]:

- Carnegie Mellon University Institute for Green Oxidation Chemistry (USA)
- Centre for Green Chemistry University of Monash (Australia)
- Center for Green Manufacturing University of Alabama (USA)
- Center for Sustainable and Green Chemistry (DK)
- Chemical Process Engineering Research Institute Centre for Research & Technology (Greece)
- Göteborg University's Centre for Environment and Sustainability (SE)
- Green Chemistry Centre of Excellence at York (UK)
- Green Chemistry Network Centre (Delhi University, India)
- Greek network of Green Chemistry (Greece)

- Institute of Applied Catalysis, A research network for catalysis (UK)
- Institute for a Sustainable Environment, University of Oregon (USA)
- NSF Science and Technology Center for Environmentally Responsible Solvents and Processes, University of North Carolina; Chapel Hill (USA)
- Queen's University Ionic Liquid Laboratories (QUILL) Queen's University of Belfast (UK)
- The Clean Technology Research Group, University of Nottingham (UK)
- University of Leicester, Leicester Green Chemistry Group (UK)
- University of Leeds, Leeds Cleaner Synthesis Group (UK)
- University of Notre Dame Energy Center (Indiana, USA)
- Green Chemistry Network of Spain REDQS (ES)

6.2
Areas of Green Chemistry

In order to achieve the best possible results, the G8 Countries aim to exploit Green Chemistry in relation to the need of each individual country. Thus, each country requires Green Chemistry to be involved in solving specific questions: Latin America to the exploitation of renewable resources, the Arab Regions to water quality and treatment, the Far East Countries to anti- and de-pollution issues. Adoption of the Green Chemistry principles is now possible because present European technology has the capacity to build new protocols for manufacturing molecular species.

Besides, despite the fact that more than ten years have passed since the seven research areas in green/sustainable chemistry were identified by the OECD, these thematic spheres remain the undisguised focus of Green Chemistry research [13].

A short report on each thematic area highlighting their advances and future research follows.

---

34) The list reports, in alphabetical order, only some of the numerous research institutions and organizations on Green Chemistry due to lack of space.

6.2.1
Alternative Feedstocks

The synthesis and manufacture of any chemical substance begins with the choice of a starting material. In many cases, the selection of a starting material can be the most significant factor in determining the impact of the synthesis on the environment. Certainly, a first-level assessment of any starting material must be whether or not the substance itself poses a hazard in terms of toxicity, accident potential, possible ecosystem damage, or in another form. In this context, it is understandable that the feasibility and benefits of using bio-based, as opposed to petroleum-based, starting materials have been actively investigated in both academia and industry. To ensure a high degree of product safety for consumers and the environment, renewable resources have been shown to have advantages when compared to petrochemical-derived raw materials and can, therefore, be regarded as being the preferred source of raw material in Green Chemistry. Moreover, the current high prices for petroleum and natural gas have spurred the chemical industry to examine alternative feedstock for the production of commodity chemicals. In this area, over the last 30 years, alternatives to conventional petroleum and natural gas feedstock have been developed, in particular by the exploitation of biomass.

Biomass is biological material derived from living, or recently living organisms, such as wood, carbohydrates, waste and gas (Figure 6.6). Biomass energy is derived from several distinct energy sources: refuse, wood, waste and landfill gases [14].

There are a number of technological options available to make use of a wide variety of biomass types as a renewable energy source. Conversion technologies may release the energy directly, in the form of heat or electricity, or may convert it to another form, such as liquid biofuel or combustible biogas [15]. The most

Figure 6.6 Biomass is biological material derived from living, or recently living organisms, such as wood and animal waste. (Please find a color version of this figure in the color plates.)

conventional application of biomass still relies on direct incineration. Currently, the New Hope Power Partnership is the largest biomass power plant in North America.[35]

The 140 MW facility uses sugar cane fiber and recycled urban wood as fuel to generate enough power for its large milling and refining operations as well as to supply renewable electricity to nearly 60 000 homes.

The facility reduces dependence on oil by more than one million barrels per year, and, by recycling sugar cane and wood waste, preserves landfill space in urban communities in Florida. A biomass power plant's size is often determined by biomass availability in its close proximity as transport costs of the (bulky) fuel play a key factor in the plant's economics. However, biomass can be converted to other usable forms of energy by turning the raw materials, or feedstocks, into a usable form such as transportation fuels. These are produced from biomass through biochemical or thermochemical processes and they include ethanol, methanol, biodiesel, biocrude, and methane.

- Ethanol is the most widely used biofuel today [16]. Brazil has the largest and most successful bio-fuel programs in the world, involving production of ethanol fuel from sugarcane, and it is considered to have the world's first sustainable biofuels economy. In the United States alone, more than 1.5 billion gallons are added yearly to gasoline as an oxygenate, to improve vehicle performance and reduce air pollution. Ethanol is produced from the fermentation of sugar by enzymes produced from specific varieties of yeast. Traditional fermentation processes rely on yeasts that convert six-carbon sugars to ethanol using a process similar to brewing beer. Ethanol made from cellulosic biomass materials or other agricultural feedstock is called second generation bioethanol. Ethanol can be used in its pure form (neat), as a blend with gasoline, or as a fuel for fuel cells.

- Methanol also can be used as a transportation fuel. Currently methanol is produced using natural gas, but it can also be produced from biomass through a two-step thermochemical process. First, the biomass is gasified to produce hydrogen and carbon monoxide. These gases are then reacted to produce methanol. Methanol can be used in its pure form (neat), as a feedstock for the gasoline additive methyl *tert*-butyl ether (MTBE), or as fuel for fuel cells.

- Biodiesel is a renewable diesel fuel substitute that can be made by chemically combining any natural oil or fat with an alcohol (usually methanol). Any vegetable oils, animal fats, and recycled cooking greases can be transformed into biodiesel and there are many different ways to do this. Biodiesel can be used neat or as a diesel additive and is typically used as a fuel additive in 20% blends (B20) with petroleum diesel in compression ignition (diesel) engines. Other blend levels can be used, depending on the cost of the fuel and the desired benefits.

---

35) Presentation of the plant http://www.psc.state.fl.us/utilities/electricgas/RenewableEnergy/Cepero-OCFC.pdf.

- Methane is the major component of compressed natural gas. Methane, in a blend of other gases, can be produced from biomass by a biochemical process under anaerobic digestion conditions.

However, it should also be considered that there are several issues in the replacement of petroleum by biomass feedstocks that include impurities, variability of feedstock composition, distributed supply, scalability and pathways for breakdown of cellulose. Although some large-scale chemical production occurs as a by-product of fuel production, widespread use of biomass feedstocks for commodity chemical manufacture will require sustained research and development in a variety of fields, such as plant science, microbiology, genomics, catalysis, and chemical separation technologies.

Another example of an alternative feedstock is lignin. Lignin is a complex chemical compound most commonly derived from wood and it is one of the most abundant organic polymers. In 1998, a German company, Tecnaro, developed a process for turning lignin into a substance, called Arboform [17]. When lignin is combined with resins and flax it forms a bio-plastic mass that looks and feels like wood and can be used to make several products such as furniture, toys, loudspeakers and even car interiors. Most significantly, Arboform is biodegradable and its raw material lignin is available in abundance, making it an environmentally friendly material that can potentially save significant natural resources. When the item is discarded, it can be burned just like wood. At the present time arboform costs €2.50 per kilogramme. The inventors Jürgen Pfitzer and Helmut Nägele have been awarded the European Inventor Awards 2010 for their studies.

Further, currently there are several ongoing investigations for seeking high value application of lignin for Green Chemistry, in particular in the fields of carbon fibers, aromatic chemicals, polymer resins and antioxidants.

Box 6.4 Light as emerging feedstock

Light is another emerging feedstock in a broad sense, a safe alternative to toxic catalysts in many synthetic transformations. In addition to utilizing UV light, the most renewable and environmentally ideal energy source is sunlight. In this regard, the quote (given roughly a century ago during a conference in New York) by Giacomo Ciamician – the founder of photochemistry – is particular pertinent [18]: "On the arid lands there will spring up industrial colonies without smoke and without smokestacks; forests of glass tubes will extend over the plants and glass buildings will rise everywhere; inside of these will take place the photochemical processes that hitherto have been the guarded secret of the plants, but that will have been mastered by human industry which will know how to make them bear even more abundant fruit than nature, for nature is not in a hurry but mankind is".

Although it appeared (and still is) futuristic, we now know that many of these former fictions can be realized and applied. It must be stated that photochemistry is already largely used in several laboratory and industrial applications (for example in the synthesis of benzyl halides). However, photocatalytic systems that are able to operate effectively and efficiently, not only under UV light but also under the most environmentally ideal energy source, sunlight, are yet to be established.

6.2.2
Use of Innocuous Reagents

As in the selection of a starting material, the selection of a reagent must include an evaluation to identify what the hazards associated with a particular reagent are. This evaluation should include an analysis of the reagent itself, as well as an analysis of the synthetic transformation associated with the use of that reagent (i.e., to determine product selectivity, reaction efficiency, separation needs, etc.).

In order to evaluate the hazards inherent to the use of a certain reagent several issues have to be address:

6.2.2.1 Less Hazardous Reagent

First, an investigation should be undertaken to determine if alternative reagents are available that are either more environmentally benign themselves or are able to carry out the necessary synthetic transformation in a more environmentally benign way. In order to answer this question alternative reagents must be identified and any hazardous properties that they possess must be compared with those associated with the reagent originally selected. One example of an innocuous reagent is dimethyl carbonate (DMC) [19].

DMC is an environmentally benign substitute of phosgene[36] in carboxymethylation reactions and of dimethyl sulfate (DMS)[37] and methyl halides[38] in methylation reactions. Reported toxicity and ecotoxicity data classify DMC as both a nontoxic and environmentally benign chemical [20]. DMC does not produce inorganic salts. In fact, the leaving group, methyl carbonate, decomposes giving only methanol and CO_2 as by products.

36) In chemistry, phosgene is a very important building block able to provide the carbonyl function in many classes of organic compounds. It is also a versatile reagent since it is employed in selective chlorocarbonylation, chlorination, dehydration and carbonylation reactions. However, the handling of phosgene, which is a gas, needs special attention. Due to its highly toxic nature (it was used as a chemical weapon during World War I), the use of phosgene gas, either on a small scale in the laboratory or on a large scale in industry poses several risks due to both storage and transportation issues. Besides, the need to replace phosgene also depends on the fact that its production involves large amounts of chlorine as raw material and results in the production of halogenated by-products.

37) Dimethyl sulfate is an extremely toxic compound, it is absorbed through the skin, mucous membranes, and gastrointestinal tract. There is no strong odor or immediate irritation (apart from eye irritation) to warn of lethal concentration in air. Delayed toxicity allows potentially fatal exposures to occur prior to development of any warning symptoms. Symptoms may be delayed 6–24 h. Moreover, its hydrolysis products, monomethyl sulfate and methanol, are environmentally hazardous. In water, the compound is ultimately hydrolyzed to sulfuric acid and methanol.

38) Methyl halides are generally toxic as well as possibly carcinogenic. Breathing methyl iodide fumes, for instance, can cause lung, liver, kidney and central nervous system damage. It causes nausea, dizziness, coughing and vomiting. Prolonged contact with skin causes burns. Massive inhalation causes pulmonary edema.

In this context, Scheme 6.1 shows the methylation of phenol by methyl halide (CH$_3$X) and DMS to give anisole, and the alkoxycarbonylation of an alcohol by phosgene (COCl$_2$).

$$PhOH + (CH_3)_2SO_4 + NaOH \longrightarrow PhOCH_3 + NaCH_3SO_4 + H_2O$$

$$PhOH + CH_3I + NaOH \longrightarrow PhOCH_3 + NaI + H_2O$$

$$2ROH + COCl_2 + 2NaOH \longrightarrow ROCOOR + 2NaCl + 2H_2O$$

Scheme 6.1 Methylation and alkoxycarbonylation using DMS, CH$_3$I and COCl$_2$ (under batch conditions).

DMC is able to perform the same reactions, using a catalytic amount of base and producing only methanol and CO$_2$ as by-products (Scheme 6.2). Moreover, no inorganic salts have to be disposed of and, therefore, it can also be used in continuous-flow synthesis.

$$PhOH + CH_3OCOOCH_3 \xrightarrow{\text{cat. base}} PhOCH_3 + CO_2 + CH_3OH$$

$$ROH + CH_3OCOOCH_3 \xrightarrow{\text{cat. base}} ROCOOCH_3 + CH_3OH$$

Scheme 6.2 Methylation and methoxycarbonylation using DMC.

DMC is classified as a flammable liquid, does not smell (methanol-like odor) and does not have irritating or mutagenic effects by either contact or inhalation. Therefore, it can be handled safely without the special precautions required for the poisonous and mutagenic methyl halides and DMS, and extremely toxic phosgene. DMC is also widely used for its many applications. In fact, recent research indicates DMC as an oxygenated fuel [21] additive (due to the high percentage of oxygen in the molecule) of gasoline or diesel oil to replace the methyl-*tert*-butyl ether (MTBE). DMC can reduce the surface tension of diesel boiling range fuels leading to an improved (diesel) fuel with better injection delivery and spray. This and other applications led to an enormous effort in the investigation of low-cost synthesis of DMC.

Besides these applications, DMC is considered green because:

1) It is produced according to a green synthesis (see Box 6.5)
2) It is nontoxic
3) It produces no inorganic waste when utilized in synthesis
4) It led to unexpected and even surprising reaction pathways

Concerning point 4, DMC has a very selective behavior, reacting with different nucleophiles (such as primary amines, CH$_2$ acidic compounds, phenols, etc.) acting as an alkylating or carboxymethylating agent. In fact, DMC, as an electrophile, has three reactive centers that can interact with nucleophiles: the

Box 6.5   Green synthesis of menthol

An example of Green Chemistry applied to fragrances is the synthesis of menthol (Scheme B5.1). Menthol is an organic compound made synthetically or obtained from peppermint or other mint oils. Menthol has local anesthetic and counterirritant qualities, and is widely used to relieve minor throat irritation, but it also has applications in perfumery and in some beauty products such as hair-conditioners. As with many widely-used natural products, the demand for menthol greatly exceeds the supply from natural sources, thus most of the menthol used in industry is made synthetically. In particular, menthol is manufactured as a single enantiomer (94% ee) on a scale of 3000 tons per year by Takasago International. This process involves an asymmetric synthesis reaction developed by a team led by Ryoji Noyori and it is green and highly efficient [31].

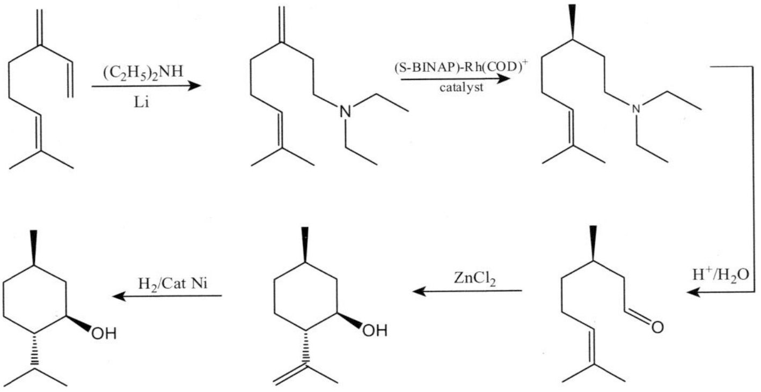

Scheme B5.1   Green synthesis of menthol.

carbonyl, and two methyl groups. Such centers can be classified according to the Hard–Soft Acid–Base (HSAB) theory [22]: the carbonyl group is the harder electrophile, as a result of its polarized positive charge and sp^2 hybridization (thus it reacts preferentially with harder nucleophiles); the two methyl groups represent softer electrophiles, thanks to their sp^3 orbital and their saturated carbon atom, which has a weaker positive charge (thus it reacts preferentially with softer nucleophiles).

Numerous investigations have verified the compliance of reactivity of ambident nucleophiles and electrophiles with the HSAB theory [23]. Many ambident nucleophiles are known, but few ambident electrophiles have been studied. As a result of its ambident electrophilic character, DMC can be used either for carboxymenthylation or methylation reactions, both with high selectivity (up to 99.9%).

It can act as an efficient carboxymethylating agent (as a phosgene substitute) for a wide string of nucleophiles. For example the carboxymethylation of amines to carbamates, which has great industrial importance in the synthesis of urethane [24], aromatic polycarbonates and isocyanate [25]. Therefore, the areas in which

DMC is used as an actual or potential phosgene substitute correspond to the main areas of phosgene industrial applications. Carbamates are very useful compounds, widely used in the synthesis of pesticides, fungicides and herbicides, pharmaceuticals, cosmetics and polyurethanes, in addition to being employed as a protecting group. DMC has been successfully used for the methylation of arylacetonitriles and methyl aryl acetates at the α position. In fact, the reaction of CH_2 acidic compounds such as arylacetonitriles, aryl acetates, aryloxyacetic esters, sulfones, sulfoxides, and lactones with DMC is highly selective, as it yields the sole monomethyl derivative (Scheme 6.3) [26]. Regardless of the high temperature and the great excess of alkylating agent (DMC is also the solvent of the reactions), at complete conversion of the substrate, selectivity for the monomethylated product is often >99%.

$$ArCH_2X + CH_3OCOOCH_3 \quad \xrightarrow[180\ °C]{K_2CO_3} \quad ArCH(CH_3)X + CO_2 + CH_3OH$$

Scheme 6.3 Monoalkylation of nitriles, esters and sulfones, X = CN, $COOCH_3$, SO_2R, SO_2Ar, etc.

Scheme 6.3 refers to monoalkylation of nitriles, esters, and sulfones. This reaction has an industrial relevance, since $ArCH(CH_3)COOH$ are well know anti-inflammatory agents.

The reason for the selectivity in monomethylation of these compounds is not immediately evident. Isolation of intermediates and a detailed kinetic study showed that the reaction mechanism does not imply a simple nucleophilic substitution [27] (Eq. 6.1).

$$\underset{\textbf{1}^-}{\overset{\overset{\displaystyle H}{|}}{\underset{\underset{\displaystyle X}{|}}{ArC^-}}} + CH_3OCOOCH_3 \quad \xrightarrow{\quad\times\quad} \quad \underset{\textbf{4}}{ArCH(CH_3)X} + CH_3OCOO^- \qquad (6.1)$$

Instead, monomethylation derives from an unusual reaction pathway that involves the reactivity of anion $\textbf{1}^-$ and anion $\textbf{2}^-$, according to two consecutive nucleophilic displacements: the first one follows a $B_{Ac}2$ mechanism, while the second occurs through a $B_{Al}2$ mechanism, according to the HSAB theory.

Thus, **4** is produced through a series of consecutive pathways, all of them being very selective. Scheme 6.4 accounts for such behavior: the reaction proceeds through the carboxymethylation species **2**, which afterward reacts with the methyl of DMC.

In summary, while anion $ArCH^-X$ does not give $ArC(CH_3)_2X$, anion $ArC^-(COOCH_3)X$ also does not allow the formation of $ArC(COOCH_3)_2X$.

We can assert that anions $\textbf{1}^-$ and $\textbf{2}^-$ give different compounds since they have different soft/hard character. Their difference in hardness provides a reason for the discrimination observed between the two electrophilic centers of DMC. The hard nucleophile $\textbf{1}^-$ attacks only the carbonyl of DMC (Scheme 6.4), while the anion of the product **2⁻** is a softer nucleophile thus it selectively produces the methyl derivative (Scheme 6.4). The change in hardness/softness of the anion, due to the presence of the carboxymethyl group, is enough to significantly alter the reactivity of the DMC molecule. Since hard–soft and soft–hard interactions are inhibited, neither double methylation nor double carboxymethylation takes place.

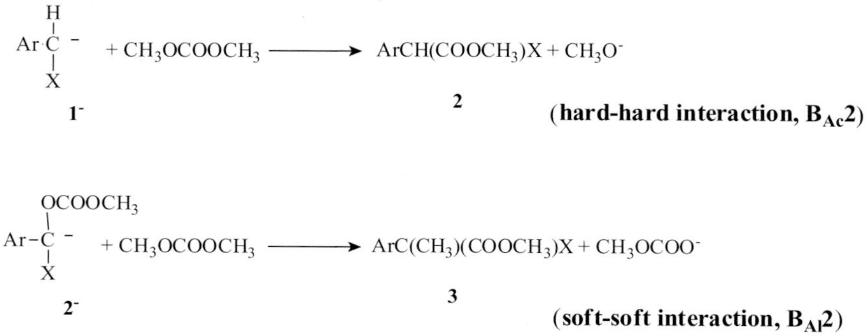

$$Ar\underset{X}{\overset{H}{\underset{|}{\overset{|}{C}}}}^{-} + CH_3OCOOCH_3 \longrightarrow ArCH(COOCH_3)X + CH_3O^-$$

$$\mathbf{1^-} \qquad\qquad\qquad\qquad\qquad\qquad \mathbf{2} \qquad\qquad \textbf{(hard-hard interaction, B}_{Ac}\textbf{2)}$$

$$Ar\underset{X}{\overset{OCOOCH_3}{\underset{|}{\overset{|}{C}}}}^{-} + CH_3OCOOCH_3 \longrightarrow ArC(CH_3)(COOCH_3)X + CH_3OCOO^-$$

$$\mathbf{2^-} \qquad\qquad\qquad\qquad\qquad\qquad \mathbf{3} \qquad\qquad \textbf{(soft-soft interaction, B}_{Al}\textbf{2)}$$

Scheme 6.4 Mechanism of the monoalkylation of nitriles, esters, and sulfones.

DMC has also emerged as a methylating agent in organic synthesis [28]. Even though its reactivity is lower than widely used methyl halides and dimethyl sulfate it has the great advantage of being less toxic.

Finally, in recent years DMC has also been used for intramolecular cyclization for the synthesis of heterocycles with applications in cosmetics and fragrances [29].

An emblematic example is (–)-norlabdane oxide that represents one of the preferred synthetic compounds with desirable ambergris-type odor and it is commercially available under various trade names (notably as amberlyn, ambroxan, ambrofix, ambrox or amberoxide). (–)-Norlabdane oxide is industrially synthesized by cyclization of the related diol, amberlyn diol, in acidic conditions. This reaction leads to a mixture of ambroxan (ca. 60%) and by-products deriving from the concurrent elimination reaction [30]. Under acidic conditions this reaction is not environmentally friendly and leads to a partial racemic mixture of products. Recently, DMC was used for efficient cyclization of the amberlyn diol in a short time span and quantitative yield (Scheme 6.5). Noteworthy is the fact that the reaction maintains the chiral integrity of the starting material.

6.2.2.2 Generate Less Waste

An important consideration and benefit associated with the use of a particular reagent is whether it is responsible for the generation of more or less waste than other reagents. The amount of waste either generated or eliminated, however, cannot be the only consideration. The type of waste generated must also be assessed. Just as not all chemical products are equal in terms of their hazard, neither are chemical waste streams. Waste streams therefore must also be assessed for any hazard properties that they possess. The way the problem is approached obviously includes recycling and reuse, but focuses mostly on prevention, and

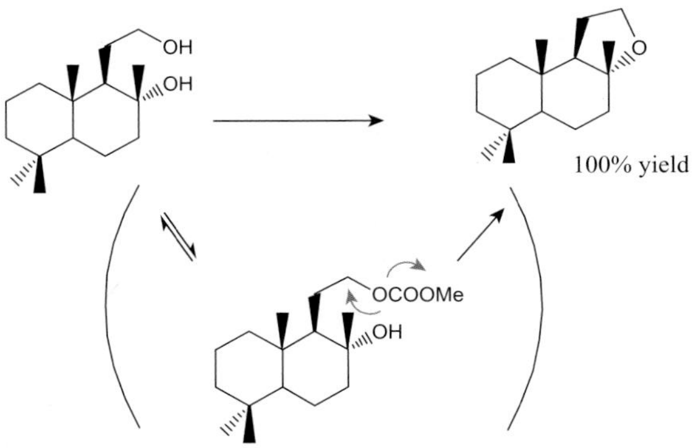

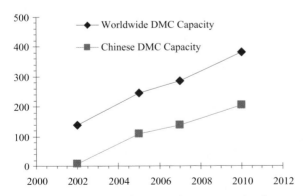

Scheme 6.5 Green synthesis of Amborxan by DMC chemistry (base, DMC, 90 °C, 3 h). The reaction mechanism is given in brackets.

Box 6.6   Green production of dimethyl carbonate – A case study

DMC was, for a long time, produced from phosgene and methanol. In this synthesis, HCl was an unwanted side product. However, since the mid-1980s DMC is no longer produced from a phosgene pathway. Nowadays, the industrial procedure to DMC – developed and recently industrialized principally in China (Figure B6.1) – does not use any chlorine, but consists of the cleavage of cyclic carbonates (Scheme B6.1).

Currently, several catalysts are under investigation for the synthesis of the cyclic carbonate which is an important green reagent and intermediate for the synthesis of DMC [32].

The current industrial applications of DMC include: polycarbonates (53%), coating and paints (29%), agrochemicals (12%), electrolyte solvents (2%), pharmaceuticals and cosmetics (5%).

Figure B6.1   Production of DMC in China compared with DMC world production.

Continued

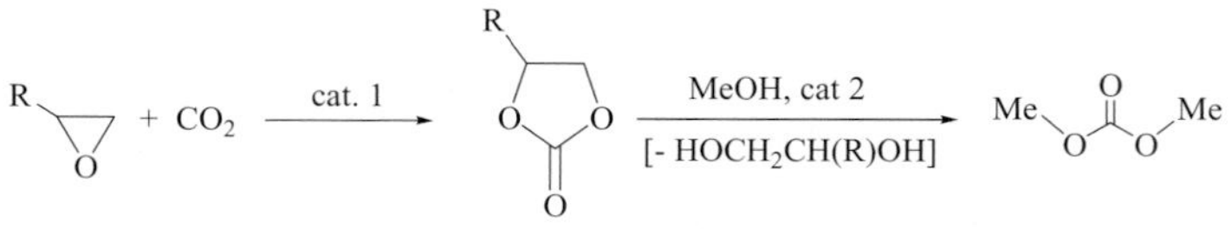

Scheme B6.1 Insertion of CO_2 into an epoxide and cleavage of the resultant cyclic carbonate. Step 1. Catalyst: MgO, CaO. Step 2. Catalyst: zeolites exchanged with alkali and/or alkaline earth metal ions. R = H and CH_3.

therefore reduction, of waste production. The idea of high conversion efficiency in a chemical process is expressed in the concept of atom economy (see metrics) postulated by Trost [33].

These considerations explain why oxidation reactions involving oxygen and hydrogen peroxide have been an outstanding priority in the last 20 years. For green oxidation reactions we refer to oxidations that use atmospheric oxygen or molecular oxygen as oxidant. They are considered green because they produce water as a by-product, they require the use of nontoxic solvents (water or CO_2) and mild reaction conditions. From these observations, it is clear that oxygen is the ideal oxidant [34] to be used due to the high active oxygen percentage content (theoretically 100%).[39] However, oxidations using air as a reagent are difficult to control and intrinsically nonselective when selectivity is very often a crucial parameter. Besides, very few reactions have been found where both atoms of oxygen can be transferred to the substrate; more often O_2 acts as an oxidant with 50% of active oxygen content leading to the formation of 1 equivalent of water. For these reasons, hydrogen peroxide is a more practical oxidant (active oxygen content 47%); it produces water as the only by-product, and a very high selectivity can be obtained. However, hydrogen peroxide used for fine chemical production, can undergo radical decomposition to water and oxygen [35] (catalase reaction). Therefore, there is a great effort to develop systems able to selectively activate oxygen and hydrogen peroxide for oxidative transformations. In this context, both homogeneous and heterogeneous catalysis play a key role. Oxidation reactions are critical to pharmaceutical, petrochemical, and agricultural industries. Several examples of how environmentally benign oxidants such as molecular oxygen, hydrogen peroxide or N_2O can be activated on heterogeneous catalysts have been reported [36]. Direct oxidation of isoprenol, β-picoline and benzene are chosen as examples for continuous gas phase processes, and oxidation of cyclopentanone, limonene, pinene, and propylene as examples of semi-continuous or batchwise processes in the liquid phase [37].

A foremost example of chemistry that produces less waste is represented by zeolites [38]. Zeolites are crystalline aluminosilicates of group IA and group IIA elements, such as sodium, potassium, magnesium and calcium. They have a

39) The active oxygen content is the mass amount of oxygen transferred to the substrate with respect to the total mass of oxidant (i.e., H_2O_2 47%; O_3 33.3% etc.).

three-dimensional framework of tetrahedra AlO_4 and SiO_4; each AlO_4 tetrahedron in the framework bears a net negative charge, balanced by a cation. Some zeolites are found in nature as minerals, many others are synthesized by industry or in the laboratory. Zeolites have many applications. The first major use for zeolites was in the purification of water [39]. Water can be softened by passing it through a zeolite, with pores that incorporate calcium and magnesium ions rending the water softer. This same type of zeolite is being increasingly used in place of polluting phosphate chemicals in laundry detergents [40]. Zeolites are also used in agriculture [41], their pores can be filled with potassium or ammonium ions, fertilizer or other micronutrients. The use of zeolite catalysts in the production of organic (fine) chemicals has appeared as a major new direction [42]. The main advantage of these materials is that their pore size, shape and properties can be modeled according to the needs of the reaction to be conducted, and to the substrate used (obviously this is not as straightforward as it seems) [43]. This improves the energy-efficiency of many industrial processes, especially in the hydrocarbon industry. It also removes the need to use other potentially polluting catalytic alternatives. This has led to numerous applications and patents in the industry (i.e., production of phenol by an alternative to the cumene process) [44]. For instance, the hydrogen form of zeolites (prepared by ion-exchange) are powerful solid-state acids [45], and can facilitate a host of acid-catalyzed reactions, such as isomerization, alkylation, and cracking since, due to the thermal stability of their structure, they can be used at high temperature.

6.2.2.3 High Conversion and Selectivity

Utilizing a reagent that is more selective means that more of the starting material is going to be converted into the desired product. On the other hand, high product selectivity does not always translate into high product yield (and less waste generated). As reported by Sheldon, both high selectivity and high conversion must be achieved in order for a synthetic transformation to generate little or no waste [46]. Utilizing highly selective reagents can mean that separation, isolation, and purification of the product will be significantly less difficult. Since a substantial portion of the burden to the environment that chemical manufacturing processes incur often results from separation and purification processes, highly selective reagents are very desirable in green chemistry.

6.2.2.4 Catalyst

One of the most important aspects in the use of a benign reagent is the substitution of antiquated stoichiometric methodologies with cleaner catalytic alternatives. Indeed, a major challenge in chemicals manufacturing in general is to develop processes based on H_2, O_2, H_2O_2, CO, CO_2 and NH_3 as the direct sources of H, O, C and N. Catalytic hydrogenation, oxidation and carbonylation are good examples of highly atom efficient, low-salt processes. The generation of copious amounts of inorganic salts can similarly be largely circumvented by replacing stoichiometric mineral acids, such as H_2SO_4, and Lewis acids and stoichiometric bases, such as NaOH, KOH, with recyclable solid acids and bases, preferably in catalytic amounts.

A large number of industrial processes are based on the use of inorganic or mineral acids. While many of these processes are catalytic, some require stoichiometric amounts of Lewis acid (e.g., acylation using $AlCl_3$). Isolation of the product necessitates neutralization steps to remove the acid, resulting in enormous quantities of hazardous waste, with the cost of disposal of this waste often outweighing the value of the product. In fact, Lewis-acid catalyzed reactions are of great interest because of their unique reactivities and selectivities and mild reaction conditions used [47]. A wide variety of reactions using Lewis acids have been developed, and they have been applied to the synthesis of natural and unnatural compounds. However, in most of the reactions, Lewis acids have to be used in stoichiometric amount. On the other hand, an interesting class of Lewis acids are lanthanide triflates $Ln(OTf)_3$ [48]. They are stable and work as Lewis acids in water. Investigations conducted in this field indicated that not only $Ln(OTf)_3$ (Ln) La, Ce, Pr, Nd, Sm, Eu, Gd, Tb, Dy, Ho, Er, Tm, Yb, Lu) but also scandium (Sc) and yttrium (Y) triflates were shown to be water-compatible Lewis acids. Thus, these rare-earth metal triflates [$RE(OTf)_3$] have been regarded as new types of Lewis acids. Many useful reactions are catalyzed by rare-earth metal triflates in aqueous media [49]. In most cases only catalytic amounts of the triflates are required to complete the reactions. Furthermore, rare-earth metal triflates can be recovered easily after reactions and reused without loss of activity [50]. An example of their use is the catalytic acylation of phenols with acid anhydrides using $Sc(OTf)_3$ [50].

One way to significantly reduce the amount of waste is to substitute traditional acids and Lewis acids with recyclable solid acid catalysts (i.e., heteropolyacids) [51]. Heteropolyacids are largely used for oxidation processes due to their low toxicity and high acidity. Heteropolyacids have been used in a variety of acid-catalyzed reactions such as esterification, etherification, hydration of olefins and dehydration of alcohols [52]. Recently, Keggin-type heteropolyacids have been used in multiphase conditions in a range of processes that is, preparation of heterocycles, protection/deprotection of organic functional groups, and oxidation processes, as well as, conversion of 2,6-dimethylphenol to 2,6-dimethyl-1,4-benzoquinone, and selective oxidation of sulfides to sulfoxides with hydrogen peroxide [53].

In any case, it must be mentioned that if a catalyst can be used, it should be used in "catalytic amount". If a reagent can be utilized and yet not consumed in the process it will require less material to continuously effect the transformation. This implies that catalysis has to be as efficient (not only effective) as possible, involving a high turnover.

The Practical Elegance in Synthesis in Green Chemistry Catalysis   Professor Ryoji Noyori (born September 3, 1938) won the Nobel Prize in Chemistry in 2001 with William S. Knowles for the study of chirally catalyzed hydrogenations and with K. Barry Sharpless for their study on chirally catalyzed oxidation reactions (Sharpless epoxidation).

Noyori believes strongly in the power of catalysis and of Green Chemistry. In a recent article he argues for the pursuit of "practical elegance in synthesis: that is chemical synthesis must be intellectually logical and technically truly efficient"

[54]. According to Noyori, every reaction in a multi-step synthesis should proceed with a high atom economy, and the overall synthesis must be accomplished with a low E-factor. In fact he states that "chemists today are asked to develop perfect chemical reactions that proceed with 100% yield and 100% selectivity without forming any waste products. Molecular catalysis, together with traditional hetero-geneous catalysis, contributes significantly to the realization of this goal."

A clear example of practical elegance in synthesis is the green oxidation of cyclohexene for the synthesis of adipic acid (Scheme 6.6). Noyori proved that if a mixture of cyclohexene and H_2O_2 in the presence of small amounts of Na_2WO_4 and methyl-(trioctyl)ammonium hydrogensulfate as phase-transfer catalyst is stirred at 75–90 °C, adipic acid is obtained directly as shiny, colorless, analytically pure crystals in a high yield. This procedure is much more environmentally benign than the commonly used oxidation of a cyclohexanol–cyclohexanone mixture with nitric acid [55].

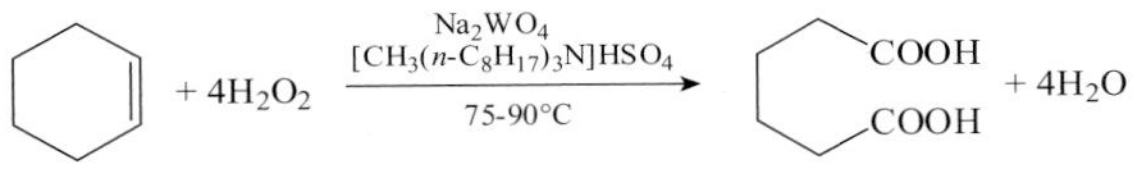

Scheme 6.6   Green oxidation of cyclohexene for the synthesis of adipic acid.

6.2.3
Employing Natural Processes

Biocatalysis has emerged as an important tool in the industrial synthesis of bulk chemicals [56], pharmaceutical and agrochemical intermediates, active pharma-ceuticals, and food ingredients [57]. The potential applications of biocatalysis for the synthesis of chemicals, are highlighted by several industrial processes that are operated by several manufacturers such as Merk [58], BASF,[40] DSM,[41] Lonza [59], Roquette[42] and Cognis.[43] These industries employ enzymes for the synthesis of medium- to high-priced compounds that cannot be produced equally well using a chemical approach (Figure 6.7).

An example of biocatalysis employed for synthetic natural pathways is the bacte-rial fermentation to produce lactic acid from corn starch or cane sugar. The lactic acid so obtained is then used as the starting material to achieve polylactic acid (PLA) [61]. PLA has a wide range of applications, such as woven shirts, microwav-able trays, hot-fill applications and engineering plastics. PLA is currently used in

40) For more details on BASF biocatalysts see http://www.basf.com/group/corporate/en/products-and-industries/biotechnology/index.

41) For further details see http://www.dsm.com/en_US/downloads/dpp/DSM_PharmaChem_SP.pdf.

42) For further details see the website: http://www.roquette.com/delia-CMS/p2/article_id-3548/topic_id-1136/index.html.

43) For further details see the website: http://www.cognis.com/company/Businesses/Care+Chemicals/Green+Chemical+Solutions/.

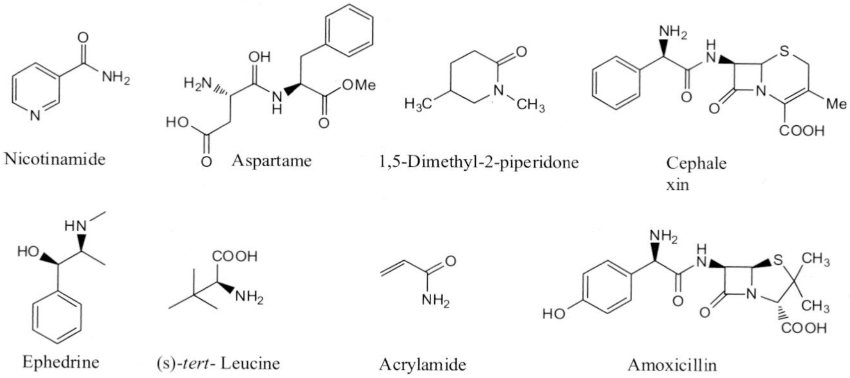

Figure 6.7 Examples of API manufactured using biocatalysts (by many steps) [60].

a number of biomedical applications, such as sutures, stents, dialysis media and drug delivery devices.

In this sense, the use of biosynthesis, biocatalysis and biotech-based chemical transformations can make an important contribution to green chemistry for both efficiency and selectivity. The range of reactions that can be carried out with microorganisms and the range of microorganisms that have already been isolated is enormous. Thus, much effort goes into the selection of new enzymatic activities. Biotech-based chemical transformations have high efficiency and selectivity; are carried out in water at ambient temperature and pressure; do not require tedious protection and deprotection of functional groups; shorten reaction sequences with fewer steps and remove the need for organic solvents. Another great advantage of biocatalysis in industry is the reduction of waste.

Enzymes often represent almost zero waste for companies as they can be reused, and once they reach the end of their service life they can be discarded through conventional waste streams. Conversely, the chemical catalysts that enzymes often replace are heavy metals, which are tightly regulated and in many cases hazardous to the environment and human health. On the other hand, a common problem in biocatalysis is that many of the desired substrates have very low solubility in water, and the catalytic activity for most enzymes is significantly reduced by the addition of even small quantities of organic solvents.

Traditionally, this has limited the use of biocatalysts, but now several examples have shown that biocatalysts can be evolved to function in nonaqueous solvents, allowing their use with water-insoluble substrates. In fact, for a biocatalyst to be effective in an industrial process, it must be subjected to improvement and optimization, and in this respect the directed evolution of enzymes has emerged as a powerful enabling technology. Currently, large-scale industrial applications of enzyme catalysis include the thermolysin-catalyzed synthesis of the low-calorie sweetener aspartame, the synthesis of semi-synthetic β-lactam antibiotics with the use of acylases, acrylamide and nicotinamide [60]. The enzymes most utilized include lipases and other esterases (for ester formation including transesterification; aminolysis and hydrolysis of esters); proteases (ester and amide

Box 6.7   Isosorbide: an example of green chemistry applied to renewables

Cyclic ethers in the form of anhydro sugar alcohols have many applications in industry, in particular in the food industry and in the therapeutic field and are employed as monomers for polymers and copolymers. Such anhydro sugar alcohols are derivatives of mannitol, iditol, and sorbitol. In particular isosorbide, an anhydro sugar alcohol derived from sorbitol, is useful as a monomer in the manufacture of polymers and copolymers, especially polyester polymers and copolymers. Isosorbide is obtained by dehydration of sorbitol by an acid-catalyzed reaction that leads to different anhydro-compounds, but also to polymer-like products. Due to the industrial relevance of this substrate, the one-pot cyclization of D-sorbitol to isosorbide has been greatly investigated.

Recent studies demonstrated that DMC can be used as an efficient dehydrating agent in the cyclization of D-sorbitol to isosorbide in up to quantitative yield (Scheme B7.1) [29]. The main difference between this synthesis and the acid-catalyzed reaction normally used for the synthesis of isosorbide is that the DMC-mediated reaction incorporates two important concepts of the sustainable chemistry: it uses green reagents and a renewable as starting material. This renders this reaction extremely appealing for industrial exploitation.

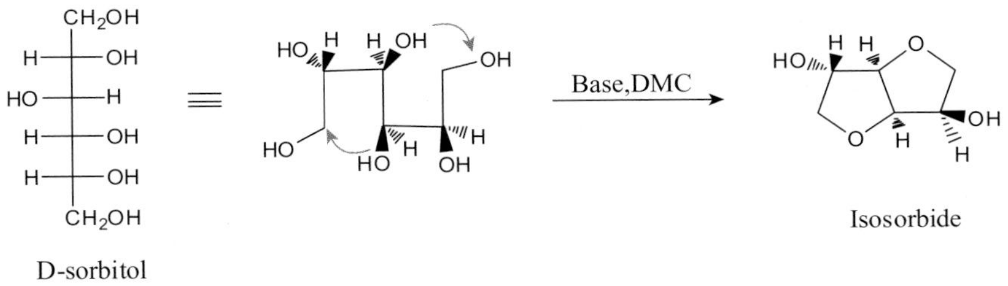

Scheme B7.1   One-step synthesis of isosorbide by DMC chemistry.

The synthesis of isosorbide represents a one-pot double cyclization reaction. Additionally, the four chiral centers were not affected by the reaction, as easily occurs in the present industrial processes that utilize the acidic-promoted cyclization.

hydrolysis, peptide synthesis); nitrilases and nitrile hydratases; hydrolases (hydrolysis of epoxides, halogenated compounds, and phosphates; glycosylation) and oxidoreductases (e.g., enantioselective reduction of ketones). However it is worth mentioning that another drawback of biocatalysis is that enzymes are very specific, thus they can only be used for one reaction at a time. In fact it is impossible at the moment to realize the synthesis of a natural product or of an active pharmaceutical ingredient and Intermediates (API) in one step by biocatalysis.

6.2.4
Use of Alternative Solvents

The use of solvents in every day laboratory work and in the chemical industry is ubiquitous [62]. Solvents are often supposed to disappear at the end of the reaction, nevertheless, they are part of the process and, consequently, they must be treated

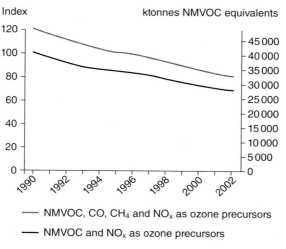

Figure 6.8   An example of non-methane volatile organic compounds (NMVOc) reduction during the last 20 years in Europe.[44]

and disposed of (or eventually recycled). Nowadays, with increasing regulatory pressure focusing on solvents, there is significant attention being paid to green alternatives to traditional solvents, which is perhaps the most active area of Green Chemistry research [62]. In fact, solvents account for the vast majority of mass wasted in syntheses and processes. Moreover, traditional solvents pose several serious issues to human health, being toxic, flammable, and/or corrosive, as well as to the environment due to their volatility and solubility, which has caused enormous air, water and soil pollution over the years. Halogenated solvents, such as carbon tetrachloride, perchloroethylene, and chloroform, have been implicated as potential and/or suspect carcinogens, while other classes of solvents have demonstrated neurotoxicological effects. However, the direct toxicity to humans is only one aspect of the hazards that solvents possess. The use of certain volatile organic compounds (VOCs) as solvents and in other applications has generated great concern about their ability to elevate atmospheric ozone levels (Figure 6.8).

Other substances used as solvents have also been found to possess significant global warming potential and are thought to contribute to the overall greenhouse gas loading in the environment. In an effort to address all these concerns for health and environment, chemists started to search for safer solutions.

As an example, DMC-derivate solvents have recently been demonstrated as efficient alternative solvents for varnishes. The varnish containing the new DMC-based solvent (specifically 2(2methoxy-2ethoxy)ethyl methyl carbonate)shows no toxicity and an improved filming performance [63].

44) For further details consults the Source EEAReport 2005 at the website: http://
www.eea.europa.eu/publications/state_of_environment_report_2005_1.

In this context, some of the main areas of research on alternative solvents include solventless systems, aqueous applications, supercritical fluids, ionic liquids and reduced hazard organic solvents.

- *Solventless Reactions.* Whenever feasible, the best solution would be to avoid the use of solvent since including a supplementary element in a chemical reaction always require extra energy consumption to remove it at the end of the process [64].

- *Water.* The increased focus on water in synthetic organic chemistry during the past few decades has resulted in a large number of reactions that can now be performed successfully in an aqueous medium [65]. Among these reactions are allylation reactions, the aldol condensation, the Michael addition, the Mannich reaction, indium-mediated allylation and Grignard-type additions, hydroformylation [66] and the benzoin condensation. In some reactions the properties of water have even led to improved results, thanks to the hydrophobic effect and easier separation, as many organic substances are not water soluble. An example of the application of water in a chemical process as a green approach to waste minimization is that of phase-transfer catalysis (PTC) [1]. This key synthetic methodology – developed before the concept of Green Chemistry had been formulated – utilizes water as the solvent and is applied in, and applicable to, a great variety of reactions in which inorganic and organic anions or carbenes react with organic substrates. It makes use of a heterogeneous two-phase system – one phase (water) being a reservoir of reacting anions or base for generation of organic anions, whereas organic reactants and catalysts are located in the second, organic phase. The reacting anions are continuously introduced into the organic phase in the form of lipophilic ion-pairs with lipophilic cations supplied by the catalyst. PTC can be carried out in liquid–liquid, solid–liquid, and gas–liquid conditions. The latter (gas–liquid phase-transfer catalysis, GL-PTC) is of importance in Green Chemistry because it is performed under continuous-flow conditions [67]. Transforming batch reactions into continuous-flow processes is a challenge for chemical engineers, but results in harmful reactions being avoided, easier reaction control and the volume in which the reaction takes place being greatly reduced.It must be mentioned that the use of water in an industrial processes can also lead to water contamination, which is energy intensive to clean.

- *Supercritical Fluids (SCFs).* SCFs can be obtained from water, carbon dioxide, methane, methanol, ethanol or acetone, to mention but a few. They include substances which have been simultaneously heated and compressed above their critical points. In particular, in the last 20 years there has been incredible growth in research involving the use of carbon dioxide as an environmentally benign solvent for chemical reactions and polymerizations, both in academia and in industry. CO_2 is a nontoxic, nonflammable, and inexpensive solvent [68]. While CO_2 is a gas under ambient conditions, its liquid and supercritical states are easily attained by compression and heating. Both liquid and $scCO_2$

have a tunable density (and dielectric constant) that increases with increasing pressure or decreasing temperature. Many small molecules are soluble in CO_2, including high-vapor-pressure solvents such as methanol, acetone, and tetrahydrofuran, numerous vinyl monomers, and azo- and peroxy-initiators. Moreover, $scCO_2$ is also widely used because it does solubilize H_2 and O_2. Water and ionic compounds are insoluble, as are most polymers. The two methodologies described by Howdle for dissolving ionic/polar species in $scCO_2$ led to a broadening of the range of applications for supercritical solvents. $ScCO_2$ has found a wide range of industrial applications, the first and most cited being the decaffeination of coffee beans.

- *Ionic Liquids.* Another class of solvents investigated as possible green solvents are ionic liquids, which offer alternatives to conventional molecular solvents for many synthetic transformations [69]. These solvents are often fluid at room temperature, and consist entirely of organic ionic species; they have no measurable vapor pressure, and hence can emit no VOCs. Their uses and applications have been pioneered only recently by Seddon. An interesting application of IL is in their use in cellulose processing [70]. Making cellulosic fibers by dissolving the so-called pulp involves the use, and subsequent disposal, of great volumes of various chemical auxiliaries. However, IL can greatly simplify these processes, serving as solvents that are nearly entirely recyclable. BASF is currently investigating the properties of fibers spun from an IL solution of cellulose in a pilot plant [71].

In this sense Green Chemistry entails the use of alternative green solvents that are nontoxic while preserving (or eventually improving) the efficiency of the synthesis in comparison to classical organic solvents while maintaining the same reaction conditions.

6.2.5
Design of Safer Chemicals

In the last 20 years, chemists have put enormous effort into designing chemicals with various applications ranging from medicines and cosmetics to materials and molecular machines. However, their work has demonstrated a quite surprising lack of interest in taking hazards into consideration in the design process.

The design of safer chemicals is a process that utilizes an analysis of the chemical structure to identify what part of a molecule is providing the characteristic or property that is desired and what part of the molecule is responsible for the toxicity or hazard. Once this information has been ascertained, it is possible to maintain efficiency of function while minimizing the hazard. The goal of designing safer chemicals can be achieved through several different strategies (i.e., computational studies), the choice of which is largely dependent on the amount of information that exists on the particular substance.

The greatest potential to design a safer chemical, in terms of toxicity or other hazards to human health and the environment, is in cases where a mechanism of action is known. Simply stated, if the pathway toward toxicity is known, and if any step within that pathway can be prevented from occurring, then the toxic endpoint will be avoided.

Although mechanisms of action may be unknown, there are often detailed correlations, by way of structure–activity relationships, that can be used to design a safer chemical. As an example, if it is known that the methyl-substituted analogue of a substance has a high toxicity, and that the toxicity decreases as the substitution moves from ethyl to propyl, and so on, it would be reasonable to increase the alkyl chain length to design a safer chemical. Even when the reason for the influence of alkyl chain length on toxicity is unknown, an empirical structure–activity relationship of this kind offers a powerful design tool.

Another important approach for the design of safer molecules is the elimination of toxic functional groups. If there is little information about the specific variations in a chemical's toxicity with structural modification or in the mechanism by which that toxicity is produced, the assumption that certain reactive functional groups will react similarly within the body or in the environment is often a good one. The assumption is especially good if there are data on other compounds in the chemical class that demonstrate a common toxic effect. Here, the design of a safer chemical could possibly proceed by removing the toxic functionality which defines the class. In some cases, this is not possible because the functionality is intrinsic for the desired properties of the molecule. In such a case, there are still options, such as masking the functional group as a nontoxic derivative form and only releasing the parent functionality when and where necessary.

Finally, if through the above methods, the structural feature of the molecule that needs to be modified in order to make it less hazardous cannot be identified, there is still the option of making the substance less bioavailable. If the substance is unable, due to structural design, to reach the target of toxicity, then it is in effect innocuous. This can be achieved through a manipulation of the water-solubility/lipophilicity relationship that often controls the ability of a substance to pass through biological membranes such as skin, lungs, or the gastrointestinal tract. The same principle applies to designing safer chemicals for the environment, such as replacements for ozone-depleting substances. In the past, it was often the goal of the chemist to design substances which were robust and could last as long as possible. This philosophy has resulted in a legacy of wastes, persistent toxic and bio-accumulative substances, and lingering toxic waste sites. Nowadays, it is known that it is more desirable to avoid substances that persist indefinitely in the environment or in landfill, and to replace them with those that are designed to degrade after use. Polymeric materials, for instance, should have no negative effect on the environment during their production, utilization or disposal. Therefore, the design of safer chemicals cannot be limited to hazards associated with the manufacture and use of the chemical, but also to its disposal, that is, its full life cycle.

6.2.6
Developing Alternative Reaction Conditions

The use of alternative reaction conditions has experienced great development in the last 20 years. Energy sources such as UV light, microwaves or ultrasound can be used in a controlled way to increase the efficiency of a chemical reaction, thus making it more eco-friendly.

Microwave chemistry, the science of applying microwave irradiation to chemical reactions, for instance, has been widely investigated since 1986. Microwaves will generally heat any material containing mobile electric charges, such as polar molecules in a solvent or conducting ions in a solid. As a results microwave heating has benefits over conventional ovens: that is, reaction rate acceleration, milder reaction conditions, higher chemical yield, lower energy usage and different reaction selectivity. Microwave heating is very attractive for chemical applications and has become a widely accepted non-conventional energy source for performing organic synthesis. A large number of examples of reactions have been described in organic synthesis: solvent-free reactions, cycloaddition reactions, the synthesis of radioisotopes, fullerene chemistry, polymers, heterocyclic chemistry, carbohydrates, homogeneous and heterogeneous catalysis, medicinal and combinatorial chemistry and Green Chemistry [72].

Acoustic waves are also considered alternative reaction conditions to mechanical milling. In particular, acoustic milling is generated between two disks counter-rotating at 30 000 rpm with a gap as small as 200 mm. Materials are introduced in the center of the disks, and transformed to nanometric powders with no crushing. This process is considered to be energetically greener and costs are claimed to be 10 times less [13].

6.2.7
Minimizing Energy Consumption

Chemistry and energy are two concepts that are strictly linked. In particular, in recent years, a great deal of investment has been made both by the American and European governments in order to promote Clean Energies development at different levels (Figure 6.9). The design of chemical transformations can reduce the required energy input in terms of mechanical, thermal, and other energy inputs, and the associated environmental impacts of excessive energy usage. In many aspects, design for energy minimization is inherently coupled with the design for material efficiency. For instance, when utilizing new solvents such as $scCO_2$, often the separation, a process which requires significant energy input, is also greatly increased. Furthermore, if a synthetic transformation is developed using a catalytic system rather than a stoichiometric process, the activation energy required for the conversion to occur is significantly lowered.

The chemical industry accounts for high energy consumption and CO_2 emissions. The industry typically consumes 25–30% of the total energy used annually

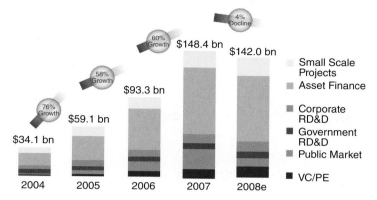

Figure 6.9 Total global new investment in clean energy, 2004–2008, US$ billions.[45] (Please find a color version of this figure in the color plates.)

by the entire manufacturing sector. The two categories of plastic manufacturing and basic organic compounds represent more that one half of chemical industry energy use. This energy relates to the energy of the molecular bonds and to some distinct manufacturing procedures, such as distillation, crystallization, separation, and so on.

The potential energy savings in the chemical industry are enormous. Improving atom economy, utilizing less hazardous reagents, reducing waste, adopting intelligent energy activation and alternative separation procedures will allow the relationship between industry and energy to be reassessed.

A convincing example is represented by the chlorine-based industry. Chlorine stands as "an iconic molecule" for industrial production. Starting from a chlorine anion, Cl_2 can easily be obtained by electrolysis. Many intermediates are produced starting from Cl_2 ($AlCl_3$, $SnCl_4$, $SOCl_2$, $COCl_2$, $TiCl_4$, $POCl_3$, $ZnCl_2$, $SiCl_4$, PCl_3, PCl_5, etc.), which in turn are starting reagents and catalysts for the production of numerous common everyday goods (Figure 6.10). Thus, each compound is the starting point of a chain leading to essential chemical derivatives.

Through a chain of chemical derivatives and relatively easily obtained compounds and intermediates, such molecules have utilized the intrinsic energy available through the use of chlorine primarily produced via electrolysis. More than 20 million tonnes of chlorine and co-products caustic soda and hydrogen are produced each year at about 80 plants across Europe, mostly (about 95%) via electrolysis-based techniques (chlor-alkali industry). Chlorine production is extremely energy intensive; recent data reported a decreasing consumption trend in Europe from 2001 to 2007. Estimates for the global warming potential (GWP) resulting from chlorine use and the primary energy consumed by the chlorine industry

45) Source: World Economic Forum, January 2009, http://www.weforum.org/pdf/climate/Green.pdf.

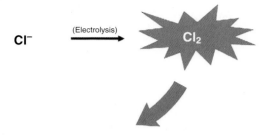

AlCl₃ , SnCl₄ , TiCl₄ , SiCl₄ , ZnCl₂
PCl₃ , POCl₃ , PCl₅ , COCl₂ , SOCl₂, etc.

Figure 6.10 Synthesis of chlorine derivatives by electrolysis.

Table 6.1 Comparing CO_2 emissions in cement, iron, steel and chloro alkali production.

	Chloro-alkali manufacturing	Cement	Iron and steel
Kg CO_2/Kg	1.5	0.95	1.7
World production (2008)	Europe 20 Mtons China 21 Mtons	2.3 Btons	1.2 Btons

in Europe are 0.29% of the total GWP and 0.45% primary energy consumption (Table 6.1).[46]

So, besides their (eco-)toxicity, a major concern with chlorine derivatives is the large amount of energy necessary for their production. This is why chlorinated molecules may have both a direct (as greenhouse gases) and indirect (energy production) impact on climate change at a global level.

The substitution of compounds where "chlorine is used in the making", means that Cl_2 can be avoid as a primary energetic source; this, however, makes chemistry "without chlorine" considerably more difficult and illustrates why it has never been taken on before.

It must be stated that chemists have always worked with the intent to substitute S_N2 with $B_{Ac}2$ following a chlorine-free idea. However, the search for chlorine-free chemistry will not imply a drastic change in the industrial chemical processes, it is more an evolution of production pathways driven by industrial needs related to the modern market.

Furthermore, reduced dependence on chlorine electrolysis will increase energy saving and will produce a smaller CO_2 footprint. Decreasing CO_2 emissions is a crucial step for containing industrial environmental pollution. Figure 6.11 clearly shows the impact of the chemical industry on CO_2 production.

46) For data and more details see http://www.eurochlor.org.

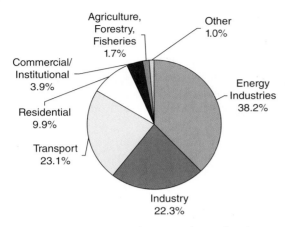

Figure 6.11 CO_2 emissions by sector (shares of total CO_2 emissions: 2007).

The whole of Europe produces 5.21 trillion t CO_2eq per year and has a primary energy consumption of about 55 trillion GJ per year. In Table 6.1 the primary energy consumption and GWP of three different materials are shown.

An approximate estimate of the share of environmental impact caused by the use of chlorine in the state-of-the-art production ranges between 20 and 55% of the whole production chain, beginning from the extraction of the raw materials down to the manufacturing of the chlorine-derived products. This environmental impact potential can be reduced by the development of chlorine-free production routes.

From a scientific and technical point of view, many scientists were, and are, devoted to a progressive substitution and prevention of the use of halogens (principally chlorine) according to Green Chemistry principles. A poignant example is in the field of Friedel–Crafts acylations and hydroxyalkylation. These reactions produce chemicals – such as aromatic ketones, alkenyl ketones and hydroxyalkyl derivatives – that are either valuable intermediates or end active ingredients in an extensive range of high value-added products within pharmaceuticals, flavors, fragrances, and fine chemicals. However, Friedel–Crafts "catalysts" are frequently halogenated Lewis acid activating compounds, $AlCl_3$, $TiCl_4$, $SnCl_4$, BF_3, $ZnCl_2$, which present significant health, safety and environmental challenges for the batch processing typically used in smaller companies [73]. The acylating processes also suffer from the production of copious amounts of aqueous effluent, generated when the products are released from activating agent/product complexes by hydration with water. The costs associated with overcoming these problems are increasing, and challenge the position of traditional acylation routes as economically viable processes for chemical companies. Some major chemical companies use liquid HF process systems, but the capital and running costs exceed the capabilities of small scale manufacturers.

To avoid the production of chlorine by-products during acylation reactions, three classes of alternative reagents can be utilized: carboxylic acids, carboxylic acids anhydrides and mixed carboxylic acid anhydrides. In this regard, a great variety of solid acid catalysts can be studied to perform the acylation reaction with carboxylic acids and carboxylic acid anhydrides and with some mixed anhydrides (in particular acyl-carbonates). The different classes of catalysts to be studied range from clays, zeolites, metal oxides, acid-treated (mixed) metal oxides and heteropolyacids [74].

6.3
Metrics in Green Chemistry

Metrics in Green Chemistry are very important indicators of environmental issues and pollution associated with chemicals manufacturing. Metrics evaluation is often required for the assessment of the operations of a process plant, as they indicate human toxicity, toxicity pathway and ecotoxicity.

The most common green metrics for the evaluation of a chemical process are:

- the *effective mass yield*, defined as the percentage of the mass of desired product relative to the mass of all non-benign materials used in a synthesis [75];

- the *E Factor*, defined as total waste per product; which is an output oriented indicator, defined as waste per mass unit of the product (Σwaste [kg] / product [kg]) [76];

- *the atom economy* (AE), describing the conversion efficiency of a chemical process in terms of atoms involved. In an ideal chemical process, the amount of starting materials or reactants equals the amount of all products generated and no atom is wasted. Thus atom economy can be written as: % atom economy = (molecular weight of desired product/molecular weight of all reactants) $\times$ 100 [77];

- the *mass index* (MI), which is an input oriented indicator, defined as the mass of all raw materials used for the synthesis per unit mass of the purified product (Σraw materials [kg] / product [kg])

Green metrics provide information already in the design phase of a chemical process and indicate consumptions measured as material and energy flows, and waste or toxic release emissions. Thus, by their application, chemists can improve their awareness of environmental issues related to new products as well as to existing ones. For example, utilizing these metrics, some comparative evaluations have been carried out to define the "greener" methylating agent among dimethyl carbonate, dimethyl sulfate and methyl iodide. The data collected demonstrate that DMC is the more eco-friendly reagent [78].

As evidence of the important role played by these green indicators, in 1998 an Ecometrics workshop was held in Switzerland with the participation of academia, industry and decision makers in order to discuss the need of metrics for the environment and to evaluate the best indices to address environmental issues [79]. The key role of the metrics was confirmed during the First International Conference on Green and Sustainable Chemistry held in Japan in 2003. During this meeting, the life cycle approach [80] was also recognized as a method to be taken into account when conducting the environmental assessment of a material.

In 2005, J. Andraos, a professor at York University (Canada) investigating the application of reaction metrics in organic reactions and total syntheses, proposed a formalism to unify the metrics used for chemical reactions and introduced the *stoichiometric factor* (SF), making use of algorithms to calculate reaction metrics and to compare the green performances [81]. In this context, in order to identify a sustainable threshold, Andraos also proposed that reactions considered to be green are characterized by a minimum atom economy of 61.8% coupled with high yield and high solvent recovery and run under stoichiometric conditions [82].

A method enabling the comparison of different chemical reactions in terms of potential environmental impact and identifying the critical phases of a synthesis process was developed by Metzeger and Eissen. This method, called EATOS (environmental assessment tool for organic synthesis), envisages two metrics as tools: the *mass index* and the *E factor*.

By means of EATOS software, other environmental indices are calculated that also consider the weighting factors [83] such as input material prices, for risk of the R-phases and for output materials toxicity and ecotoxicity. The same authors have also introduced in the EATOS tool the cost index. This tool has been utilized by various researchers to evaluate, the "greenness" of pyrazole derivates [84], of photochemistry [85] and of functionalization of heterocyclics [86]. This software, elaborating data relative to reagents, solvents, auxiliary materials, products and secondary products, provides an evaluation graph, useful for the understanding of which phase has more environmental impact and for comparing different processes. Finally, it must be mentioned that, despite all the work so far conducted in the field of green metrics, due to the complexity of a chemical process, the debate on the best metrics to be used and the threshold for deciding the "greenness" of reactions still remains open.

6.4
Conclusions and Future Perspectives

Since 1990, Green Chemistry has gained ever increasing importance in organic synthesis, inspiring scientists in academia, industry, and research institutes around the world. The results so far achieved are encouraging: a new "generation" of reactions that avoid the use of toxic and dangerous chemicals and/or waste have been developed, new eco-compatible solvents have been discovered, compounds that perform comparably or better than those already in existence, but that are biodegradable, have been produced and energy requirements have been reduced.

Substainability of Chemical Industry:

Where we are...   ...where we want to go

Figure 6.12 From the current industrial scenario to green industrial chemistry [13]. (Please find a color version of this figure in the color plates.)

However, the road is still open and big challenges for scientist remain around the corner.

In this prospect the future of Green Chemistry relies upon three open roads: new renewable feedstock, new synthetic pathways and new products.

Regarding the field of renewable feedstock, the recent BP oil drilling disaster in the Gulf of Mexico which began on the night of April 20, 2010 with a tragic explosion that claimed 11 lives, and that now has become the largest environmental disaster in US history, teaches us a hard lesson: fossil fuel utilization has several dangerous repercussions. Their constant depletion leads to continuous prospecting for oil in ever more isolated and/or difficult to reach sites. Furthermore, their use results in a large amount of CO_2 emissions and environmental damage. In this regard, the use of renewable feedstocks is a necessity that chemists must work on, both to ensure the energy needs of future generations and, most importantly, to preserve a green future for our children and grandchildren (Figure 6.12).

New synthetic pathways is also a field with high competition that will employ the talents of many of the next generation of organic and inorganic chemists. It is important to emphasize the huge efforts needed in pursuing synthetic pathways that mimic natural processes in order to avoid emissions into the environment of products that nature is not able to take in and degrade. As chemistry is fast advancing, the sooner more chemists start working in this direction the better it will be for humankind.

Finally, new products should also be taken into account. In fact, the modification of existing products according to human needs should be achieved using Green Chemistry as a focal reference point. The new products must be intrinsically secure since they are made for consumers that is, for us. In this context, solvents surely have great importance. Chemists must use solvents that not only are environmental friendly but that also aid the efficiency of the reaction being worked on.

Chemists must also be aware that to develop Green Chemistry following the principles of Green Chemistry is not enough. In fact, they should be focused on exploiting new green synthetic routes that are as good as or better than the ones currently used by the industry. This means that the economical aspects related to a reaction that is, reagents employed, energy consumed, and waste disposal should also and foremost be kept in mind.

In this sense, Green Chemistry has several essential targets to achieve in the near future, targets that can be achieved only by a strong connection between fundamental research and industry. Nowadays, industry has the skills to work for the welfare of people and to ultimately demonstrate that chemistry is an essential support for the development and evolution of humankind.

References

1 Starks, C.M. and Liotta, C. (1978) *Phase Transfer Catalysis*, Academic Press, New York; Montanari, F., Landini, D., and Rolla, F. (1982) *Top. Curr. Chem.*, **101**, 147; Dehmlow, E.V., and Demlow, S. (1983) *Phase Transfer Catalysis*, II edn, Verlag Chemie, Weinheim; Dehmlow, E.V., and Dehmlow, S.S. (1983) *Phase Transfer Catalysis*, Verlag Chemie, Weinheim.

2 Carson, R. (1962) *Silent Spring*, Mariner books.

3 Anastas, P.T. and Farris, C.A. (eds) (1994) *Benign by Design: Alternative Synthetic Design for Pollution Prevention* (ACS Symposium Series), American Chemical Society, Washington.

4 Anastas, P.T. and Williamson, T.C. (eds) (1996) *Green Chemistry: Designing Chemistry for the Environment* (ACS Symposium Series), American Chemical Society, Washington; Anastas, P.T., and Williamson, T.C. (eds) (1998) *Green Chemistry: Theory and Practice*, Oxford University Press, New York.

5 Tundo, P., Anastas, P., Black, D., Breen, J., Collins, T., Memoli, S., Miyamoto, J., Polyakoff, M., and Tumas, W. (2000) *Pure Appl. Chem.*, **72** (7), 1210.

6 Anastas, P.T. and Warner, J.C. (eds) (1998) *Green Chemistry: Theory and Practice*, Oxford University Press, New York, p. 30.

7 Anastas, P. and Eghbali, N. (2010) *Chem. Soc. Rev.*, **39**, 301.

8 The meeting was sponsored by IUPAC and co-sponsored by ACS, EPA and UNESCO and led to the publication of Tundo, P. and Anastas, P.T. (eds) (2000) *Green Chemistry: Challenging Perspectives*, Oxford University Press, Oxford.

9 Jun, B., Jie, J., Zengwei, Y., and Huang, J. (2000) *Circular Economy: An Industrial Ecology Practice Under the New Development Strategy in China*, Center for Environmental Management & Policy, Nanjing University; Elizabeth C. Economy, 2004. China's Environmental Challenges. http://www.cfr.org/pub7391/elizabeth_c_economy/congressional_testimony_chinas_environmental_challenges.php (accessed 22 September 2004); China-EU workshop, March 17th-18th, 2010, Yong Xing Gordon Hotel, Beijing, China.

10 Aricò, F. and Tundo, P. (2007) *Chem. Int.*, **29**, 5.

11 Kennet, M. (2008) Introduction to Green Economics, in Harvard School Economics Review.

12 Tundo, P., Maggiorotti, P., and Cici, M. (2002) Awards for Green Products and Processes, ISBN 8888214004.

13 Jenck, J.F., Agterberg, F., and Droescherc, M.J. (2004) *Green Chem.*, **6** (11), 544.

14 Priyadarsan, S., Annamalai, K., Sweeten, J.M., Holtzapple, M.T., and Muhktar, S. (2005) *Proceedings of the Combustion Institute*, **30**, 2973–2980; Ragauskas, A.J., Williams, C.K., Davison, B.H., Britovsek, G., Caimey, J., Eckert, C.A., Frederick, W.J., Jr, Hallet, J.P., Leak, D.J., Liotta, C.L., Mielenz, J.R., Murphy, R., Templer, R., and Tschalpliski, T. (2006) *Science*, **311**, 484.

15 Lynd, L.R., Cushman, J.H., and Wyman, C.E. (1991) *Science*, **251**, 1318; McKendry, P. (2002) *Bioresour. Technol.*, **83**, 37; Mosier, N., Wyman, C., Dale, B.E., Elander, R., Lee, Y.Y., Holtzapple, M., and Ladisch, M.R. (2005) *Bioresour. Technol.*, **96**, 673.

16 Datar, R.P., Shenkman, R.M., Cateni, B.G., Huhnke, R.L., and Lewis, R.S. (2004) *Biotechnol. Bioeng.*, **86**, 587. Some more details on "Towards Sustainable Production and Use of Resources: Assessing Biofuels, United Nations Environment Programme", http://www.unep.fr/scp/rpanel/pdf/Assessing_Biofuels_Full_Report.pdf (accessed October 2009).

17 Nägele, H., Pfitzer, J., Eisenreich, N., Eyerer, E., Elsner, P., and Eckl, W. (2000) US 6,509,397 B1 to Fraunhofer-Gesellschaft zur Förderung der angewandten Forschung e.V., München (Germany); Nägele, H., Pfitzer, J., Inone, E., Eyerer, P., Eisenreich, N., and Eckl, W. (2000) WO/2000/027924 to Tecnaro Gesellschaft Zur Industriellen Anwendung Nachwachsender Rohstoffe.

18 Ciamician, G. (1912) *Science*, **36**, 385.

19 Tundo, P. and Selva, M.M. (2002) *Acc. Chem. Res.*, **35**, 706; Selva, M. and Tundo, P. (2006) *J. Org. Chem*, **71**, 1464; Tundo, P., Perosa, A., and Zecchini, F. (eds) (2007) *Methods and Reagents for Green Chemistry*, chapter 4, John Wiley & Sons.; Tundo, P. and Esposito V. (2008) *Green Chemical Reactions*, chapter 10, Springer, 77–103.

20 Tundo, P. and Anastas, P.E. (2000) Dimethylcarbonate: an answer to the need for safe chemicals, in *Green Chemistry: Challenging Perspectives* (ed. F. Rivetti), Oxford University Press, p. 201; Sweet, D.V. (ed.) (1986) *Registry of Toxic Effects of Chemical Substances*, vol. 2, U.S. Department of Health and Human Services, Washington, DC, p. 186.

21 Petroleum Energy Center Report (1999) http://www.pecj.or.jp/japanese/report/e-report/99F.2.1.1-e.pdf (accessed February 1999).

22 Pearson, R.G. (1963) *J. Am. Chem. Soc.*, **85**, 3533; Pearson, R.G. and Songstad, J. (1967) *J. Am. Chem. Soc.*, **89**, 1827; Pearson, R.G. (1988) *J. Am. Chem. Soc.*, **110**, 7684–7690; Chattaraj, P.K., Less, H., and Parr, R.G. (1991) *J. Am. Chem. Soc.*, **113**, 1855–1856; Rauk, A., Hunt, I.R. and Keay, B.A. (1994) *J. Org. Chem.*, **59**, 6808–6816.

23 Fleming, I. (1991) *Frontier Orbitals and Organic Chemical Reactions*, John Wiley & Son, Ltd, West Sussex, UK, pp. 66–73.

24 Tundo, P., Grego, S., Rigo, M., and Paludetto, R. (2008) EP 08172275.3 to DOW Chemical.

25 Anastas, P., Black, D., Breen, J., Collins, T., Memoli, S., Miyamoto, J., Polyakoff, M., and Tumas, W. (2000) *Pure Appl. Chem*, **72** (7), 1207; Tundo, P., Rossi, L., and Loris, A. (2005) *J. Org. Chem.*, **70** (6), 2219; Tundo, P., Bressanello, S., Loris, A., and Sathicq G. (2005) *Pure Appl. Chem.*, **77** (10), 1719; Trotta, F., Tundo, P., and Moraglio G. (1987) *J. Org. Chem.*, **52**, 1300; Tundo, P., Rosamilia, A.E., and Aricò, F. (2010) *J. Chem. Educ.*, **87** (11), 1233–1235.

26 Selva, M., Marques, C.A., and Tundo, P. (1994) *J. Chem. Soc. Perkin Trans. 1*, 1323; Loosen, P., Tundo, P., and Selva, M. (1994) U.S. Patent 5,-278,533; Bomben, A., Marques, C.A., Selva, M., and Tundo, P. (1995) *Tetrahedron*, **51**, 11573; Bomben, A., Selva, M., and Tundo, P. (1997) *J. Chem. Res. Synop*, (12), 448; Tundo, P., Trotta, F., and Moraglio G. (1990) Italian Pat. 20159A/90C.

27 Tundo, P., Selva, M., and Bomben, A. (1999) *Org. Synth.*, **76**, 169.

28 Rahmathullah, S., Hall, J.E., Bender, B.C., McCurdy, D.R., Tidwell, R.R., and Boykin, D.W. (1999) *J. Med. Chem.*, **42**, 3994.

29 Pattison, D.B. (1957) *J. Am. Chem. Soc.*, **79** (13), 3455–3456; Bevinakatti, H.S.,

Newman, C.P., Ellwood, S., Tundo, P., and Aricò, F. (2009) WO2009010791 (A2) to Imperial Chemical Industry, ICI (now to Givaudan and Croda).

30 Davey, P.N., Payne, L., and Sidney, T. (1998) US 5821375; Barton, D.H., Parekh, S.I., Taylor, D.K., and Tse, C. (1994) US Patent 5463089.

31 Tani, K., Yamagata, T., Akutagawa, S., Kumobayashi, H., Taketomi, T., Takaya, H., Miyashita, A., Noyori, R., and Otsuka, S. (1984) *J. Am. Chem. Soc.*, **106**, 5208–5217.

32 Mélendez, J., North, M., and Pasquale, R. (2007) *Eur. J. Inorg. Chem.*, 3323; Clegg, W., Harrington, R., North, M., and Pasquale, R. (2010) *Chem. Eur. J.*, **16**, 6828–6843. DOI: 10.1002/chem.201000030; North, M. and Pasquale, R. (2009) *Angew. Chem. Int. Ed.*, **48**, 2946.

33 Trost, B.M. (1995) *Angew. Chem. Int. Ed. Engl.*, **34** (3), 259.

34 Goti, A. and Cardona, F. (2008) Hydrogen peroxide in green oxidation reactions: recent catalytic processes, in *Green Chemical Reactions* (eds P. Tundo and V. Esposito), Springer, Dordrecht, The Netherlands, pp. 191–212.

35 Fita, I, *et al.* (1985) *J. Mol. Biol.*, **185**, 21.

36 Hoelderich, W.F. and Kollmer, F. (2000) *Pure Appl. Chem.*, **72** (7), 1273; Schuster, H., Rios, L.A., Weckes, P.P., and Hoelderich, W.F. (2008) *Appl. Catal. A Gen.*, **348** (2), 266; Laufer, MC., Hinze, R., Hoelderich, WF., Bonrath, W., and Netscher, T. (2009) *Catal. Today*, **140**, 105.

37 Hoelderich, W.F. and Kollmer, F. (2000) *Pure Appl. Chem.*, **72** (7), 1273.

38 Kulprathipanja, S. (2010) *Zeolites in Industrial Separation and Catalysis*, Wiley-VCH Verlag GmbH, Weinheim.

39 Moreno, N., Querol, X., Ayora, C., Pereira, C.F., and Janssen-Jurkovicova, M. (2001) *Environ. Sci. Technol.*, **35**, 3526.

40 Johnson, L.B. (2005) US6893632 B2; US2001/31220 A1; US6440415 B1; US2002/197247 A1; US6893632 B2; Ebihara, F. and Watano, S. (2003) *Chem. Pharm. Bull.*, **51** (6), 743.

41 Quimby, P.C., Birdsall, J.L., Caesar, A.J., Connick, W.J., Boyette, C.D., Caesar, T.C., and Sands, D.C. (1994) to The United States of America as represented by the Secretary of the (Washington, DC). Appl. No.: 08/039,679.

42 Ballini, R., Bigi, F., Gogni, E., Maggi, R., and Sartori, G. (2000) *J. Catal.*, **191**, 348; Srivastava, R., Iwasa, N., Fujita, S.-I., and Arai, M. (2008) *Chem. Eur. J.*, **14**, 9507.

43 Hegedues, A., Hell, Z., and Potor, A. (2006) *Synth. Commun.*, **36**, 3625; Bonino, F., Damin, A., Bordiga, S., Selva, M., Tundo, P., and Zecchina, A. (2005) *Angew. Chem. Int. Ed.*, **44** (30), 4774; Kim, S.-S., Pinnavaia, T.J., and Damavarapu, R. (2008) *J. Catal.*, **253** (2), 289; Tachikawa, T., Yamashita, S., and Majima, T. (2010) *Angew. Chem. Int. Ed.*, **49** (2), 432.

44 Tanabe, K. and Hölderich, W.F. (1999) *Appl. Catal. A*, **181**, 399.

45 Gounder, R. and Iglesia, E. (2009) *J. Am. Chem. Soc.*, **131** (5), 1958.

46 Sheldon, R.A. (2000) *Pure Appl. Chem.*, **72** (7), 1233.

47 Schinzer, D. (1989) *Selectivities in Lewis Acid Promoted Reactions*, Kluwer Academic Publishers, Dordrecht; Yamamoto, H. (ed.) (2000) *Lewis Acids in Organic Synthesis*, Wiley-VCH Verlag GmbH, Weinheim.

48 Kobayashi, S. (1994) *Synlett*, 689; Marshman, R.W. (1995) *Aldrichim. Acta*, **28**, 77.

49 Kobayashi, S., Sugiura, M., Kitagawa, H., and Lam, W.W.-L. (2002) *Chem. Rev.*, **102**, 2227–2302.

50 Hofle, G., Steglich, W., and Vorbruggen, H. (1978) *Angew. Chem. Int. Ed. Engl.*, **17**, 569; Vedejs, E. and Diver, S.T. (1993) *J. Am. Chem. Soc.*, **115**, 3358; Vedejs, E., Bennet, N.S., Conn, L.M., Diver, S.T., Gingrass, M., Lin, S., Oliver, P.A., and Peterson, M.J. (1993) *J. Org. Chem.*, **58**, 7286.

51 Ratton, S. (1997) *Chem. Today*, March/April, 33; Sheldon, R.A. (2000) *Science*, **287**, 1636; Sheldon, R.A. (2005) *Green Chem.*, **7**, 267.

52 Simoes, M.M.Q., Conceicao, C.M.M., Gamelas, J.A.F., Domingues, P.M.D., Cavaleiro, A.M.V., Cavaleiro, J.A.S., Ferrer-Correia, A.J.V., and Johnstone, R.W.A. (1999) *J. Mol. Catal. A Chem.*,

144, 461–468; Wang, J., Yan, L., Li, G., Wang, X., Ding, Y., and Suo, J. (2005) *Tetrahedron Lett.*, **46**, 7023; Heravi, M.M., Zadsirjan, V., Bakhtiari, K., Oskooie, H.A., and Bamoharram, F.F. (2007) *Catal. Commun.*, **8**, 315; Nagaraju, P., Pasha, N., Sai, P., Prasad, S., and Lingaiah, N. (2007) *Green Chem.*, **9**, 1126; Park, D.R., Park, S., Bang, Y., and Song, I. (2010) *Appl. Catal. A Gen.*, **373**, 201; Rao, P.S.N., Venkateswara, K.T., Said Prasad, P.S., and Lingaiah, N. (2010) *Catal. Commun.*, **11**, 547; Romanelli, G., Autino, J., Vázquez, P., Pizzio, L., Blanco, M., and Cáceres, C. (2009) *Appl. Catal. A Gen.*, **352**, 208.

53 Kaczorowska, K., Kolarska, K., Mitka, K., and Kowalski, P. (2005) *Tetrahedron*, **61**, 8315–8327; Villabrille, P., Romanelli, G., Quaranta, N., and Vázquez, P. (2010) *Appl. Catal. B Environ.*, **96**, 379; Villabrille, P., Romanelli, G., Vázquez, P., and Cáceres, C. (2008) *Appl. Catal. A Gen.*, **334**, 374.

54 Noyori, R. (2005) *Chem. Commun.*, **14**, 1807.

55 Sato, K., Aoki, M., and Noyori, R. (1998) *Science*, **281**, 1646.

56 Fahrenkamp-Uppenbribk, J. (2002) *Science*, **297**, 798.

57 Davies, I.W. and Welch, C.J. (2009) *Science*, **325**, 701.

58 Shultz, C.S. and Krska, S.W. (2007) *Acc. Chem. Res.*, **40** (12), 1320.

59 Leresche, J.E. and Meyer, H.-P. (2006) *Org. Process Res. Dev.*, **10**, 572.

60 Shoemake, H.E., Mink, D., and Wubbolts, M.G. (2003) *Science*, **299**, 1694.

61 Martin, O. and Averous, L. (2001) *Polymer*, **42**, 6209; Gruber, P. and O'Brien, M. (2001) *Biopolymer online*, **6**, Chapter 8, *Polylactides:NatureWorks® PLA*, June 2001. 59.

62 Marcus, Y. (ed.) (1999) *The Properties of Solvents*, John Wiley & Sons, Inc., New York.

63 Tundo, P., Riva, L., and Mangano, R. (2008) PCT/IB2008/003409; (2009) IPN # WO 2009/147469 A1.

64 Kerton, F.M. (ed.) (2009) *Alternative Solvents for Green Chemistry*, RSC Green Chemistry Book Series, Royal Society of Chemistry, p. 23; Tanaka, K. (ed.) (2003)

Solvent-Free Organic Synthesis, Wiley-VCH Verlag GmbH, Weinheim, Germany; Cave, G.W.V., Raston, C.L., and Scott, J.L. (2001) *Chem. Commun.*, 2159.

65 Li, C.-J. and Chan, T.-H. (2007) *Comprehensive Organic Reactions in Aqueous Media*, 2nd end, John Wiley & Sons, Inc., Hoboken, NJ; Kerton, F.M. (2009) *Alternative Solvents for Green Chemistry*, RSC Green Chemistry Book Series, Royal Society of Chemistry.

66 Wiebus, E. and Cornils, B. (1994) *Chem. Ing. Technol.*, **66**, 916; Kohlpaintner, C.W., Fischer, R., and Cornils, B. (2001) *Appl. Catal. A Gen.*, **221** (1–2), 219–225; Paganelli, S., Ciappa, A., Marchetti, M., Scrivanti, A., and Matteoli, U. (2006) *J. Mol. Catal. A Chem.*, **247**, 138–14; Paganelli, S., Zancheta, M., Marchetti, M., and Mangano G. (2000) *J. Mol. Catal. A Chem.*, **157**, 1–8.

67 Tundo, P. (ed.) (1991) *Continuous-Flow Methods in Organic Synthesis*, E. Horwood Pub., Chichester, UK.

68 Arai, Y., Sako, T., and Takebayashi, Y. (2002) *Supercritical Fluids, Springer Series in Materials Processing*, Springer, New York.

69 Earle, M.J. and Seddon, K.R. (eds) (2002) *Clean Solvents: Alternative Media for Chemical Reactions and Processing – Ionic Liquids: Green Solvents for the Future*, ACS Symposium Series, American Chemical Society, p. 819; Plechkova, N.V. and Seddon, K.R. (2007) *Methods and Reagents for Green Chemistry – Ionic Liquids: "Designer" Solvents for Green Chemistry*, John Wiley & Sons, Inc., Hoboken, p. 105; Chiappe, C. and Pieraccini D. (2004) *J. Phys. Org. Chem.*, **18** (4), 275.

70 Swatloski, R.P., Spear, S.K., Holbrey, J.D., and Rogers, R.D. (2002) *J. Am. Chem. Soc.*, **124**, 4974.

71 Hermanutz, F., Gähr, F., Massonne, K., and Uerdingen, E. (2006) Oral presentation at The 45th Chemiefasertagung, Dornbirn, Austria, September 20th–22nd. More details at http://www.basionics.com/en/ionic-liquids (accessed 13 August 2008).

72 De la Hoz, A., Dıaz-Ortiz, A., and Moreno, A. (2005) *Chem. Soc. Rev.*, **34**,

164–178; Strauss, C. and Trainor, R. (1995) *Aust. J. Chem.*, **48**, 1665; Varma, R.S. (1999) *Clean Prod. Process.*, **1**, 132–147; Varma, R.S. (2002) *Advances in Green Chemistry: Chemical Syntheses Using Microwave Irradiation*, Astra Zeneca Research Foundation, Kavitha Printers, Bangalore, India; Bose, A.K., Manhas, M.S., Ganguly, S.N., Sharma, A.H., and Banik, B.K. (2002) *Synthesis*, 1578; Nuchter, M., Ondruschka, B., Bonrath, W., and Gum, A. (2004) *Green. Chem.*, **6**, 128.

73 Olah, G.A. (1973) *Friedel-Crafts Chemistry*, Wiley-VCH Verlag GmbH, New York.

74 Sartori, G. and Maggi, R. (2006) *Chem. Rev.*, **106**, 1077–1104; Cardoso, L.A.M., Alves, W., Jr, Gonzaga, A.R.E., Aguiar, L.M.G., and Andrade, H.M.C. (2004) *J. Mol. Catal. A Chem.*, **209**, 189; Klisakova, J., Cerveny, L., and Cejka, J. (2004) *Appl. Catal. A Gen.*, **272**, 79; Chiche, B., Finiels, A., Gauthier, C., and Geneste, P. (1986) *J. Org. Chem.*, **51**, 2128; Tauster, S. (1987) *J. Acc. Chem. Res.*, **20**, 389.

75 Hudlicky, T., Frey, D.A., Koroniak, L., Claeboe, C.D., and Brammer, L.E., Jr (1999) *Green Chem.*, **1**, 57.

76 Sheldon, R.A. (1992) Organic synthesis; past, present and future *Chem. Ind. (Lond.)*, 903–906.

77 Trost, B.M. (1991) *Science*, **254**, 1471.

78 Selva, M. and Perosa, A. (2008) *Green Chem.*, **10**, 457.

79 Biswas, G., Clift, R., Davis, G., Ehrenfeld, J., Förster, R., Jolliet, O., Knoepfel, I., Luterbacher, U., Russell, D., and Hunkeler, D. (1998) *Int. J. LCA*, **3** (4), 184.

80 Yasui, I. (2003) Conference report. *Green Chem.*, **5**, G70–G77.

81 Andraos, J. (2005) *Org. Process Res. Dev.*, **9**, 149.

82 Andraos, J. (2005) *Org. Process Res. Dev.*, **9**, 404.

83 Eissen, M. and Metzger, J.O. (2002) *Chem. Eur. J.*, **8** (16), 3580.

84 Corradi, A., Leonelli, C., Rizzuti, A., Rosa, R., Veronesi, P., Grandi, R., Baldassari, S., and Villa, C. (2007) *Molecules*, **12**, 1482.

85 Protti, S., Dondi, D., Fagnoni, M., and Albini, A. (2009) *Green Chem.*, **11**, 239.

86 Ravelli, D., Dondi, D., Fagnoni, M., and Albini, A. (2008) *Appl. Catal. B Environ.*, **79** (4), 368.

7

Water: Foundation for a Sustainable Future

Maya A. Trotz, James R. Mihelcic, Omatoyo K. Dalrymple, Arlin Briley, Ken D. Thomas, and Joniqua A. Howard

7.1
Introduction

> "Water is so familiar, we generally consider it be a rather bland fluid of simple character. However, ... if scientists had discovered it in recent times, it would undoubtedly have been classified as an exotic substance." [1]

Water is the foundation of our current and future economic, societal, and environmental prosperity. It is a critical component of agricultural and industrial production (including electric power generation), household consumption, hygiene and sanitation, recreation, ecosystem maintenance and various processes at the earth's surface [2]. Water also plays a vital role as a material in society. For example, when compared to other material flows in the Unite States, water usage is comparable on a mass basis with other material flows such as aggregates, industrial materials, recycled metals, and primary metals. Water is also unique because it can serve as a transport medium for waste and feedstocks, and it comprises a portion of many final products. Ecosystems are also dependent on water and they provide a tremendous amount of services to humans. Importantly, much of the world's poor (those living on less than 1 or 2 US$ per day) depend on ecosystems for their economic livelihood [3].

Water has received much more attention at the global level since the 1972 United Nations Conference on the Human Environment added the environment to the list of global problems. For example, at the 2002 World Summit on Sustainable Development, world leaders reaffirmed the principles of sustainable development adopted at the Earth Summit ten years earlier and also adopted the Millennium Development Goals (MDGs) [3]. The MDGs are an ambitious agenda for reducing poverty and improving lives based on global agreements made at the Millennium Summit in September 2000. Many of the MDGs are related to health and thus indirectly related to water and sanitation. In particular, Goal 7 (i.e., ensure environmental sustainability) seeks to not only integrate the principles of sustainable development into country policies and programs and reverse the loss

The Chemical Element: Chemistry's Contribution to Our Global Future, First Edition.
Edited by Javier Garcia-Martinez, Elena Serrano-Torregrosa
© 2011 Wiley-VCH Verlag GmbH & Co. KGaA. Published 2011 by Wiley-VCH Verlag GmbH & Co. KGaA.

Table 7.1 Location of 35 million km³ of total freshwater found on Earth (data from [4]).

Location	Percent of world's freshwater
Glaciers and permanent snow cover	68.7
Groundwater	30.1
Lakes	0.26
Soil moisture	0.05
Atmosphere	0.04
Marshes and swamps	0.03
Biological water	0.003
Rivers	0.006

of environmental resources but also reduce by half the proportion of people without access to safe drinking water and achieve significant improvement in the lives of at least 100 million slum dwellers. Given the magnitude of delivering sanitation and sufficient amounts of safe water to billions of the global population, in December 2003, the United Nations General Assembly proclaimed the period 2005–2015 the International Decade for Action "Water for Life". This ten-year period officially started on World Water Day (March 22, 2005).

The total volume of the world's water is estimated to be 1.386 billion km³.	However, only 35 million km³ of total freshwater is found on Earth.

Table 7.1 shows the locations of the world's freshwater reserves. Note that, as shown in Table 7.1, over two thirds (68.7%) of the world's freshwater is not readily available, being present as glaciers and permanent snow cover.

Globally, 3800 km³ of water are withdrawn every year and 2100 km³ of this volume are consumed. Consumed water is evapotranspirated or incorporated into products or organisms. The remaining 1700 km³ of water that are not consumed are typically returned to local water bodies, usually as wastewater that is generated primarily from domestic and industrial sources. This large volume of water returned to local water systems may not be available for easy reuse though, depending on its next use and, importantly, whether it has been contaminated and/or treated prior to discharge. In addition, the treatment of wastewater consumes materials and energy [3]. Fresh water represents a small fraction of the Earth's total water resources, and fresh accessible water is scarce. In addition, water is not evenly distributed geographically [2]. Individuals and governments must use and manage this precious resource in a more intelligent and sustainable manner for current and future generations.

Table 7.2 summarizes ranges of concentrations of natural chemicals and environmental pollutants found in different natural sources of water. Major cations in water include K^+, Na^+, Ca^{2+}, and Mg^{2+}. Major anions include SO_4^{2-}, Cl^-, HCO_3^-, and

Table 7.2 Range of chemical concentrations found in natural waters (from [5]. Reprinted with permission of John Wiley & Sons, Inc.).

Substance	Rain, fog	Lakes, rivers	Groundwater	Oceans
Trace metals (e.g., Pb, Cu, Hg, Zn)	$0.01–100\,\mu g\,L^{-1}$	$0.001–10\,\mu g\,L^{-1}$	$0.1–10^6\,ng\,L^{-1}$	$0.01–100\,ng\,L^{-1}$
Organic pollutants (e.g., PCBs, pesticides, solvents)	$1–5000\,ng\,L^{-1}$	$0.1–500\,ng\,L^{-1}$	$0.001–10^6\,ng\,L^{-1}$	$0.001–10\,pg\,L^{-1}$
Major ions (Ca^{2+})	$0.1–20\,mg\,L^{-1}$	$1–120\,mg\,L^{-1}$	$1–120\,mg\,L^{-1}$	$800\,mg\,L^{-1}$
Major ions (Cl^-)	$0.05–10\,mg\,L^{-1}$	$0.1–30\,mg\,L^{-1}$	$0.1–50\,mg\,L^{-1}$	$35\,000\,mg\,L^{-1}$

CO_3^{2-}. Concentrations of H^+ and OH^- are not typically significant as most natural waters have pH values that range between 6 and 8. Many inorganic and organic chemicals find their way into natural waters. The number of chemicals detected in water is quite staggering and should not be surprising since approximately 70 000 chemicals are commonly sold in commerce this century. Unfortunately, many of these chemicals that find their way into water pose a risk to ecosystems and human health.

Table 7.3 lists several sources of commonly used water and some issues associated with the source. Surface water (e.g., rivers, streams, lakes) is widely used as a source of water because of its high flowrate and availability. However, it is easy to contaminate, especially with pipe discharges and non-point source runoff from agricultural and urban areas. This is why treatment of surface water usually consists of several treatment processes to remove particulate and dissolved substances that make their way into surface water. In contrast, groundwater is naturally filtered by the subsurface that can remove particulate matter and other pollutants; however, it has lower flows compared to surface waters and the renewal times of groundwater can be very long if it is extracted in a nonsustainable manner.

Having access to an appropriate amount of water is also important. Water scarcity occurs when there is insufficient water to satisfy normal human requirements. Vörösmarty et al. [6] estimate that nearly 2 billion people suffer from severe water scarcity. Furthermore, they report that of the one billion additional people expected to face water scarcity by the year 2025, 20% will be associated with direct effects of climate change. A water footprint determines the water required to support human activities and readers can calculate their personal and country's water footprint at: http://www.waterfootprint.org/.

Figure 7.1 provides water footprints for several countries around the world. The following eight countries (in order of consumption) are responsible for 50% of the global water footprint: India, China, United States, Russia, Indonesia, Nigeria, Brazil, and Pakistan. On a per capita basis, for 1997–2001, the United States had

Table 7.3 Sources of fresh water and some issues associated with the source (adapted from
[3]. Reprinted with permission of John Wiley & Sons, Inc.).

Source of water	Comments
Surface water	High flows, easy to contaminate, relatively high suspended solids (TSS), turbidity, and pathogens. In some parts of the world, rivers and streams dry up during the dry season.
Groundwater	Lower flows but has natural filtering capacity that removes suspended solids (TSS), turbidity, and pathogens. May be high in dissolved solids (TDS) including Fe, Mn, Ca, and Mg (hardness). Difficult to clean up after contaminated. Renewal times can be very long.
Rainwater	Used for drinking water, hygiene, irrigation, and sanitation (e.g., flushing toilets). Can also be directed into rain gardens and also low impact development techniques that promote groundwater recharge and minimize surface runoff.
Reclaimed and Reused Water	Technically feasible. Currently used for irrigating agricultural crops, landscaping, groundwater recharge, and potable water. Includes decentralized use of greywater (defined as the wastewater produced from baths and showers, clothes and dishwashers, and lavatory sinks).
Seawater	Energy intensive so costly compared to other sources. Disposal of resulting brine must also be considered. Multistage distillation and reverse osmosis are the two technologies most commonly used, accounting for approximately 87% of worldwide desalination capacity.

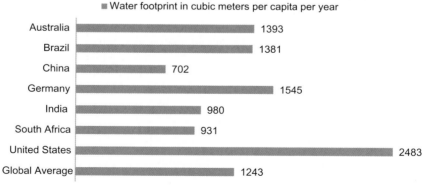

Figure 7.1 Water footprints of several countries along with the global average (1997–2001)
(data from [7]).

the highest footprint: 2483 m³/capita-year. In comparison, for the same period the global water footprint was 1243 m³/capita-year [3].

7.2
Water Pollution and Water Quality

Pollution occurs when a chemical or material is discharged to the environment at levels leading to a negative impact on human health and/or the environment. Chemical pollutants are discharged to water from point sources and nonpoint sources. Point sources are stationary locations such as pipes, and nonpoint sources include runoff from land and atmospheric deposition [3].

7.2.1
Biochemical Oxygen Demand

Biochemical oxygen demand (BOD) is a measure of a wastewater's ability to consume oxygen and is related to the presence of chemicals and materials in the water that consume oxygen from chemical and/or biological processes. For example, organic carbon can be biologically degraded by microorganisms, and, in the process, consume oxygen. Nitrogen in the form of ammonia can be oxidized to nitrate by bacteria as well, and, in the process, consume oxygen. BOD measurements can also be used to determine if natural waters are negatively impacted by human discharges.

As shown in Table 7.2, pollutants can be either organic or inorganic. Pollutants such as nitrogen, phosphorus, organic matter and suspended solids are discharged to the rivers of the world in the tens of millions of tons per year (Figure 7.2). Their discharge is strongly related to human population, technology, and affluence. Wastewater discharges can include domestic and industrial wastewater, agricultural runoff, erosion of land, and runoff from paved surfaces that make up the infrastructure of our built environment. The strength of organic wastes in a water sample can be measured by BOD.

Figure 7.2a shows the BOD discharged to global waterways in 1995 (in mega tons per year) for agricultural, domestic (i.e., households), and industrial sectors. The figure also provides expected future discharges in 2010 and 2020. Figure 7.2b also shows the agricultural nitrogen loading to our waterways for OECD and non-OECD countries for similar time periods. Note, in Figure 7.2 the large contribution of BOD and nitrogen pollution emitted to water from the agricultural sector, a major contributor to poor water quality around the world.

7.2.2
Nutrients (Nitrogen and Phosphorus)

Nitrogen and phosphorus are two of the three major nutrients required for the growth of plants and animals (potassium is the third). Because there is limited

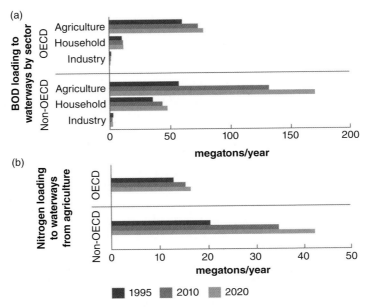

Figure 7.2 (a) Annual BOD loading (mega tons) into global waterways (in OECD and non-OECD countries) for agricultural, domestic, and industrial sectors for the year 1995 and estimated for 2010 and 2020. (b) Nitrogen loading (mega tons) for OECD and non-OECD countries for agricultural sectors into global waterways for the year 1995 and estimated for 2010 and 2020 (data from [4] and [3]. Reprinted with permission of John Wiley & Sons, Inc.).

availability of forms of nitrogen and phosphorus readily available for growth in many soils, modern agriculture depends on manufactured fertilizers. Though nitrogen and phosphorous are essential, their improper use and release leads to ecosystem degradation.

In 1840, German scientist Justus von Liebig showed that sulfuric acid reacted with insoluble phosphates, like bone, to produce a more bioavailable phosphorus. In 1842, Englishman John Bennett Lawes received a patent for phosphorus production from coprolites (fossilized animal dung) using sulfuric acid. Later discoveries of easily accessible phosphorus bearing minerals led to mining of areas like Florida where the desired ore is found about 4–15 m below the earth's surface and is about 3–6 m thick. Reaction of the ore with sulfuric acid produces phosphoric acid which is then used to make fertilizers. For example, ammonia is reacted with phosphoric acid to produce the fertilizer diammonium phosphate (DAP) which is easily taken up by plants. In Florida, approximately 4500 kg of phosphogypsum is produced per 900 kg of phosphoric acid and, unlike natural gypsum $(CaSO_4(s))$ prevalent in construction materials, radioactivity due to radium limits its useful reuse. In addition to radiation, Florida ore also contains 2–4% fluoride which

Table 7.4 Nutrients found in human urine and feces (data from [9]).

Nutrient	Human Urine kg/person-year	Human Feces kg/person-year	% Nutrient in Urine
Nitrogen	4.0	0.5	89
Phosphorus	0.4	0.2	67

negatively impacts local atmosphere and agriculture. China, the United States and Morocco produced 67% of the phosphate rock in the world in 2009, however, Morocco has the largest known reserves, of the order of six times that of the United States or 40% of the world's share. Despite existing reserves, the world is consuming phosphorus faster than it is replenished geologically [8] and this is forcing new advances in agricultural practices, including fertilizer production, to ensure food security.

The annual global production of phosphate rock is approximately 40 million tons. This requires the mining of 140 million tons of ore. The use of the resulting phosphate is broken down as: 80% for fertilizer, 12% for detergent, 5% for animal feed, and 3% for specialty applications such as fire retardants. Though phosphorus is the eleventh most abundant element on earth, a dwindling reserve of high-grade ore combined with annual increases in demand, associated with increased global population and affluence, suggest the depletion of current economically available reserves will occur in 60 to 130 years. In addition, it is estimated that the current abundance of inexpensive phosphorus will be exhausted within the first half of this century.

Table 7.4 shows the mass of phosphorus and nitrogen discharged every year by an individual human. The values in Table 7.4 are broken down by the amount of the two nutrients found in urine and feces. Note the large percentage of these nutrients found in human urine; the nitrogen content in urine is estimated to be as high as 89%, 67% for phosphorus, and 75% for potassium [9]. The International Fertilizer Industry Association (IFIA) [10] reports the 2007 worldwide "Apparent Consumption" of phosphate (P_2O_5) was 54 428 metric tons of which 40 million metric tons was for agricultural use [11]. Hence, for a population of 6 billion people the amount of phosphorus produced per year by humans represents approximately a quarter of that used by the fertilizer industry. The lack of management of this valuable nutrient resource is occurring at a time when readily-available phosphorus is running out globally and in spite of the environmental impact of mining.

Nitrogen gas (N_2O) comprises 78% of the world's atmosphere, but must be "fixed" to make it chemically reactive and biologically available. Nitrogen fixation, the reaction of $N_2(g)$ with carbon, hydrogen and/or oxygen, occurs through biological processes on land and in the ocean, lightning bolts in the atmosphere, fertilizer production, industrial combustion, and biomass burning [2].

Table 7.5 Major nitrogen species involved in global nitrogen cycle.

Species	Nitrogen oxidation state	Examples of reactions	Notes
NO_3^-	+5	$NO_3^- + 2e^- + 2H^+ \leftrightarrow$ $NO_2^- + H_2O$	A predominant form of N taken up by organisms.
NO_2^-	+3	$NH_3 + CO_2 + 1.5O_2$ $+ Nitrosomonas \rightarrow$ $NO_2^- + H_2O + H^+$ $NO_2^- + CO_2 + 0.5O_2$ $+ Nitrobacter \rightarrow NO_3^-$	Nitrification is used in wastewater treatment plants to convert toxic N in the chemical form of ammonia to nitrate and nitrite.
N_2O	+1	In stratosphere: $N_2O \rightarrow NO_x = NO + NO_2$ $NO + O_3 \rightarrow NO_2 + O_2$ $O + NO_2 \rightarrow NO + O_2$ net: $O + O_3 \rightarrow 2O_2$	Important greenhouse gas and catalyst for ozone destruction [13].
N_2	0	$N_2(g) + O_2(g) \leftrightarrow 2NO(g)$ $NO(g) + O_3(g) \leftrightarrow NO_2(g) + O_2(g)$ $NO_{2(g)} + OH^- \leftrightarrow H^+ + NO_3^-$	Comprises 78.082 % of atmosphere by volume. Must be fixed to be made biologically available.
NH_3	−3	$NH_3 + 2O_2 \leftrightarrow NO_3^- + H^+ + H_2O$ $NH_4^+ + NO_2^- \leftrightarrow N_2 + 2H_2O$	NH_4^+ is a predominant form of N taken up by organisms. Denitrification is used to eliminate N from wastewater.

The Haber process is used by industry to manufacture ammonia from nitrogen and hydrogen gases in the presence of a catalyst. The resulting ammonia is used directly as a fertilizer and combined to formulate other inorganic N compounds. Biological N fixation provides about 90–130 Tg N/year on the continents and human activities have resulted in the fixation of an additional 150 Tg N/year [2, 12]. Fixation due to human activities includes cultivation of crops like legumes. Once fixed, nitrogen cycles through a complex biogeochemical cycle and some of the major species are listed in Table 7.5. Two important microbially induced processes responsible for transformation of fixed nitrogen in both natural (e.g., soils) and engineered processes (e.g., wastewater treatment) are nitrification and denitrification. Nitrification entails the oxidation of nitrogen in ammonia to nitrite and nitrate and denitrification refers to the conversion of nitrate back to nitrogen gas. Denitrification results in fluxes of >291 Tg N/year from the oceans and land [2].

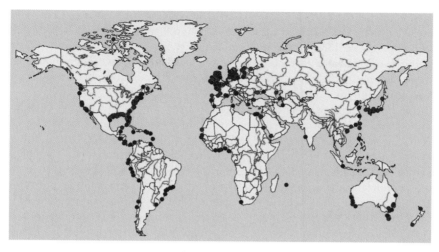

Figure 7.3 Coastal eutrophic and hypoxic zones (sites identified by Selman *et al.* 2008. Figure from [3]. Reprinted with permission of John Wiley & Sons, Inc.).

Human impact on the nitrogen cycle is a growing area of research and of concern since the process involves a potent greenhouse gas (e.g., the Global Warming Potential of $N_2O(g)$ is 298 times that of $CO_2(g)$) [3] and also involves a nutrient source with implications for water quality. Excessive input of nutrients such as nitrogen and phosphorus can cause major problems in water systems, such as eutrophication. In this process, the nutrient enrichment results in increases in organic matter due to excessive growth of plants and algae. Subsequent decay of this organic matter consumes oxygen and this leads to the loss of biodiversity and death of aquatic life.

Figure 7.3 shows the 400 coastal areas in the world that currently experience some form of nutrient enrichment. Of these, 169 are reported to experience decreased oxygen levels in the water column. These areas are referred to as "dead zones" where the oxygen concentration in the water is decreased below $2\,mg\,L^{-1}$, either on a seasonal or long-term basis. This has a large negative impact on the food chain of aquatic ecosystems, which then has an adverse impact on regional and global fisheries along with the productivity of marine mammals [3].

One of the largest dead zones in the world is located in the Gulf of Mexico. The Gulf of Mexico, the ninth largest body of water in the world, is bordered by the United States to the north and east, Mexico to the west and south, and the island of Cuba to the southeast. The Gulf region covers approximately 600 000 square miles. The Mississippi River drains 41% of the lower 48 states of the United States and 15% of North America. It has been documented that 65% of the nutrient input to the Gulf of Mexico dead zone is carried by the Mississippi River and originates in agricultural areas that grow a large amount of corn. These areas not only provide food for a growing population but are now also viewed as a source of production of biofuel materials. Besides applying excessive amounts of fertilizers, farmers in these locations have drained up to 80% of natural wetlands which would have served as natural buffers to runoff containing nitrogen and phosphorus. Other nitrogen and phosphorus inputs into the watershed are associated with municipal wastewater and industrial discharges, urban runoff, and atmospheric deposition associated with fossil fuel combustion [3].

Nutrient loadings of water bodies can occur from atmospheric inputs which, in some cases, can be a large percentage of the total loading. For example, 25% of the nitrogen input to Chesapeake Bay (Maryland, US) is attributable to the burning of fossil fuels in the generation of electricity and for transportation. Air (being almost 80% N_2 and about 20% O_2) can form NO_x when subjected to conditions of high temperature and pressure, such as in an internal combustion engine. The exhaust gases are expelled to the atmosphere, where by various pathways the NO_x combines with water and eventually is deposited on land or in a water body.

The area of the Gulf of Mexico off the southern US coast is currently suffering another water quality threat, the Deepwater Horizon oil leak, considered the worst environmental disaster in the US. Named for the floating oil platform operated by oil giant British Petroleum, the disastrous crude oil leak began in April 2010 when an explosion occurred and the ensuing fire could not be contained, causing the rig to sink in 1500 m deep water. The rate of crude oil escaping into Gulf waters from the remaining drilling pipes and equipment is uncertain, but even the most conservative official estimate (12 000 barrels per day) [14] indicates that a minimum of 30 million gallons (114 000 m³) was released. Open sea oil slick burning efforts plus oil capture operations removed 20%, at best, (6.29 million gallons or 23 800 m³) of the oil from the ecosystem in the first 50 days of the disaster [15]. At the time of writing, greater amounts of oil are successfully being captured, but the prospect for complete sealing of the hole was still quite uncertain.

In addition to the oil, over 4243 m³ (1 121 000 gallons) of chemical dispersants, including Corexit 9500, were applied at the surface and injected onto the leaking well head [15]. The dispersant has both a hydrophobic end (binds with the oil fraction) and a hydrophilic end (binds with water). Its purpose is to break up the oil into smaller particles which are denser than oil. Usually applied on the surface of an oil spill, the resulting particles sink below the surface and "mix" with the water. The Gulf of Mexico has about 2434 trillion m³ (643 million billion gallons) of water [16], and although that volume is vastly greater than that of leaked oil and Corexit 9500 combined to date, the damage to wildlife habitat is already evident from the effects of surface oil alone.

The future impacts of the oil/dispersant particles below the surface of the water are largely unknown. Although the Material Safety Data Sheet for Corexit 9500 states that the chemical poses a low environmental risk [17], the US Environmental Protection Agency ordered the use of less toxic dispersants [18]. In addition to the direct physical and chemical impact on the Gulf ecosystems, bacterial degradation of the contaminants will also lead to oxygen consumption. Coupled with the nutrient inputs to the Gulf from the Mississippi river, this oil spill will undoubtedly expand the area's dead zones.

7.2.3
Global Cycling of Carbon in Water

The global cycling of carbon is closely linked to the oxygen cycle through the processes of photosynthesis and respiration [3]. Photosynthesis is the main source term in the oxygen cycle. It is also the origin of the organic carbon that is eventually converted to carbon dioxide in the carbon cycle. Photosynthesis can be described as:

$$CO_2 + H_2O + \Delta \rightarrow C(H_2O) + O_2$$

Here Δ represents the Sun's energy and $C(H_2O)$ represents organic carbon. There is a positive free-energy change in photosynthesis; therefore, the reaction requires the input of energy from the Sun. The chemical energy that is stored by photosynthesis is subsequently made available for use by organisms through respiration. Respiration is the major sink term in the oxygen cycle and is responsible for the conversion of organic carbon to carbon dioxide in the carbon cycle. It can be described as:

$$C(H_2O) + O_2 \rightarrow CO_2 + H_2O + \Delta$$

Photosynthesis is carried out by plants and some bacteria and respiration is carried out by all organisms, including those that photosynthesize. The interplay of photosynthesis and respiration plays a key role in maintaining the oxygen levels required by water-based ecosystems [3].

Related specifically to our water resources, climate change is expected to: (i) increase water availability in moist tropics and high latitudes, (ii) decrease water availability and increase drought in the mid-latitudes and semi-arid low latitudes, and (iii) expose hundreds of millions more people around the world to increased water stress. Climate change is also expected to have an adverse impact on ecosystems, food production, coasts, and human health, all which depend on water [3].

Figure 7.4 shows the carbon cycle under natural conditions and those conditions modified by human activities. Figure 7.4 indicates that the natural carbon cycle maintains relatively constant reservoirs of carbon in the air with balanced transfers among compartments. Human carbon emissions add to the atmospheric reservoir at rates that are not balanced through uptake by the land and oceans. The result is an increase in atmospheric $CO_2(g)$ concentrations which is linked to global climate change (global warming effects and increased ocean acidification). Carbonate (H_2CO_3, HCO_3^-, CO_3^{2-}) chemistry is extremely important in aquatic systems as it regulates changes in pH due to its buffering capacity (ability to consume protons without major changes in pH). The bicarbonate ion, HCO_3^-, is the major carbonate species in most natural waters and its average concentration in the oceans is $122\,mg\,L^{-1}$

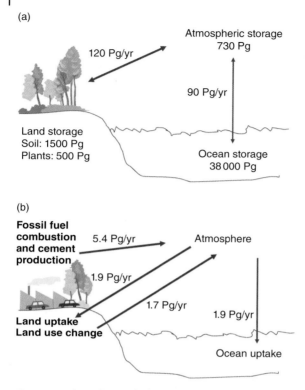

Figure 7.4 The carbon cycle: (a) under natural conditions and (b) as modified by human perturbation. The mass unit petagram (Pg) is 10^{15} grams, or a billion metric tons; from [3]. Reprinted with permission of John Wiley & Sons, Inc.

[2]. Dissolution of $CO_2(g)$ into oceans releases protons and can be represented by the following reaction with water (H_2O) to release a proton (H^+) and form a bicarbonate ion (HCO_3^-):

$$CO_{2(g)} + H_2O \leftrightarrow H^+ + HCO_3^-$$

7.2.4
Turbidity and Pathogens

Human (and animal) waste contains a large amount of pathogens, nutrients, and particulate matter. The municipal wastewater that is collected in sewers and at on-site treatment systems (e.g., septic systems) also contains large amounts of these pollutants. To provide an idea of the number of pathogens in wastewater, there can be millions of disease-causing microorganisms found in just 100 mL of untreated municipal wastewater. These organisms are usually associated with fecal matter (not with the urine). In addition, the concentration of suspended solids (i.e., particulate matter) in an average municipal wastewater collection system in the United States may be several hundred $mg\,L^{-1}$. This concentration may be larger in countries that use less water (which dilutes the wastewater stream). Particulates cause the water to be turbid, and may contain organic matter and thus exert BOD. They may also contain other pollutants or pathogens. Turbidity measures the optical clarity of water and is used to determine the quality of water samples. It is caused by the scattering and absorbance of light by suspended particles in the water [19].

The World Health Organization (WHO) defines safe drinking water as water that "does not represent any significant risk to health over a lifetime of consumption, including different sensitivities that may occur between life stages" [20]. Their published guidelines for drinking water quality are available online [20]. The WHO recommends that at a minimum, *Escherichia coli* (*E. coli*), thermotolerant (fecal) coliforms, and chlorine residuals (where there is chlorination) be monitored in community water supply systems. They recommend that there be zero *E. coli* organisms in a 100-mL sample. Total coliforms is not considered an appropriate indicator for fecal contamination by the WHO because many bacteria of no significance occur naturally, especially in tropical waters. In regard to turbidity, water with a turbidity of less than 5 NTU is usually esthetically acceptable to a user. The WHO [20] has proposed no health-based guidelines for turbidity, however, they recommend that median turbidity should be less than 0.1 NTU for effective disinfection [19] (the unit of NTU is commonly used to measure turbidity of water).

7.2.5
Arsenic and Fluoride

Arsenic is a Group 15 element that is the 51st most abundant element in the earth's crust. It occurs in over 245 minerals. It concentrates in sedimentary rocks like arsenopyrites, igneous rocks containing iron oxide and volcanic glass and aluminosilicate minerals [21]. It was a common ingredient in insecticides (e.g., lead arsenate, Paris Green, calcium arsenate), herbicides, and dessicants for cotton. All of these applications resulted in widespread contamination of agricultural lands and soil and watersheds surrounding manufacturing plants [22, 23]. Arsenic is now used mainly as a wood preservative, with other applications in the paint, dye, drug, and semi-conductor industries. In addition to these anthropogenic sources, various geochemical phenomena naturally release arsenic from mineral formations resulting in high aqueous arsenic concentrations ($>50\,\mu g\,L^{-1}$) in many groundwaters around the world. Arsenic can exist in both organic and inorganic forms (+5, +3, 0, −1, −2 oxidation states) and the most prevalent are +3 (arseneous acid, arsenite), +5 (arsenic acid, arsenate), and −3 (arsenide).

Table 7.6 lists the acidity constants (pK_a) of the main inorganic ions of the +5 and +3 oxidation state, commonly referred to as the arsenate or arsenite anions, respectively. The pK_a value is defined as the pH value where an acid (e.g., H_3AsO_4) and its conjugate base (e.g., $H_2AsO_4^-$) exist at equimolar concentrations.

Table 7.6 Acidity constants (pK_a) at $I = 0$ and 25 °C, for arsenic species.

Oxidation state	Reaction	pK_a
+5	$H_3AsO_4 = H^+ + H_2AsO_4^-$	$pK_{a1} = 2.24$
	$H_2AsO_4^- = H^+ + HAsO_4^{2-}$	$pK_{a2} = 6.94$
	$HAsO_4^{2-} = H^+ + AsO_4^{3-}$	$pK_{a3} = 12.19$
+3	$H_3AsO_3 = H^+ + H_2AsO_3^-$	$pK_{a1} = 9.29$
	$H_2AsO_3^- = H^+ + HAsO_3^{2-}$	$pK_{a2} = 12.13$
	$HAsO_3^{2-} = H^+ + AsO_3^{3-}$	$pK_{a3} = 13.40$

Values taken from NIST database [24].

Comparison of the pK_a values of the arsenate and arsenite systems indicates that for most natural waters (pH values between 6 and 8), $H_2AsO_4^-$ and $HAsO_4^{2-}$ dominate arsenate speciation and the uncharged H_3AsO_3 species dominates arsenite speciation. This difference in charge plays an important role in the fate of the species in the environment (e.g., arsenate bonds to charged mineral oxide surfaces and is immobilized whereas arsenite remains soluble) and in the design of technologies to remediate arsenic-contaminated waters.

The routes of human exposure to arsenic include inhalation, ingestion and sorption through the skin. Each of these routes of exposure provides unique challenges when attempting to reduce the adverse impacts of exposure. Most of the recognized chronic health effects of arsenic exposure come from studies of human populations exposed to considerably high concentrations of arsenic in drinking water in Taiwan [25–28], Argentina [29], Bangladesh and West Bengal [30]. These studies mainly found high incidences of arsenical dermatosis, blackfoot disease, bladder, kidney, and lung cancers, though a list of other health effects like pulmonary diseases have also been linked to arsenic exposure. The International Agency for Research on Cancer (IARC) currently classifies arsenic as a group 1 chemical: one that is carcinogenic to humans. The drinking water standard for arsenic set by the World Health Organization, the United States, Canada, Britain and other European countries is $10 \, \mu g \, L^{-1}$.

The widespread presence of high levels of arsenic in the groundwaters of West Bengal and Bangladesh is currently seen as the worst environmental catastrophe ever as over 70 million people are at risk of suffering from chronic arsenic exposure. In these two areas, approximately 20 million groundwater wells exist, many of which provide water to small villages that desperately need affordable treatment technologies for the high arsenic concentrations of the water. These wells were dug over 20 years ago under a program sponsored by the United Nations International Children's Emergency Fund (UNICEF) and it was not until 1993, after symptoms of chronic arsenic exposure appeared, that the extent and severity of the problem was realized. Only a small percentage of the wells have been tested to date and unfortunately over 50% of them have arsenic concentrations higher than $10 \mu g \, L^{-1}$. In Bangladesh, for example, arsenic concentrations range from 10 to greater than $1000 \mu g \, L^{-1}$ [31].

Exposure to high concentrations of arsenic through well water also continues in Inner Mongolia, northeastern Taiwan, and Hanoi [26, 28, 32]. High arsenic concentrations have also been found in the ground water and surface water in the Unites States, especially in the south central and western parts of the country [21, 33–35]. The $10 \mu g \, L^{-1}$ rule became effective in the United States in 2006, impacting approximately 13 million Americans, many of whom depend on small water supply systems. Given the costs associated with reducing population exposure to high levels of arsenic in drinking water, disparate levels of intervention exist around the world and millions continue to suffer as affordable solutions are unattainable.

Fluoride (F^-) is the ionic form of fluorine. It is a Group 17 element of the periodic table that is the 13th most abundant element in the earth's crust. It is found in many minerals like fluorite, $CaF_2(s)$, and complexes with many cations owing to its similarity to hydroxide ions. Hydrofluoric acid (HF) has a pK_a value of 3.17 and in natural waters F^- dominates speciation. Chronic exposure to fluoride results in dental and skeletal fluorosis, both of which are detrimental to the bones. Elevated fluoride concentrations in drinking water has led to fluorosis in countries like India, China, Tanzania, Mexico, Argentina, and South Africa, and researchers believe that many more countries, especially in Northern Africa, have high groundwater fluoride levels [36]. The WHO guideline is $1.5 \, mg \, L^{-1}$ and, in some places, fluoride is actually added to drinking water (fluoridation) because of its ability to reduce dental cavities.

As mentioned earlier during the discussion on phosphorus, fluoride represents up to 4% of the ore mined in Florida. Scrubbers are used to remove fluoride-containing gases during phosphoric acid manufacture which leads to the final production of industrial grade silicon fluorides (e.g., sodium fluorosilicate Na_2SiF_6). These silicon fluorides are used at water treatment plants. Despite the beneficial use of the fertilizer industry by-product, there is some concern that the industrial grade fluoride is not as effective and that toxic by-products like arsenic further threaten human health.

7.2.6
Global Cycling of Mercury

Mercury is a natural (e.g., volcanoes) and anthropogenic (e.g., mined ore) contaminant that causes deleterious human health effects, including impaired mental function, neurological disorders and kidney damage. It is a transition metal commonly found in three oxidation states (Hg(0), Hg(I), and Hg(II)) and is unique because its elemental state can vaporize at room temperature and it can form methylmercury (CH_3HgCl or CH_3HgOH and represented as MeHg), the most toxic form known. Bioaccumulation and biomagnification of methyl mercury occurs to a relatively high extent in aquatic systems [37–39], making fish consumption the leading route of human exposure. As a result, US EPA recommended surface and drinking water limits are $12\,\text{ng}\,\text{L}^{-1}$ and $2000\,\text{ng}\,\text{L}^{-1}$ total inorganic mercury, respectively, and many states in the United States and countries in the world issue fish consumption advisories, especially targeting pregnant women.

In 1995, estimated global anthropogenic contributions to atmospheric mercury emissions were 2200 metric tons/year with power plants being the largest contributor [40] and artisanal gold mining next with a reported 300 metric tons/year [40, 41]. Of the 3386 metric tons of mercury produced per year, batteries, chlor-alkali processes and small scale gold and silver mining account for 32%, 24% and 20% of the market [42]. There is great variation in these estimates because of difficulties with accounting and lack of enforcement and regulations in some countries.

Various physical, chemical and biological processes influence mercury speciation and transfer between soil, water and air, and Morel *et al.* [43] provide a good review. The residence time of mercury in the atmosphere is about a year, giving it enough time to be transported around the world, hence the term "global pollutant". Atmospheric deposition has been identified as the main route of aquatic mercury contamination in non-mining areas of temperate and cold regions where the bulk of scientific research has been conducted to date. Difficulties still remain in understanding global inputs to localized systems due to lack of data and understanding of complex processes governing mercury transformation and emission. Recent calls for the global elimination of mercury production testify to the harmfulness of this element, however, it does not eliminate the existence of significant quantities of mercury already present in our biosphere. Figure 7.5 summarizes the biogeochemical processes governing mercury cycling in the environment.

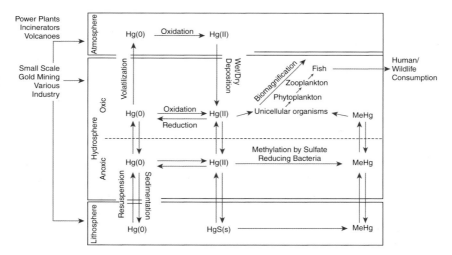

Figure 7.5 Major processes in the cycling of mercury. MeHg represents CH_3HgCl or CH_3HgOH, Hg(II) represents species like $HgCl_2$, $Hg(OH)_2$, HgClOH in the oxic environment and HgNOM, $Hg(HS)_2$ in the anoxic environment. Adapted from [43].

Different chemical (e.g., sorption, precipitation, photoinduced volatilization) and biological (methylation or demethylation) processes influence the speciation of mercury in aquatic systems and extensive research has been done using both model and real conditions to understand them. The few examples given next just barely capture the complexities. Mercury (II) sorbs to common sediment constituents [44–46] and this is influenced by conditions like pH, the presence of ligands (e.g., chloride, sulfate, phosphate, natural organic matter (NOM)), and other heavy metals (e.g., Ni(II), Pb(II)). Methylation and demethylation of mercury can occur via abiotic and microbial pathways [47]. Researchers found that dissolved organic matter (DOM) influenced the abiotic, photoinduced methylation rates of mercury [48]. Recent studies on mercury volatilization (to Hg(0)) found that the presence of NOM decreased volatilization in aqueous solutions, but that mercury volatilization in real lake samples was significant in sunlight [49].

7.2.7
Emerging Chemicals of Concern

Emerging chemicals of concern are characterized by relative newness in production, use, and appearance in the environment. They started showing up in the environment, including the influent of publicly-owned drinking water and wastewater treatment facilities over the last two decades. In some cases, the occurrence in the environment and potential impact may have long been suspected, but only recently validated by modern analytical equipment. The presence of these compounds is most often not monitored or their use is not generally regulated. However, they are thought to have adverse ecological and human health effects once released into the environment [50]. Most of what is known about the environmental impact of such chemicals is still "emerging" from ongoing research. Their detection and removal from water is becoming an important area of research, especially in anticipation of stricter regulations as new information on their transport, fate, and effects is released.

Emerging chemicals constitute a fairly large group of complex compounds and include pharmaceuticals (human and veterinary medicine), natural and synthetic hormones, personal care products, sanitizers and cleaning products, and surfactants, among others. Most of these chemicals were designed to offer improvements in common household conveniences, industry, medical treatment, and agriculture, for example. Pharmaceutical compounds which have been detected in wastewater and the environment include human and veterinary antibiotics (for example, sulfamethoxazole, erythromycin, and roxithromycin); prescription analgesic drugs such as codeine and anti-epileptic drugs including carbamazepine; and non-prescription drugs such as acetaminophen and ibuprofen. The environmental effects of these chemicals include the development of antibiotic-resistant microbes in water treatment processes and in the aquatic environment [51, 52], and the potential increased toxicity of chemical combinations and metabolites [53].

A particularly worrisome group of emerging chemicals is endocrine disrupting compounds (EDCs) which can upset normal endocrine functions even at a low concentration ($\mu g \, L^{-1}$ and $ng \, L^{-1}$ range), due to their steroid-like structures. Adverse effects on developmental and reproductive systems of animals and humans have been reported [54–56].

Major emerging chemicals of concern, due to their endocrine disrupting properties, global use, presence and potential harm to aquatic life, are caffeine, DEET (*N,N*-diethyl-*meta*-toluamide), bisphenol A and β-estradiol. Table 7.7 summarizes some of the attributes of these chemicals. 1,3,7-trimethylxantine (i.e., caffeine) is widely consumed globally with average consumption of about 70 mg per person per day. The major source of caffeine in domestic wastewater comes from unconsumed coffee, tea, sodas, or discarded medication. Caffeine is highly soluble in water at $13 \, g \, L^{-1}$ and has a low volatility. DEET is commonly found as an active ingredient in many insect repellent products and is used by approximately 33% of the United States' population annually [57]. Its potential as a wildlife endocrine disruptor [57, 58] has pushed Brazil to regulate concentrations of DEET in visitors' insect repellents in certain parts of the rainforest. Bisphenol A, which was investigated in the 1930s as a synthetic estrogen, is used industrially for polycarbonate plastic and epoxy resins. Effluents from facilities that manufacture epoxy and polycarbonate plastics and elution from the products containing it are suspected to be the major source of this contaminant in the environment [59]. β-Estradiol, one of the most potent EDCs, is excreted through feces and released from sewage treatment plants [60].

Understanding the chemistry of emerging chemicals of concern is very important. The fate of contaminants in the aquatic environment is strongly influenced by their biodegradability and physical and chemical properties. The potential for bioremediation of some compounds exists, however, the biodegradability of many emerging chemicals is still a major challenge, largely because they have stable heterocyclic and/or conjugated polycyclic structures (Table 7.7). In addition, pharmaceutical compounds are bioactive and may interact with microbes even at very low concentrations of just a few $ng \, L^{-1}$. These interactions may result in antibiotic-resistant organisms or by-products which have similar or worse impacts than the parent compounds. Apart from being bioactive, pharmaceuticals and natural and synthetic hormones tend to be hydrophilic because they are designed to stimulate physiological responses in humans and animals. This makes them highly mobile in water.

Conventional water and wastewater treatment processes are ineffective at removing trace EDCs. A number of advanced treatment options are available and these include membrane filtration technologies, advanced oxidation processes (AOPs), and hybrid systems. The advanced oxidation processes have the benefit of degrading the compounds, but the analysis of by-products is important because they have the potential to be toxic. In that sense, toxicity assays will determine the final efficiency of the treatment process for this class of pollutants.

Table 7.7 Selected emergent chemicals of concern.

Structure	Chemical formula	Molecular weight (g)	Chemical name	Common name	CAS number	Typical use
	$C_6H_{10}O_2N_4$	194.2	1,3,7-trimethylxantine	Caffeine	58-08-2	Stimulant
	$C_{12}H_{17}NO$	191.3	N-N-diethyl toluene	DEET	134-62-3	Insect repellent
	$C_{18}H_{24}O_2$	272.3	Estra-1,3,5[10], 7-triene-3,17diol	17β-estradiol	50-82-2	Reproductive hormone
	$C_{15}H_{16}O_2$	228.3	2,2-bis-(4-hydroxyphenyl) propane	Bisphenol A	80-05-7	Plasticizer

7.3
Water Treatment Technologies

7.3.1
Point of Use Treatment and Advanced Oxidation Processes

Point of use treatment refers to technology that is employed at the point of use, usually at the household level, for the removal of various contaminants from drinking water.

Household-based ceramic water filters are seen as an affordable, accessible technology for effectively removing bacteria, especially thermotolerant coliforms, thereby reducing diarrheal incidences. The pore size of the ceramic material allows water to flow through, but blocks many bacteria and particles with particles sizes around 0.1 to 1 μm.

Figure 7.6 depicts a household ceramic filter which sits in a larger receptacle [19]. Contaminated water is placed in the inner, ceramic container which restricts the passage of bacteria and dirt through its pores. The cleaner water flows into the larger receptacle where it can be accessed via the tap. Some filters are coated or impregnated with silver which kills the bacteria and inhibits mold and algae growth. The inner side of the ceramic filter requires cleaning with soap, water and a soft brush to unclog pores.

Solar Water Disinfection (SODIS) is another point of use approach to inactivating bacteria through exposure to sunlight [61]. Contaminated water is placed in transparent plastic (PET) bottles which are placed on rooftops and exposed to

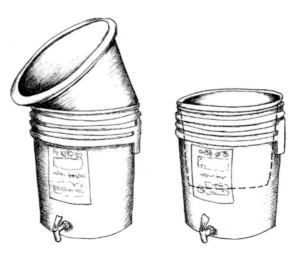

Figure 7.6 Example of ceramic clay pot filter that is used to filter water. A water storage container is covered by the clay pot to store the filtered water (Figure from [19]. Reprinted with permission of Linda A. Phillips).

sunlight for a given timeframe (usually 6 h) that depends on the weather conditions (e.g., cloud cover). Additional chemicals, pre-filtration, and use of solar collectors, improve the effectiveness of SODIS [62, 63]. The process of disinfection is believed to be through the processes of pasteurization and advanced oxidation.

The point of use treatment methods mentioned above are gaining popularity around the world as countries work to meet the Millennium Development Goals (MDGs) by 2015. This is because some governments and nongovernmental organizations have decided that it is more cost effective to treat water at the household level with technologies mentioned above than by development of more centralized and larger scale community-managed water supply systems that improve collection of water at a nearby or faraway source. The emergence of new chemicals of concern, like those described in Section 7.2.7, requires more advanced treatment options in the absence of policies to curb their manufacture, use and release.

Advanced oxidation processes (AOP) refers to the oxidation of organic and inorganic materials through the use of chemical reagents like ozone (O_3), hydrogen peroxide (H_2O_2), oxygen (O_2), and air, under well-controlled conditions that produce powerful hydroxyl radical (OH·) oxidants. The radicals then oxidize the contaminant of interest, many times, to less innocuous substances, including complete mineralization to CO_2 and H_2O.

Wastewater treatment facilities are already using a combination of UV with H_2O_2 for both disinfection and removal of toxic organics. Fenton's Reagent, discovered by British chemist Henry John Horstman Fenton in the 1890s, contains hydrogen peroxide and a dissolved iron catalyst (e.g., $FeSO_4$) that is used to oxidize contaminants or waste waters. When used as an AOP, ferrous iron (Fe^{2+}) is oxidized by hydrogen peroxide to ferric iron (Fe^{3+}), a hydroxyl radical and a hydroxy anion (OH^-) and Fe^{3+} is reduced back to Fe^{2+}, a peroxide radical (OOH·) and a proton (H^+):

Iron II oxidation: $Fe^{2+} + H_2O_2 \rightarrow Fe^{3+} + OH\cdot + OH^-$

Iron III reduction: $Fe^{3+} + H_2O_2 \rightarrow Fe^{2+} + OOH\cdot + H^+$

Photocatalysis is an AOP that combines UV irradiation with a catalyst to promote free radical formation. Extensive literature exists on photocatalysts, their applications and limitations [64, 65], since it is a promising remediation technology for environmental contaminants. The most widely studied catalyst is titanium dioxide which has been shown to degrade a range of organic contaminants like pesticides and aromatic compounds in the presence of visible or UV light and Bhatakhande *et al.* summarize the various photocatalytic investigations on both organic and inorganic transformations [64].

One of the challenges of creating an economically viable technology lies in the solid/liquid separation after treatment, since the photocatalytic particles tend to be fine powders, especially in the case of titanium dioxide. Methods to increase particle size include coating of titanium dioxide on larger substrates like magnetite which has properties that can be beneficial during the solid/liquid separation step [66]. Researchers have also coated flat surfaces of glass through which the contaminated solution flows while being irradiated and have explored heavy metal doping agents like silver and iron that can increase the efficiency of the catalyst and catalyze reactions at higher wavelengths in the visible range [67, 68].

7.3.2
Membranes

Membranes treat water through a physiochemical technique that uses pressure or vacuum to move water across the membrane. The process allows the passage of water through the membrane barrier but does not allow passage of chemicals and particles. By 2007, over 20000 membrane treatment plants were operating worldwide. There are several membrane processes that can remove pathogens, including viruses (approximate size 0.01–0.1 µm), bacteria (approximate size 0.1–10 µm), and Protoza such as *Giardia* and *Cryptosporidium* (approximate size 1–10 µm) [69].

Membranes are comprised of natural or synthetic materials and are typically manufactured in flat sheets, tubular form, or as fine hollow fibers. Most membranes manufactured for drinking water treatment are made of a polymeric material because they are less expensive [69]. The most common configurations seen at water treatment plants are hollow-fiber, spiral wound, and cartridges. Hollow-fiber membranes consist of hollow-fiber membrane material. Spiral wound membranes consist of a flat sheet membrane material wrapped around a central collection tube. Cartridges consist of a flat sheet membrane material that is often pleated to increase the surface area [69]. Synthetic membrane materials may be composed of polyamide, polysulfone, acrylonitrile, polyethersulfone, Teflon, nylon, and polypropylene polymers. Modified natural materials used to construct membranes consist of cellulose acetate, cellulose diacetate, cellulose triacetate, and a blend of both di- and tri-acetate materials [3]. Table 7.8 provides some information on how material type impacts performance and operation.

7.3.3
Arsenic

There are many different mitigation strategies for the remediation of arsenic-contaminated waters and Table 7.9 summarizes those evaluated by the US EPA for small drinking water systems; defined as a system serving 10000 or fewer people with average water demand normally less than 1.4 million gallons per day. This list is by no means exhaustive, but provides an idea of the range of options available for arsenic removal, including the provision of alternative sources of water, like bottled water.

Given that As(III) exists mainly as the uncharged H_3AsO_3 species, reduced inorganic As(III) (arsenite) is usually converted to As(V) (arsenate) to facilitate

Table 7.8 How material used to manufacture a membrane impacts performance and operation (information obtained from [69]).

Material property	Comment
membranes constructed of polymers	Reacts with oxidants commonly used in drinking water treatment; therefore, should not be used with chlorinated feed water
mechanical strength	Membranes with greater strength can withstand larger transmembrane pressure levels. This provides greater operational flexibility.
bi-directional strength	May allow cleaning operations or integrity testing to be performed from either the feed or the filtrate side of the membrane.
membrane with a particular surface charge	Used to achieve enhanced removal of particulate or microbial contaminants of the opposite surface charge due to electrostatic attraction. In addition, a membrane can be characterized as being hydrophilic (i.e., water attracting) or hydrophobic (i.e., water repelling) which influences the propensity of the material to resist fouling.

Table 7.9 Mitigation strategies for the remediation of arsenic-contaminated waters. [70].

- Non-treatment and treatment minimization
  - Source abandonment and provision of clean water
  - Seasonal use
  - Blending of waters
  - Sidestream treatment
- Enhancement of existing treatment processes
  - Enhanced coagulation/filtration
  - Enhanced lime softening
  - Iron/manganese filtration

- Treatment (full stream or side stream)
  - Ion exchange
  - Sorption (activated alumina; iron-based sorbents)
  - Coagulation-assisted microfiltration (CMF)
  - Coagulation-assisted direct filtration (CADF)
  - Oxidation/filtration
- Point-of-use treatment
  - Sorption (activated alumina; iron-based sorbents)
  - Reverse osmosis

removal. Chlorine, permanganate, and ozone are examples of oxidizing agents that can convert As(III) to As(V) and that are usually applied at the beginning of the treatment process:

Chlorine oxidation: $H_3AsO_3 + OCl^- \leftrightarrow H_2AsO_4^- + H^+ + Cl^-$

Permanganate oxidation: $3H_3AsO_3 + 2MnO_4^- \leftrightarrow 3H_2AsO_4^- + H^+ + 2MnO_2 + H_2O$

Ozone oxidation: $H_3AsO_3 + O_3 \leftrightarrow H_2AsO_4^- + H^+ + O_2$

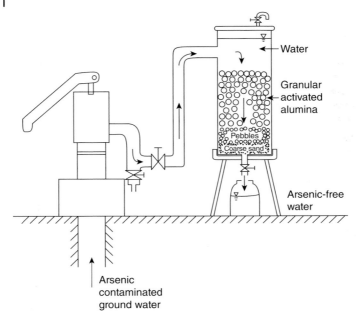

Figure 7.7 Small-scale system installed at a hand pump to treat arsenic-contaminated community water supplies (Figure from [19]. Reprinted with permission of Linda A. Phillips. Adapted from drawing provided by Dr. Arup SenGupta and others at Lehigh University).

During the removal process, for example in the case of sorption technologies, the sorbent material (activated alumina for example) surface forms a complex with the arsenic species. The charged As(V) species tend to complex with the surface more and are less affected by other water constituents like sulfate, silicate and phosphate.

Figure 7.7 illustrates a community sorption-based treatment process to treat arsenic-contaminated water where contaminated water is delivered via a hand pump to a bed filled with activated alumina sorbent particles which will remove the arsenic and provide cleaner water which is collected at the outflow of the bed. As arsenic sorbs to the activated alumina surface, pH changes can occur and an additional step may be needed to adjust the pH of clean water. This pH shift is less common with the iron-based sorbents. The sorbents can be regenerated by pH adjustment (to a pH where arsenic is less favored) or by addition of a chemical that effectively competes for surface sites (e.g., carbonate).

Spent sorbent poses potential concerns, especially in communities with limited disposal facilities, since the material can be a concentrated source for further arsenic contamination. In the United States, a toxicity characteristic leaching procedure (TCLP) is used to determine whether spent sorbents are hazardous. Based on these tests, results have shown that this is not the case and thus the materials can be placed in non-hazardous landfills. Ghosh *et al.* [71], however, argue that the TCLP tests do not replicate landfill conditions and are inappropriate for determining disposal of arsenic-laden sorbent particles from water treatment processes and that these sorbents should be disposed of in Class 1 landfills which are lined and equipped to accept hazardous waste.

7.3.4
Water Reuse

Wastewater is now being viewed as a renewable resource from which water, materials (e.g., nutrient fertilizers, biomaterials), and energy can be recovered [72]. Water conservation and water reclamation along with water reuse are all methods that can be used to decrease water demand. In addition, water treatment technology is currently so advanced that domestic wastewater can be converted to drinking water, or for use in other residential, agricultural, commercial, and industrial capacities. Globally, the volume of water being reused is increasing rapidly. One challenge though is matching the supply of reused water with quality and demand. For example, the water needs of agriculture are typically located far away from centralized wastewater treatment plants located in urban areas. In this case, it might not make sense to construct a long piping system and then use energy to pump reclaimed water back to rural areas. In this situation, smaller decentralized satellite reclamation processes can be employed that extract water and nutrients from the collection system farther upstream closer to agricultural users [3].

There is also great interest in recovering chemicals and materials from wastewater. For example, the solids generated during the wastewater treatment process (termed biosolids) are valued as a soil amendment in many parts of the world for their high content of nutrients and organic carbon. There is also progress being made in obtaining struvite ($MgNH_4PO_4(s)$) from the wastewater solids treatment processes. However, there is concern (and associated regulations) for application of biosolids because many inorganic and organic chemicals accumulate in the solids.

In addition, there is heightened interest to recover nutrients from source-separated urine [73, 74]. This is because, as mentioned in an earlier section (Table 7.4), urine contains the majority of nitrogen and phosphorus found in human wastes. In addition, it is relatively free of pathogens, unlike the fecal solids [19, 74]. In addition, the World Health Organization [75] states that recycled nutrients from human wastes has the potential to decrease poverty through: (i) improved household food security and nutritional variety, which reduce malnutrition; (ii) increased income from sale of surplus crops (the use of excreta and greywater may allow cultivation of crops year-round in some locations); and (iii) money saved on fertilizer, which can be put toward other productive uses. It is also estimated that if all urine was collected from one person, it would provide enough nutrients to fertilize 300–400 m² of crops per year [76]. Figure 7.8 shows a four-step process where urine can be collected for application to agricultural crops. This collection

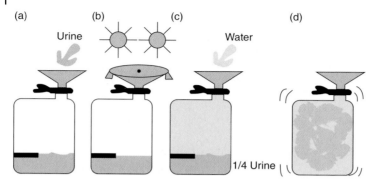

Figure 7.8 Four-step process to collect urine to use as agricultural fertilizer. (a) Fill 20-L can 25% full with human urine (in this case 5 L of urine); (b) let urine stand for 2 days to ensure destruction of the pathogen, *Schistosoma Haematobium*; (c) fill remainder of container with water (in this case 15 L); (d) mix and apply directly to crops (reprinted with permission from [74]). (Please find a color version of this figure in the color plates.)

Table 7.10 K_{so} values for calcite, hydroxyapatite (HAP), and struvite at 25 °C.

Mineral	Dissolution Reaction	K_{so}
Calcite	$CaCO_3(s) \leftrightarrow Ca^{2+} + CO_3^{2-}$	$10^{-8.48}$
HAP	$Ca_5PO_4OH(s) \leftrightarrow 5Ca^{2+} + 3PO_4^{3-} + OH^-$	$10^{-57.5}$
Struvite	$MgNH_4PO_4(s) \leftrightarrow Mg^{2+} + NH_4^+ + PO_4^{3-}$	$10^{-13.15}$

and application method has been successfully applied in Mali, Africa and many other parts of the world [74].

Research is also being done on waterless urinals to overcome maintenance issues caused by precipitate formation. When the urea in the urine decays pH normally increases. Udert *et al.* [77] studied precipitation in conventional toilets and waterless urinals and saw pH increase from 7 to ~9 as the amount of hydrolyzed urea (a reaction mainly catalyzed by the enzyme urease found in urea positive bacteria) increased. Table 7.10 provides the solubility products for the three main precipitates: calcite, hydroxyapatite (HAP) and struvite. In general, calcite is the main precipitate found in conventional toilets, struvite is primarily formed in waterless urinals, and HAP was found to be more common in toilets that had less water dilution. Processes that manipulate the concentrations of the ions on the right-hand side of the dissolution reactions shown in Table 7.10 would limit precipitate formation. For example, ion exchange cartridges placed at the bottom of the urinal can selectively remove ions like NH_4^+ and be replaced more easily than pipes [78].

7.3.5
Carbon Sequestration

The link between global climate change and the levels of carbon dioxide (CO_2) and other greenhouse gases (GHG) in the atmosphere [79] as discussed in Section 7.2.3 and Chapter 9, have led to great interest in processes of capturing and storing CO_2, or "carbon sequestration", as a remedy for rising atmospheric CO_2 levels. Studies have shown that pre-Industrial Age atmospheric CO_2 levels were about 280 ppm_v, whereas today the level is over 380 ppm and rising [80]. In addition, CO_2 dissolved in the oceans is correspondingly higher, lowering the pH and threatening coral and many other ocean species. Carbon sequestration offers the hope of reducing emissions of CO_2 to the atmosphere, allowing natural CO_2 cycling to bring levels back down.

Carbon sequestration methods range from simple concepts to sophisticated carbon capture and storage (CCS) technologies; from inducing higher organic content of soil by altering agricultural practices to selecting porous mineral formations deep underground and injecting CO_2 at high pressure (Figure 7.9). A sequestration method must be evaluated for its capacity, economic efficiency, energy efficiency, environmental impact, and stability (preferably unfailing containment on geologic timescales). For example, setting aside land and encouraging forest growth and accumulation of woody plant matter (forms of terrestrial sequestration) require little energy input, and concurrently there are positive environmental impacts. However, the risk of forest fire returning the CO_2 to the atmosphere and

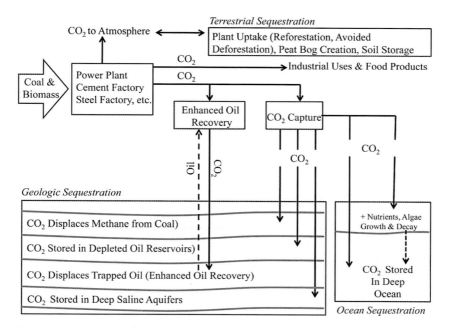

Figure 7.9 CO_2 release and sequestration pathways. Adapted from [83].

negating the sequestration effort must be considered as well. Similarly, outfitting a coal-fired power generation plant to capture the CO_2 emissions for deep well injection into a saline aquifer (geologic sequestration) may provide a high likelihood of successful long-term sequestration, but this technology is energy intensive, requiring pumps to compress and transport the gas to the formation. Ocean sequestration includes the direct injection of liquid CO_2 into the deep ocean as well as the manipulation of biological growth at the ocean surface using CO_2 and nutrients, followed by sedimentation of dead organisms in the deep ocean [81, 82].

Evaluating engineered carbon capture and storage methods is an area of very active research. The geologic carbon sequestration potential of three types of formations has been compiled for the United States and parts of Canada, for example: storage of CO_2 in unmineable coal seams; storage of CO_2 in depleted oil and gas reservoirs; and storage of CO_2 in otherwise unusable aquifers bearing saline water. These estimates indicate that all the US and Canadian CO_2 emissions from major stationary sources of CO_2 (which account for more than two thirds of anthropogenic CO_2 production, including power plants, petroleum and natural gas processing facilities, cement plants, and energy-intensive industries such as steel or fertilizer production) could be stored for over 1000 years based on current consumption levels [83]. While this is promising, geologic CCS field test data to date is too limited to know if carbon sequestration will prove to be feasible on the scale needed to achieve GHG reduction goals. The impact of leaked CO_2 on water quality in groundwater aquifers is also a concern and questions remain on risks from pressure build-up in the subsurface.

7.4
Conclusions

Water is the foundation of the world's current and future economic, societal, and environmental prosperity. As stated at the beginning of this chapter, it is a critical component of agricultural and industrial production, electric power generation, household consumption, hygiene and sanitation, recreation, and ecosystem maintenance. Population growth, land use change, climate change, increased consumption, and inefficient use have all placed additional stressors on the quantity and quality of the world's water resources. Unfortunately, chemical and microbiological pollution of water from human activities has resulted in decreased function of the water and has increased risks to ecosystems and human health. Further compounding the problem is that the subsequent treatment of contaminated waters also requires additional chemicals, materials, and energy.

Managing water for a sustainable future requires integrated approaches that include participation of all sectors of society who not only use the resource, but whose daily activities contribute to its quality and availability. Opportunities exist for transforming our world through the adoption of policies and practices that reduce or eliminate risks to water quality and quantity. Green Chemistry, defined as the design of products and processes that reduce or eliminate the use of, and generation of hazardous substances [84], minimizes risk by reducing or eliminating the hazard versus controlling exposure to a particular hazard. The future of a world that embraces Green Chemistry could be a place where the chemicals and materials used in commerce and society, and subsequently released to the environment, would not cause harm to the environment or humans who are exposed to them. Also needed are innovative technologies to supply and treat water and that match the source of water with the specific type of demand. These technologies must be combined with appropriate behavioral changes that take place at the individual, household, and community level, along with leadership provided by governments and nongovernmental organizations that take into consideration current and future generations.

References

1 Voet, D. and Voet, J. (1990) *Biochemistry*, John Wiley & Sons, Inc., New York, p. 1223.

2 Berner, E.K. and Berner, R.A. (1996) *Global Environment: Water, Air and Geochemical Cycles*, Prentice-Hall Inc., Upper Saddle River, NJ.

3 Mihelcic, J.R. and Zimmerman, J.B. (2010) *Environmental Engineering: Fundamentals, Sustainability, Design*, John Wiley & Sons, Inc., New York.

4 United Nations Educational, S., and Cultural Organization (UNESCO) and World Water Assessment Programme (UNECO-WWAP) (2003) *Water for People Water for Life: The United Nations World Water Development Report*, United Nations Educational, Scientific, and Cultural Organization (UNESCO) and Berghahn Books, Paris, France.

5 Mihelcic, J.R. (1999) *Fundamentals of Environmental Engineering*, John Wiley & Sons, Inc., New York.

6 Vörösmarty, C.J.G.P., Salisbury, J., and Lammers, R.B. (2000) Global water resources: vulnerability from climate change and population growth. *Science*, **289**, 284–288.

7 Hoekstra, A.Y. and Chapagain, A.K. (2007) Water footprints of nations: water use by people as a function of their consumption pattern. *Water Resources Management*, **21** (1), 35–48.

8 Vaccari, D.A. (2009) Phosphorus: a looming crisis. Scientific American Magazine, pp 54–59.

9 Drangert, J.O. (1998) Urine blindness and the use of nutrients from human excreta in urban agriculture. *GeoJournal*, **45**, 201–208.

10 IFIA (2010) Statistics. http://www.fertilizer.org/ifa/Home-Page/STATISTICS/Fertilizer-supply-statistics (accessed 6/6/10).

11 Huang, W.Y. (2009) Factors Contributing to the Recent Increase in U.S. Fertilizer Prices, http://www.ers.usda.gov/Publications/AR33/AR33.pdf (accessed 6/5/10).

12 Galloway, J.N. (1998) The global nitrogen cycle: changes and consequences. *Environ. Pollut.*, **102** (1), 15–24.

13 Ravishankara, A.R., Daniel, J.S., and Portmann, R.W. (2009) Nitrous oxide (N_2O): the dominant ozone-depleting substance emitted in the 21st century. *Science*, **326** (5949), 123–125.

14 U.S. Department of Interior (2010) http://www.doi.gov/news/pressreleases/Flow-Rate-Group-Provides-Preliminary-Best-Estimate-Of-Oil-Flowing-from-BP-Oil-Well.cfm (accessed 1/20/11).

15 U.S. Government (2010) http://www.deepwaterhorizonresponse.com/go/doc/2931/600283/ (accessed 7/22/10).

16 U.S. Environmental Protection Agency (2010) http://www.epa.gov/gmpo/about/facts.html (accessed 1/20/11).

17 Nalco (2010) *COREXIT® EC9500A,* MSDS; Nalco Company, Naperville, IL.

18 U.S. Environmental Protection Agency (2010) http://yosemite.epa.gov/opa/admpress.nsf/d0cf6618525a9efb85257359003fb69d/0897f55bc6d9a3ba852577290067f67f!OpenDocument (accessed 1/20/11).

19 Mihelcic, J.R., Fry, L.M., Myre, E.A., Phillips, L.D., and Barkdoll, B.D. (2009) *Field Guide in Environmental Engineering for Development Workers: Water, Sanitation, Indoor Air,* American Society of Civil Engineers (ASCE) Press, Reston, VA.

20 WHO (2006) Guidelines for Drinking Water Quality: First Addendum to Third Edition, vol. 1. Recommendations, http://www.who.int/water_sanitation_health/dwq/gdwq0506.pdf (accessed 5/17/2010).

21 Welch, A.H., Lico, M.S., and Hughes, J.L. (1988) Arsenic in ground water of the Western United States. *Groundwater,* **26** (3), 333–347.

22 Aurillo, A.C., Mason, R.P., and Hemond, H.F. (1994) Speciation and fate of arsenic in 3 lakes of the aberjona watershed. *Environ. Sci. Technol.,* **28** (4), 577–585.

23 Mariner, P.E., Holzmer, F.J., Jackson, R.E., Meinardus, H.W., and Wolf, F.G. (1996) Effects of high ph on arsenic mobility in a shallow sandy aquifer and on aquifer permeability along the adjacent shoreline, commencement bay superfund site, Tacoma, Washington. *Environ. Sci. Technol.,* **30** (5), 1645–1651.

24 Martell, A.E. and Smith, R.M. (1997) *Critical Selected Stability Constants of Metal Complexes Database. Version 4.0,* National Institute of Standards and Technology (NIST), Gaithersburg, MD.

25 Chen, S.L., Dzeng, S.R., Yang, M.H., Chiu, K.H., Shieh, G.M., and Wai, C.M. (1994) Arsenic species in groundwaters of the blackfoot disease area, Taiwan. *Environ. Sci. Technol.,* **28** (5), 877–881.

26 Tseng, W.P. (1989) Blackfoot disease in Taiwan – a 30-year follow-up-study. *Angiology,* **40** (6), 547–558.

27 Chiou, H.Y., Hsueh, Y.M., Liaw, K.F., Horng, S.F., Chiang, M.H., Pa, Y.S., Lin, J.S.N., Huang, C.H., and Chen, C.J. (1995) Incidence of internal cancers and ingested inorganic arsenic – a 7-year follow-up study in Taiwan. *Cancer Res.,* **55** (6), 1296–1300.

28 Guo, X.J., Fujino, Y., Kaneko, S., Wu, K.G., Xia, X.J., and Yoshimura, T. (2001) Arsenic contamination of groundwater and prevalence of arsenical dermatosis in the Hetao Plain Area, Inner Mongolia, China. *Mol. Cell. Biochem.,* **222** (1-2), 137–140.

29 Hopenhayn-Rich, C., Biggs, M.L., Fuchs, A., Bergoglio, R., Tello, E.E., Nicolli, H., and Smith, A.H. (1996) Bladder cancer mortality associated with arsenic in drinking water in Argentina. *Epidemiology,* **7**, 117–124.

30 Chowdhury, U.K., Biswas, B.K., Chowdhury, T.R., Samanta, G., Mandal, B.K., Basu, G.C., Chanda, C.R., Lodh, D., Saha, K.C., Mukherjee, S.K., Roy, S., Kabir, S., Quamruzzaman, Q., and Chakraborti, D. (2000) Groundwater arsenic contamination in Bangladesh and West Bengal, India. *Environ. Health Perspect.,* **108** (5), 393–397.

31 Bhattacharya, P., Chatterjee, D., and Jacks, G. (1997) Occurrence of arsenic-contaminated groundwater in alluvial aquifers from delta plains, eastern india: options for safe water supply. *Water Resour. Dev.,* **3** (1), 79–92.

32 Berg, M., Tran, H.C., Nguyen, T.C., Pham, H.V., Schertenleib, R., and Giger, W. (2001) Arsenic contamination of groundwater and drinking water in vietnam: a human health threat. *Environ. Sci. Technol.,* **35** (13), 2621–2626.

33 Frey, M.M. and Edwards, M.A. (1997) Surveying arsenic occurence. *J. Am. Wat. Works Assoc.,* **89** (3), 105–117.

34 Peters, S.C., Blum, J.D., Klaue, B., and Karagas, M.R. (1999) Arsenic occurrence in new hampshire drinking water. *Environ. Sci. Technol.,* **33** (9), 1328–1333.

35 Schreiber, M.E., Simo, J.A., and Freiberg, P.G. (2000) Stratigraphic and geochemical controls on naturally occurring arsenic in Groundwater, Eastern Wisconsin, USA. *Hydrogeology Journal,* **8**, 161–176.

36 Amini, M., Mueller, K., Abbaspour, K.C., Rosenberg, T., Afyuni, M., Moller, K.N., Sarr, M., and Johnson, C.A. (2008) Statistical modeling of global geogenic fluoride contamination in groundwaters.

Environ. Sci. Technol., **42** (10), 3662–3668.

37 Monson, B.A. and Brezonik, P.L. (1998) Seasonal patterns of mercury species in water and plankton from softwater lakes in Northeastern Minnesota. *Biogeochemistry*, **40** (2-3), 147–162.

38 Watras, C.J., Back, R.C., Halvorsen, S., Hudson, R.J.M., Morrison, K.A., and Wente, S.P. (1998) Bioaccumulation of mercury in pelagic freshwater food webs. *Sci. Total Environ.*, **219** (2-3), 183–208.

39 Kim, J.P. and Burggraaf, S. (1999) Mercury bioaccumulation in rainbow trout (*Oncorhynchus mykiss*) and the trout food web in Lakes Okareka, Okaro, Tarawera, Rotomahana and Rotorua, New Zealand. *Water Air Soil Pollut.*, **115** (1-4), 535–546.

40 UNEP (2002) Global Mercury Assessment (accessed 7/14/2006).

41 Lacerda, L.D. (1997) Global mercury emissions from gold and silver mining. *Water Air Soil Pollut.*, **97**, 209–221.

42 EPA (2006) *Epa's Roadmap for Mercury*, EPA-HQ-OPPT-2005-0013; EPA, Washington, DC.

43 Morel, F.M.M., Kraepiel, A.M.L., and Amyot, M. (1998) The chemical cycle and bioaccumulation of mercury. *Annu. Rev. Ecol. Syst.*, **29**, 543–566.

44 Sarkar, D., Essington, M.E., and Misra, K.C. (2000) Adsorption of Mercury(II) by Kaolinite. *Soil Sci. Soc. Am. J.*, **64** (6), 1968–1975.

45 Bonnissel-Gissinger, P., Alnot, M., Lickes, J.P., Ehrhardt, J.J., and Behra, P. (1999) Modeling the adsorption of mercury (II) on (hydr)oxides II: alpha-FeOOH (goethite) and Amorphous Silica. *J. Colloid Interface Sci.*, **215** (2), 313–322.

46 Mac Naughton, M.G. and James, R.O. (1974) Adsorption of aqueous mercury (II) complexes at the oxide/water interface. *J. Colloid Interface Sci.*, **47** (2), 431–440.

47 Benoit, J.M., Gilmour, C.C., Heyes, A., Mason, R.P., and Miller, C.L. (2003) Geochemical and biological controls over methylmercury production and degradation in aquatic ecosystems. *Biogeochemistry Environmentally Important Trace Elements*, **835**, 262–297.

48 Siciliano, S.D., O'Driscoll, N.J., Tordon, R., Hill, J., Beauchamp, S., and Lean, D.R.S. (2005) Abiotic production of methylmercury by solar radiation. *Environ. Sci. Technol.*, **39** (4), 1071–1077.

49 Amyot, M., Mierle, G., Lean, D.R.S., and McQueen, D.J. (1994) Sunlight-induced formation of dissolved gaseous mercury in lake waters. *Environ. Sci. Technol.*, **28** (13), 2366–2371.

50 Kolpin, D.W., Furlong, E.T., Meyer, M.T., Thurman, E.M., Zaugg, S.D., Barber, L.B., and Buxton, H.T. (2002) Pharmaceuticals, hormones, and other organic wastewater contaminants in U.S. streams, 1999–2000: a national reconnaissance. *Environ. Sci. Technol.*, **36** (6), 1202–1211.

51 Kummerer, K. (2001) drugs in the environment: emission of drugs, diagnostic aids and disinfectants into wastewater by hospitals in relation to other sources – a review. *Chemosphere*, **45** (6-7), 957–969.

52 Wise, R., Hart, T., Cars, O., Streulens, M., Helmuth, R., Huovinen, P., and Sprenger, M. (1998) Antimicrobial resistance. *BMJ*, **317** (7159), 609–610.

53 Eljarrat, E. and Barcelo, D. (2003) Priority lists for persistent organic pollutants and emerging contaminants based on their relative toxic potency in environmental samples. *TrAC Trends in Analytical Chemistry*, **22** (10), 655–665.

54 Rhind, S.M. (2005) Are endocrine disrupting compounds a threat to farm animal health, welfare and productivity? *Reproduction in Domestic Animals*, **40** (4), 282–290.

55 Roepke, T.A., Snyder, M.J., and Cherr, G.N. (2005) Estradiol and endocrine disrupting compounds adversely affect development of sea urchin embryos at environmentally relevant concentrations. *Aquat. Toxicol.*, **71** (2), 155–173.

56 Spano, M., Toft, G., Hagmar, L., Eleuteri, P., Rescia, M., Rignell-Hydbom, A., Tyrkiel, E., Zvyezday, V., and Bonde, J.P. (2005) INUENDO exposure to PCB and p, p'-DDE in European and inuit populations: impact on human sperm chromatin integrity. *Human Reproduction*, **20** (12), 3488–3499.

57 Lee, K.E., Barber, L.B., Furlong, E.T., Cahill, J.D., Kolpin, D.W., Meyer, M.T., and Zaug, S.D. (2004) *Presence and Distribution of Organic Wastewater Compounds in Wastewater, Surface, Ground and Drinking Waters of Minnesota 2000–2002*, USGS, Reston, VA.

58 Jobling, S.R., Williams, A., Johnson, A., Taylor, M., Gross-Sorokin, C.R., Tyler, C., van Aerle, R., Sanots, E., and Brighty, G. (2006) Predicted exposures to steroid estrogens in U.K. rivers correlate with widespread sexual disruption in wild fish populations. *Environ. Health Perspect.*, **114** (S-1), 32–39.

59 Suzuki, T., Nakagawa, Y., Takano, I., Yaguchi, K., and Yasude, K. (2004) Environmental fate of bisphenol A and its biological metabolites in river water and their xenoestrogenic activity. *Environ. Sci. Technol.*, **38** (8), 2389–2396.

60 Barel-Cohen, K., Shore, L., Shemesh, M., Wenzer, A., Murell, J., and Schorr, N. (2006) Monitoring of natural and synthetic hormones in a polluted river. *J. Environ. Manage.*, **78** (1), 16–23.

61 Sommer, B., Marino, A., Solarte, Y., Salas, M.L., Dierolf, C., Valiente, C., Mora, D., Rechsteiner, R., Setter, P., Wirojanagud, W., Ajarmeh, H., AlHassan, A., and Wegelin, M. (1997) SODIS – an emerging water treatment process. *J. Water Supply Res. Technol. AQUA*, **46** (3), 127–137.

62 Amin, M.T. and Han, M.Y. (2009) Roof-harvested rainwater for potable purposes: application of solar collector disinfection (SOCO-DIS). *Water Res.*, **43** (20), 5225–5235.

63 Fisher, M.B., Keenan, C.R., Nelson, K.L., and Voelker, B.M. (2008) Speeding up solar disinfection (SODIS): effects of hydrogen peroxide, temperature, pH, and copper plus ascorbate on the photoinactivation of E-coli. *J. Water Health*, **6** (1), 35–51.

64 Bhatkhande, D.S., Pangarkar, V.G., and Beenackers, A. (2002) Photocatalytic degradation for environmental applications – a review. *J. Chem. Technol. Biotechnol.*, **77** (1), 102–116.

65 Kabra, K., Chaudhary, R., and Sawhney, R.L. (2004) Treatment of hazardous organic and inorganic compounds through aqueous-phase photocatalysis: a review. *Ind. Eng. Chem. Res.*, **43** (24), 7683–7696.

66 Beydoun, D., Amal, R., Low, G.K.C., and McEvoy, S. (2000) Novel hotocatalyst: titania-coated magnetite. Activity and photodissolution. *J. Phys. Chem. B*, **104**, 4387–4396.

67 Fretwell, R. and Douglas, P. (2001) An active, robust and transparent nanocrystalline anatase TiO_2 thin film – preparation, characterisation and the kinetics of photodegradation of model pollutants. *J. Photochem. Photobiol. A*, **143**, 229–240.

68 Chen, J., Ollis, D.F., Rulkens, W.H., and Bruning, H. (1999) Kinetic processes of photocatylitic mineraliztion of alcohols on metallized titanium dioxide. *Water Res.*, **33** (5), 1173.

69 EPA (2005) *Membrane Filtration Guidance Manual*, EPA 815-R-06-009; Office of Water, Washington, DC.

70 EPA (2003) Arsenic Treatment Technology Evaluation Handbook Office of Water (4606M), http://www.epa.gov/safewater/arsenic/pdfs/handbook_arsenic_treatment-tech.pdf (accessed 5/26/10).

71 Ghosh, A., Mukiibi, M., and Ela, W. (2004) TCLP underestimates leaching of arsenic from solid residuals under landfill conditions. *Environ. Sci. Technol.*, **38** (17), 4677–4682.

72 Guest, J.S.S.S.J., Barnard, J.L., Beck, M.B., Daigger, G.T., Hilger, H., Jackson, S.T., Karvazy, K., Kelly, L., Macpherson, L., Mihelcic, J.R., Pramanik, A., Raskin, L., van Loosdrecht, M.C.M., Yeh, D., and Love, N.G. (2009) A new planning and design paradigm to achieve sustainable resource recovery from wastewater. *Environ. Sci. Technol.*, **43**, 6126–6130.

73 Maurer, M.P.W. and Larsen, T.A. (2006) Treatment processes for source-separated urine. *Water Res.*, **40** (17), 3151–3166.

74 Shaw, R. (2010) The Use of Human Urine as a Crop Fertilizer in Mali, West Africa, http://cee.eng.usf.edu/peacecorps/ (accessed 6/1/2010).

75 WHO (2006) Guidelines for the Safe Use of Wastewater, Excreta, and Greywater Volume 4: Excreta and Greywater Use in

Agriculture, http://www.who.int/water_sanitation_health/wastewater/gsuweg4/en/index.html (accessed 5/17/2010).

76 Jonsson, H., Stinzing, A.R., Vinneras, B., and Salomon, E. (2004) Guidelines on the Use of Urine and Faeces in Crop Production, p. 1–43. http://www.ecosanres.org/pdf_files/ESR_Publications_2004/ESR2web.pdf (accessed 5/17/2010).

77 Udert, K.M., Larsen, T.A., and Gujer, W. (2003) Estimating the precipitation potential in urine-collecting systems. *Water Res.*, **37** (11), 2667–2677.

78 Boyer, T. (2010) University of Florida, Gainesville, FL. Personal Communication.

79 Solomon, S., Qin, D., Manning, M., Alley, R.B.B.T., Bindoff, N.L., Chen, Z., Chidthaisong, A., Gregory, J.M., Hegerl, G.C., Heimann, M., Hewitson, B., Hoskins, B.J., Joos, F., Jouzel, J., Kattsov, V., Lohmann, U., Matsuno, T., Molina, M., Nicholls, N., Overpeck, J., Raga, G., Ramaswamy, V., Ren, J., Rusticucci, M., Somerville, R., Stocker, T.F., Whetton, P., Wood, R.A., and Wratt, D. (2007) Technical summary, in *Climate Change 2007: The Physical Science Basis. Contribution of Working Group I to the Fourth Assessment Report of the Intergovernmental Panel on Climate Change* (eds S. Solomon, D. Qin, M. Manning, Z. Chen, M. Marquis, K.B. Averyt, M. Tignor, and H.L. Miller), Cambridge University Press, Cambridge, UK and New York, pp. 19–91.

80 Easterling, D. and Karl, T. (2008) Global Warming Frequently Asked Questions, http://www.ncdc.noaa.gov/oa/climate/globalwarming.html (accessed 5/31/2010).

81 Voormeij, D.A. and Simandl, G.J. (2004) Geological, ocean, and mineral CO_2 sequestration options: a technical review. *Geosci. Can.*, **31** (1), 11–22.

82 Shih, D.C.F., Wu, Y.M., and Hu, J.C. (2010) Potential volume for CO_2 deep ocean sequestration: an assessment of the area located on Western Pacific Ocean. *Stochastic Environ. Res. Risk Assess.*, **24** (5), 705–711.

83 National Energy Technology Laboratory (2008) *Carbon Sequestration Atlas of the United States and Canada*, 2nd edn, U.S. Department of Energy (DOE), Washington, DC.

84 Anastas, P.T. and Warner, J.C. (1998) *Green Chemistry: Theory and Practice*, Oxford University Press, Oxford.

8

Facing the Energy Challenges through Chemistry in a Changing World

Gabriele Centi and Siglinda Perathoner

8.1
Introduction

Energy is a very pervasive concept and a vital part of our daily lives, because all aspects of our life depend on the availability of energy in one form or another, from the production of food and all goods required for our style of life (health care, water supply, etc.) to mobility, communication, and many other aspects [1]. Without energy, we cannot heat homes, power businesses or use transport.

Availability of energy is an indicator of social development. A rough linear relationship exists between GDP (gross domestic product) and energy consumption, even if deviations are present due to different politics on energy saving. The latter are mainly associated with taxes on energy use, and, in part also, to different attitudes toward saving energy and climate situations, but the differences are also significantly related to different degrees of development of science and technology (S&T). Energy use and its impact on the sustainable development of society is thus both an indicator of the socio-political context and of the degree of S&T development.

The use of energy requires the availability of technologies to transform energy into a form suitable for distributed use (electrical energy, for example) and of technologies/processes to produce the energy vectors (gasoline, for example) which allow one to transfer, in space and time, a quantity of energy [2]. These energy vectors should have not only suitable characteristics for their commercial use (gasoline, which is a mixture of various hydrocarbons, for example, must possess many properties, from a boiling point within a certain range to a minimum octane number–see later), but also suitable characteristics to minimize their impact on the environment upon use (gasoline specifications include a maximum content of aromatics, and particularly of benzene due to its toxicity, a maximum volatility of light components, a maximum content of sulfur, etc.). Energy is thus a vital part of our daily lives, but its use requires having

- materials, devices and processes which transform it into a form suitable for daily use, including availability at the needed space and time

The Chemical Element: Chemistry's Contribution to Our Global Future, First Edition.
Edited by Javier Garcia-Martinez, Elena Serrano-Torregrosa.
© 2011 Wiley-VCH Verlag GmbH & Co. KGaA. Published 2011 by Wiley-VCH Verlag GmbH & Co. KGaA.

- technologies which allow its efficient use and minimize the impact on the environment

Chemistry is the core science making possible this sustainable use of energy, although the role of chemistry is often hidden. A car is not considered a product of chemistry, but cars are chemistry in motion. Today's automobiles depend heavily upon chemical industry innovations to enhance their performance, safety and fuel efficiency. Polyurethane seat cushions make travel comfortable on long journeys while neoprene hoses, brake fluids, sealants, adhesives and coolants maintain vehicle performance and endurance. Seat belts and air bags, very important for car safety, both rely on polymer chemistry. The list of car components depending on chemistry is endless: tires, fuels, lubricants, paints and surface coatings, batteries, cables, dashboard, steering wheel, and so on.

In the energy area, chemistry plays a similar multi-area enabling role. We should more correctly discuss energy systems rather than energy itself. An energy system uses energy resources to produce an effect, that is, to obtain at the end of the energy chain a final use of mechanical, thermal, lighting or electric energy. It is necessary to extract energy from the initial resource and distribute it in the appropriate form and with appropriate characteristics for the final use that allow the production of the desired effect (heating, lighting, mechanical power, etc.). Figure 8.1 gives a schematic representation of this energy chain that, in reality, is a complex energy system with many interactions. There is a set of primary energy resources, processes and technologies for conversion, transport, storage and final uses of energy. Chemistry has a main role in many of the critical components of this energy chain in terms of key elements to realize sustainable energy production and use, as shown schematically in Figure 8.1.

These multiple contributions of chemistry to the energy challenge are further detailed in Figure 8.2 which outlines the main chemistry role (differentiated in contributions from material and process developments) to the four main energy issues towards sustainable energy [3]: conservation, rational use of fossil fuels, nuclear energy and use of renewable sources. Chemistry is the strategic core discipline for the conversion processes based on solar energy, but it is indispensable to save energy in both production and use. Many material production processes benefit from improved chemical reactions and from optimized processes, often

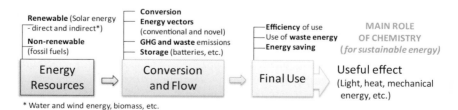

Figure 8.1 Energy chain and the role of chemistry in enabling sustainable energy production and use.

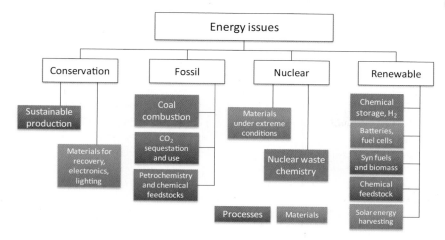

Figure 8.2 Multiple contributions of chemistry to the energy challenge; in blue contribution from material and in red from process developments; adapted from [3]. (Please find a color version of this figure in the color plates.)

through improved catalysts, and not only energy conversion from fossil sources. Improved materials are critical for energy saving and to enable more energy-efficient processes such as house climatization, high-temperature combustion, solid-state lighting and more efficient transportation. Electrical energy storage in Li ion or redox flow batteries requires multiple efforts from chemistry to design functional electrodes and storage materials. Next generation clean and energy-efficient technologies will emerge from new materials made available by emerging nanoscience and nanotechnologies [4]. Examples are materials that (i) maintain performance under extreme conditions, (ii) self-heal in harsh environments, (iii) are lighter than steel but also stronger, (iv) convert sunlight or produce light more efficiently, (v) allow the saving of energy in separation processes, (vi) permit higher energy density storage, and so on. The development in nanotechnology over the last decades has now allowed progress from the laboratory scale to commercial application, and we thus expect a boost in new materials with properties optimized to sustainable energy technology requirements.

Chemistry is thus the enabling science for the sustainable production and use of energy and plays a critical role in its many aspects, from the production of the materials needed in the various technologies to transform energy (for example, production of the semiconductors and other materials necessary in the photovoltaic cells to produce electrical energy using solar light) to the technologies/processes to produce the energy vectors (for example, refinery processes to produce fossil fuels) and the materials/technologies to use energy (storage in a distributed form – for example, batteries). In addition, chemical processes (often catalytic) are leading to a reduction in the impact on the environment of the use of energy, for example, in reducing pollutant emission in power generation and in cars.

8.2
Chemistry and the Role for Development of Society

Chemistry and catalysis play a special role in making energy available cheaply from fossil fuels and renewable resources [3, 5]. Of the about 500 ExaJoules (EJ) of global primary energy demand in 2009, only about 8% was from nuclear and hydropower and 10% from biomass combustion [6], the remainder comes essentially from fossil fuels. Although natural gas and coal also require chemical processing, we use here as an example of the role of chemistry in the development of society the petroleum (oil) case, which accounts for about 34% of the global primary energy demand. Oil cannot be used as such and requires a complex series of separation and conversion processes to prepare the products in the amount and with the characteristics required for energy market applications, that is, gasoline, diesel and jet fuel for transport, kerosene and heavy oil for energy production (heating, electrical energy), and so on [7, 8].

Oil refining, for example, the area of chemistry dealing with the transformation of the crude oil, has changed progressively to adapt to the changing society demand for energy [9]. Oil has been known for a very long time, and, even in Egyptian times, was used in small amounts for coating mummies and sealing pyramids . The Chinese also used petroleum for heating and the Bible claims that Noah used it to make his Ark seaworthy. However, the start of the industrial use of petroleum dates back to the second half of the 18th century, and reveals how chemistry, combined with science and technology, was the steering factor for the societal evolution.

Oil refinery

An oil refinery converts crude oil into various products: liquid petroleum gas (LPG), gasoline and diesel for motor fuel, jet fuel, kerosene, heating fuel oils, lubricating oils, and asphalt and petroleum coke. In a simplified view, an oil refinery comprises a first operation of distillation (often called topping), where the main fractions of petroleum are separated based on their range of boiling points. Then, a number of chemical processes of transformation (cracking, reforming, coking, etc.) and separation (vacuum distillation, solvent extraction, etc.) allow maximization of the amount and improvement of the quality of the energy products.

During the 18 and 19th centuries, for example, during the first and second Industrial Revolutions, major changes occurred in agriculture, manufacturing, mining, and transport, with consequent effects on the socioeconomic and cultural conditions. The same period coincided also with the start of industrial chemical production, which was driven by two factors: the availability of cheap raw materials and the societal needs for new products.

One of the consequences of the Industrial Revolution was the increasing demand for illumination. Whale oil was mainly used at that time in lamps and as candle wax, but around the middle of the 18th century the shortage of whales as a consequence of heavy hunting created the need for alternative oils. George Bissell first thought that petroleum might be converted into kerosene for use in lamps, and, in 1858, founded the first oil company and refinery with a capacity around five orders of magnitude lower than that of the actual oil refineries. The main product was kerosene for illumination, but the other fractions were not used. With the increased market, it was thus necessary to find applications for these unused products, and so began the commercialization of synthetic lubricants and waxes, and tar as a roofing material. The success in refined petroleum products greatly spread the technology and, by 1865, 194 refineries were already in operation.

At the beginning of the 19th century, three milestones created the conditions for a change in the refinery concept: the invention (by Lenoir) of the internal combustion engine using gasoline as fuel, the introduction of this engine in cars (by Daimler and Benz) and the mass production of cars (by Ford). Making the use of vehicles affordable to a larger public created the conditions for mass mobility. This caused the societal demand for more fuel for these cars, and thus a change in the oil refinery structure to adapt to the production of gasoline and diesel, driven also by the declining market for kerosene for illumination (the light bulb was invented by Edison in 1878). The efficiency of production of these products by distillation of the crude oil, for example, the process of separation of the different oil fractions based on the boiling point, was not high at that time, around 30% of the oil. Therefore, it was necessary to introduce new chemical technologies to convert the less desirable fractions of the oil to the market products. In 1901, in Texas, a refinery was built with a production of 100 000 barrels per day, compared to the 35 barrels per day of the first refinery only 50 years earlier. This had required huge progress in chemical technology to make available oil products at low costs. The low cost and availability led to an increasing number of ships and trains being switched from coal to oil. In the United States, between 1880 and 1920, the amount of oil refined annually jumped from 26 million barrels to 442 million barrels.

Octane number

An important parameter for gasoline is the octane number, which is a measure of the resistance of the gasoline components to autoignition in spark-ignition internal combustion engines. The octane number of a fuel is measured in a test engine, and is defined by comparison with the mixture of 2,2,4-trimethylpentane (iso-octane) and heptane which would have the same anti-knocking capacity as the fuel under test: the percentage, by volume, of 2,2,4-trimethylpentane in that mixture is the octane number of the fuel.

Around 1930, the process of cracking oil, that is, the catalytic process of breaking down heavier molecules to lighter ones suitable for gasoline or diesel, was introduced in refineries. During the consecutive years, the improvements in engines (in particular, compression) led to demand for a better quality gasoline, in particular with a higher octane number. Therefore, various new chemical processes (thermal and catalytic cracking, reforming, alkylation and isomerization, polymerization, etc. – mostly catalytic processes) were introduced in the refinery to improve both the amount and quality (combustion characteristics, volatility, stability, etc.) of the products.

The evolution of the airplane created a further need for high-octane aviation gasoline and then for jet fuel, a sophisticated form of the original product, kerosene. A further major change in the refinery structure derived from the beginning of concerns about the environmental impact of emissions from motor vehicles. Around 1970, the first crude desulfurization unit to reduce the level of sulfur in gasoline was introduced. This required the development of large-scale plants to produce the H_2 necessary for desulfurization and other refinery processes needed to improve the environmental compatibility of gasoline (lower volatile compounds emissions, lower aromatics in general and, specifically, benzene, etc.). The introduction of catalytic converters led to the need to eliminate lead compounds from gasoline since lead deposits deactivate the catalytic converters. It was then necessary to develop alternative compounds (oxygenated compounds such as methyl *tert*- butyl alcohol – MTBE) to compensate for the loss in anti-knocking properties due to the elimination of tetra-ethyl lead.

Finally, a new revolution in refinery (the fourth generation plants) will be progressively derived from the new societal demand for the reduction of the carbon footprint of fuels, with thus incentives for developing biofuels, for example, fuels derived from biomass resources [3, 4]. We can visualize this as a way to introduce a shorter path in energy production for mobile applications. Petroleum is derived from the transformation of biomass under the effect of pressure and temperature when these sediments move under the ground. This occurs on a geological timescale.

The use of biomass to produce fuels without passing through petroleum introduces a shorter path for using solar energy. In fact, plants use carbon dioxide, water and solar energy (plus some other elements) to construct biomass in a process that is complex and delicate, although occurring on a large scale. Instead of waiting for their transformation to oil and then producing from it the fuels (the rate of petroleum consumption is much larger than that of formation, and thus there is a progressive depletion, even if the reserves are still large), it is possible to use the biomass directly. However, we then need to harvest, transport and transform this biomass to produce the energy vectors (biofuels) to be used in cars and airplanes.

A future possible simplification in this complex, multistep process could derive from the genetic engineering of bacteria and microorganisms in order to adapt them to produce biofuels directly from CO_2, water and solar light. This is a great challenge being somewhat against the natural evolution of bacteria and microorganisms, because the production of fuels from living elements is not part of a self-sustained life mechanism. In addition, there will probably still be huge difficulties in recovering and purifying the biofuels, and other great technical problems will probably remain. We thus suggest that this route appears unlikely, and a strategically more appealing possibility is the direct conversion of CO_2 and water with solar light in nano-devices (artificial trees) which could produce the fuels directly [10]. This possibility will be briefly discussed later in this chapter. However, the complexity and multifaceted characteristics of the energy problem indicate that it is unlikely a single solution will be achieved and thus, at this stage, all possible solutions should be explored.

We can thus summarize with the indication that after the next step of *biorefineries*, for example, those that will use biomass resources to produce fuels, often integrated with the production of chemicals, the future will be *solar refineries*.

The concept and structure of refineries has thus progressively evolved following the societal (and market) demand to provide a solution to the needs of mobility and energy, and later of environmental compatibility and sustainability. Chemistry has provided the knowledge, technologies and materials to realize this progressive transition [3].

This first historical example evidences how the chemistry and chemical technology was always driven by the societal demand for new products (fuels) to enable the newly developed technologies (cars, airplanes, etc.). Many other components of cars besides fuels are associated with chemistry, from tires and lubricants, to car paint and coating, to dashboards and the many plastic components, and so on. The same is true for airplanes, trains, ships, and so on. Quality and security of mobility is thus highly dependent on the enabling character of chemistry in providing the solutions for the societal demand [11].

8.3
Chemistry and Sustainable Energy

Chemical science and technology play multiple key roles in sustainable energy scenarios, and particularly in the development of processes and materials for the following areas:

- Energy conversion reactions
- Energy storage and transport
- Energy use efficiency

Chemical science and technology is the basis for our technical production processes for transport, and will be the core discipline for all energy conversion processes that are based on primary solar energy [3]. Chemistry provides solutions not only in terms of new processes, but also of new materials (catalysts, for example) and solutions which allow the saving of energy and raw materials. Materials chemistry science contributes in many additional ways to energy saving and efficiency, for example, in developing all the sets of new technologies for the "Smart Energy House", the house of the future which combines quality of life with minimization of energy and water consumption, and self-production of energy [12, 13].

Multidisciplinary cooperation between chemistry (in its multiple aspects going from biotechnologies to materials science and chemical processes, or involving the different synthetic, physical and engineering branches of chemical sciences) and all the other science and technology areas are the key to progress in the challenging area of development of new solutions for sustainable energy. Often chemistry has been considered a tool to application, but it is instead a broad-based discipline that creates materials and processes within its own paradigm based upon molecular understanding of the transformation processes of matter [3].

Catalysis

Catalysis is the use of chemical substances, often solids (heterogeneous catalysis) but also metal complexes or other compounds operating in the same phase as the reactants / products (homogeneous catalysis) or based on bio-compounds such as enzymes (biocatalysis) which are able to accelerate the reaction rates and/or selectively form a product in a complex reaction network. Therefore, catalysis is a key element for industrial chemical production (over 90% of chemical processes, including those for the energy sector, use catalysts in one or more steps) and for sustainability, because they allow one to operate in a way that saves resources (raw materials, energy) and is safer.

In a simplified approach, we can distinguish three main sectors in which chemistry plays a pivotal role to address the energy challenges:

- *Energy saving:* reducing the energy and raw materials consumption by introducing new processes, materials and technologies [4, 14]. For example: (i) new processes that are more selective and/or that operate with lower energy consumption; (ii) new materials (for example, foams for improved thermal isolation, new materials to reduce consumption by cars – lighter materials in place of steel, nanostructured paints to increase the vehicle aerodynamics, tires with reduced friction, and so on, solid-state lighting, for example, LEDs, and so on.); (iii) new technologies (for example, new materials to recover electrical energy from waste low-temperature streams or save energy in the conditioning of buildings). We may also include fuel cells in this class [15], particularly low-temperature proton exchange membrane fuel cells, also known as polymer electrolyte membrane (PEM) fuel cells, for mobile applications (cars, etc.), and high-temperature solid oxide fuel cells (SOFCs) for residential applications. The main objective of fuel cells, in fact, is to improve the efficiency in converting chemical to electrical energy.

- *Sustainable use of fossil fuels,* with the introduction of new processes and technologies, which together with a further reduction in the impact of current refineries (only about half of the petroleum entering a refinery is converted to valuable products, the remainder being used for the energy needs of the refinery), allow the valorization in a sustainable approach of unconventional (or less-conventional) fossil fuels: (i) coal (for example, coal combustion integrated with emissions purification and CO_2 sequestration/use; coal to liquid (CTL)) [16], (ii) strained natural gas (NG) in remote areas, for example, the large amounts of NG which are in remote areas and for which transport by pipelines or liquefaction is not convenient (for example, by converting to liquid fuels – gas to liquid (GTL)) [17], (iii) very heavy oils, tar sand or oil shale, bitumen, and so on, for example, by developing new energy-effective membranes for O_2 separation, which is then used for the combustion of these fractions – the air separation method by distillation used currently is very energy-intensive.

- *Use of renewable energy resources* [18–21], first, solar energy by developing new (nanostructured) materials able to more efficiently harvest and convert light to electrical energy and in the long-term to use solar energy to produce solar fuels [10]. Within this area falls also that related to the use of biomass and, in general, to the development of a bio-based (energy efficient) economy [22, 23]. A critical aspect for electrical energy is the difficult storage of electrons, and therefore this area includes the intense developments related to the improvement of electron storage (Li batteries, supercapacitors, etc.) [4, 24, 25]. In general, storage and transport (out of the grid) of the electrical energy is the critical issue which determines the need for developing suitable energy vectors, where chemistry plays a central role. Other forms of renewable energy depend also in one or more aspects on the improved materials, which can be developed by chemistry. For example, a rotor blade reinforced with carbon nanotubes improves the performances of wind turbines. Further aspects will be discussed in the following sections.

In nuclear energy, chemistry contributes to the development of materials resistant under the extreme conditions present in these reactors (especially, in the new generation of fusion reactors) and in the nuclear waste reprocessing. In addition, nuclear power could be, in principle, used in a wider range of applications, from water desalination to industrial process heat and hydrogen production, although it is currently used almost exclusively to generate electrical power.

An important concept to outline is that chemistry provides knowledge and materials, which are then engineered to develop new solutions for a sustainable energy. Therefore, the progresses in chemistry often have multiple applications. An interesting example is given from the recent development in carbon nanotubes (CNT) [26]. CNT is a tubular form of carbon with a diameter as small as 1 nm and length from a few nm to microns. CNT is configurationally equivalent to a two-dimensional graphene sheet rolled into a tube (Figure 8.3).

CNT, and particularly multi-walled carbon nanotubes (MWCNT), can now be produced on the tons scale quite cheaply by a technique called chemical vapor deposition (CVD) which consists essentially in the decomposition of organic sources (hydrocarbons, alcohols, etc.) over catalyst nanoparticles. Nanotubes grow at the sites of the metal catalyst.

CNTs are exceptional strong and stiff materials, due to the covalent sp^2 bonds formed between the individual carbon atoms. In addition, they have excellent thermal and electrical conductivity properties. They thus find application in preparing composite polymers with high performance, in electrical circuits, in various advanced devices for energy conversion and storage (solar cells, fuel cells, ultracapacitors and ion batteries), in sensors, in novel catalysts and advanced materials for environmental clean-up, in novel devices, and so on [27–29]. Some of these aspects will be discussed in more detail later.

(a)

(b)

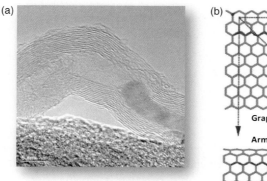

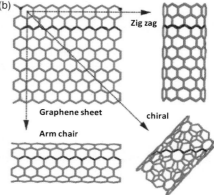

Figure 8.3 (a) Transmission electron microscopy (TEM) image of a multiwall carbon nanotube (MWCNT) which grows by chemical vapor deposition (CVD) from a catalytic nanoparticle. (b) Different modali-ties in which a graphene sheet (e.g., the basic structural element of CNTs) can be rolled to form the carbon nanotubes. (Please find a color version of this figure in the color plates.)

Carbon nanotubes and graphene

Carbon nanotubes are allotropes of carbon with a cylindrical nanostructure, and very large (up to 10^9) length-to-diameter ratio. They have unique properties, which are now exploited in nanotechnology, catalysis, electronics, optics and other fields of materials and chemistry science. Nanotubes are categorized as single-wall nanotubes (SWNTs) and multi-wall nanotubes (MWNTs).

Graphene is a one-atom-thick planar sheet of sp^2-bonded carbon atoms. It is the basic structural element of CNTs, and its direct use in advanced nanomaterials is expanding rapidly.

It is important to evidence the value chain in sustainable energy, which is exemplified in Figure 8.4 for the case of the use of CNTs in developing novel improved turbines for the production of renewable energy by wind [26]. Chemical science and technology (S&T) is the starting element in this value chain. Together with other S&T areas it leads to practical progress towards a sustainable energy and benefits for society. A large part of the manufacturing sector, which accounts for about a third of the industrial production and a significant fraction of the employment, depends for its competiveness on the capacity of the chemical industry to supply new and tailored materials. It is a strategically economic mistake (unfortunately often made) to produce these materials elsewhere (out of the country) and then import them. In a chain, breaking a step means breaking all the chain with a consequent large negative impact on society, not only in terms of the need to import new technologies, but in direct terms of higher costs and decrease in employment.

20 nm
Chemical science
Chemical technology
Synthesis CNT
Surface modification CNT
CNT + polymer composite
Rotor blades
Wind turbines
Renewable energy

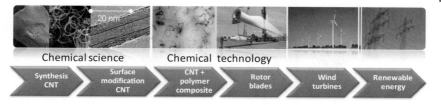

Figure 8.4 The value chain in sustainable energy: the use of CNT in wind turbines; adapted from [26]. (Please find a color version of this figure in the color plates.)

Conversion
• Wind turbines
• Fuel cell catalysts
• Efficient lighting/displays
• Solar cells
Transport
• Under-floor heating
• Windshield defroster heating
• Microwave antennas
• Electrical circuits
Storage
• Li-ion batteries
• Supercapacitors
• Hydrogen storage
Saving
• Lightweight materials
• Low rolling resistances tires
• Catalysts for energy-efficient processes
• Electrostatic coating

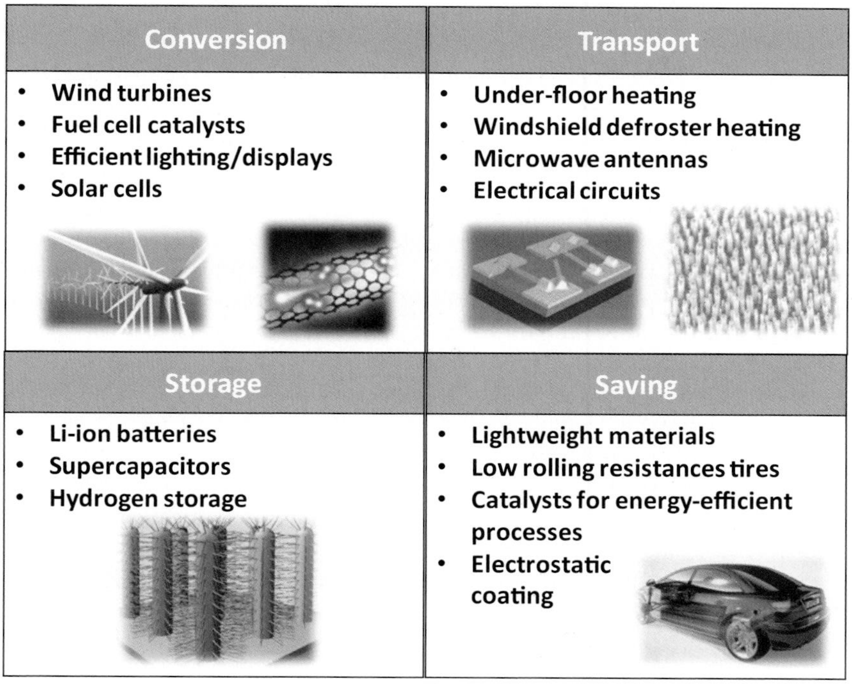

Figure 8.5 Use of carbon nanotubes in sustainable energy applications; adapted from [26]. (Please find a color version of this figure in the color plates.)

Chemical innovation for its enabling character in the entire downstream manufacture sector influences many different areas of application. Figure 8.5 shows as an example some of the applications of carbon nanotubes related to different areas of energy [26]. It is thus evident how the development of new materials by chemistry has led to a fast growth of new applications which is reflected in a significant step forward towards realizing sustainable energy. The progress in science and technology of chemistry is continuing to respond to new societal challenges. In a concise form, we can define as follows the main areas of development in chemistry to address the needs of sustainable energy:

- Need to **reduce electricity losses** due to transport and distribution. Transmission and distribution losses on average are about 6%, but this depends on the distance it needs to be transported [6]. There is often a mismatch between areas of production and consumption of electrical energy. Another critical issue is how to store electrical energy. Several energy production plants should operate at a nearly constant rate, but there are great changes in energy consumption during the day (peak consumption is about twice the average value). This requires the design and development of new materials that reduce the resistive losses in the conductor, for example materials super-conducting at higher temperature, preferably at room temperature or close to it. The DESERTEC project (http://www.desertec.org, accessed 6 January 2011) plans to construct a large number of solar power plants in the Sahara desert and then transport the produced electrical energy to Europe. This would be possible only when more efficient electrical energy long-distance transport systems will be available.

- Development of efficient **energy vectors** (and related devices and infrastructures) to transport and store energy. H_2 is a clean energy vector [30], although suffering from the problem of difficult storage (see later). Novel storage systems for H_2 (based for example, on novel nanostructured materials) [4] could be coupled to systems for producing H_2 from renewable resources. For example, photovoltaic cells could be coupled to water electrolyzers to realize devices able to store hydrogen and produce electrical energy (by fuel cells) when needed [30]. An alternative option, and in our opinion preferable, is to develop alternative energy vectors with a higher energy density [10], as discussed later. In parallel, it is necessary to develop efficient systems to store electrical energy for small mobile applications, for example, novel batteries, supercapacitors, and so on [31].

  It is necessary to push the development of *batteries for hybrid engines* and develop technologies for the safe use of hydrogen in cars. Efficient systems for chemical hydrogen storage in vehicles should be developed together with fuel cells with increased performance, for example by developing new solid polymer electrolytes and new electrocatalysts which do not use expensive and rare metals such as platinum.

- Increase the **fuel reserves** by increasing the raw materials, which can be used. There are ample reserves of natural gas (strained NG) which cannot be used currently, and it is necessary to develop chemical processes that are more efficient in transforming them to liquid products (hydrocarbons, alcohols, etc.). This technology is called GTL. Conceptually similar technologies could also convert biomass (BTL) or coal (CTL) to liquid products.

  Coal is an abundant resource, which should be *valorized in a sustainable way*. It is necessary to develop technologies for *clean coal* that could be realized by not only reducing all emitted pollutants, but also developing *effective methods of CO_2 sequestration and utilization*. The use of CO_2 can be achieved by developing catalytic processes capable of turning it into high value energy and commodities such as alcohols and hydrocarbons, polymers, and so on.

It is necessary to accelerate the introduction of the *second-generation* (and, in future, third-generation) *biofuels* that do not compete with food [4, 32–34]. It is necessary to obtain biofuels from lignocellulosic wastes with reference not only to bioethanol and biodiesel, but widening the type of products obtained (hydrocarbons, butanol, etc.). It is also necessary to choose processes, which can be well integrated with the production of bio-based chemicals [35].

Full use of biomass and integration between biofuels and bio-based products in *novel biorefineries* should be realized. An example is the development of efficient catalytic methods for converting glycerol, which is a by-product of the transesterification of vegetable oils in biodiesel production.

- The use of **solar energy** (photovoltaic, thermal and by semiconductors – see later) should encourage reduction in production costs and developing DSC photovoltaic systems with higher efficiency and greater versatility of use and integration in residential applications.

- In the field of **nuclear chemistry**, it is necessary to develop *new materials for the safe containment of radioactive waste* from nuclear power plants and that are compatible/stable with the severe conditions that are realized in the Generation IV reactors (ITER) (presence of high temperatures and neutron fluxes).

Clean coal conversion

Coal is a complex mixture of chemical compounds, mostly organic, deriving from the transformation of partly decayed plants subjected to pressure and temperature for millions of years. Coal is composed primarily of carbon along with variable quantities of other elements, chiefly sulfur, hydrogen, oxygen and nitrogen, and inorganic elements. Coal is primarily used as a solid fuel to produce electricity and heat through combustion, but is one of the largest worldwide anthropogenic sources of carbon dioxide emissions. The development in this area, together with more efficient technologies to reduce pollution emissions (SO_x and NO_x, fly ash) are thus technologies for carbon capture and sequestration (CCS).

In some countries, such as China, various processes are under development and in some cases already commercial to convert coal to liquid fuels (CTL) or to chemicals. The first step is the production of syn gas, that is, a mixture of CO and H_2, from coal by gasification:

$(Coal) + O_2 + H_2O \rightarrow H_2 + CO$

Then the CO/H_2 ratio is adjusted by the catalytic water gas shift reaction followed by removal of CO_2, typically by pressure swing adsorption (PSA): $CO + H_2O \rightleftarrows CO_2 + H_2$

Finally, the syn gas could be catalytically converted to methanol or dimethylether (DME), or to hydrocarbons by the Fischer–Tropsch process:

$CO + 2H_2 \rightarrow CH_3OH$

$2 CH_3OH \rightleftarrows CH_3OCH_3 + H_2O$

$nCO + (2n+1)H_2 \rightarrow C_nH_{2n+2} + nH_2O$
(for alkanes)

$nCO + 2nH_2 \rightarrow C_nH_{2n} + nH_2O$
(for alkenes)

8.4
Sustainable Energy Scenarios and Climate Changes

Concerns regarding greenhouse gases have been progressively turned from possibility to reality, and the fraction of the world population accessing a massive use of energy is increasing exponentially. The already significant modification of the climate due to CO_2 emissions thus calls for rapid action [36]. Chemistry can contribute to this task in multiple ways, but few of these contributions will be large compared to the scale of present emissions within short timescales (less than one decade) [2]. The fastest method is saving energy by increasing the efficiency of its use and altering human behavior. However, in the longer term it is necessary to develop alternative strategies. It is thus not possible to discuss chemistry and energy without trying to imagine a possible energy scenario [37].

The design and development of future fuels requires definition of sustainable energy scenarios and the effective scientific strategies necessary to address the interconnected energy and environment issues. It is not possible here to give a detailed discussion of these aspects, but we may briefly recall the need in the short-term for new technologies for energy saving and efficiency, and for the use of biomass (Figure 8.6). In the medium term, a better use of renewable resources is needed, including solving the issue of energy storage and transport, and finding a sustainable solution to CO_2 emissions, because the full transition to non-fossil fuels will require a longer time. Finally, in the long-term, the renewable energy scenario, based, in particular, on solar energy, will become predominant.

Therefore, alternative feedstocks for synthetic fuels, from stranded and unconventional fossil fuel reserves to green biomass uses represent the short to medium term solution, which integrates with energy saving and an increasing use of renewable energy (wind, geothermal and solar energy, particularly using photovoltaic (PV) cells). However, in a medium to long term perspective, it is necessary to push the role of solar energy to produce fuels diretly and recycle CO_2 [10, 38, 39].

Essential ingredients for future energy scenarios are, besides sustainability, the continuity in usage of existing technologies (and infrastructure) wherever possible. In fact, the investment on energy infrastructure (distribution and use) is so large that the integration of future supplies into the actual system is a key condition.

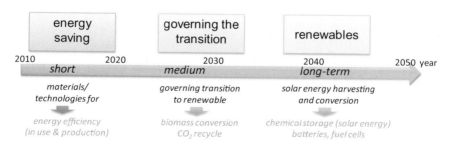

Figure 8.6 Simplified scenario for sustainable energy in relation to the use of renewable and solar energy.

Photovoltaic cells produce electrical energy, which can be put into the existing distribution system (grid). Furthermore, the investment can be modulated and can be possible even at a personal level. This enlargement of the possible investors in a new energy technology is a critical factor to accelerate the introduction of new sustainable technologies and, thus, a key element for its success. When an energy technology requires very large investments, only a few players (companies) can enter the market, but often they are the same companies holding major shares in the actual system. They have thus scarce incentive to introduce new technologies. The possibility of success of new energy technologies has thus two key characteristics:

1) Easy integration into the actual system
2) The possibility of enlarging the base of investors, due to the possibility of efficient implementation also on a small scale.

From this perspective, the large debate in the last five years on the use of hydrogen (H_2) as the future energy vector [40] has often not considered the need to change the actual energy infrastructure and thus the need for very large investment. For this reason, and for the increasing recognition of the probable impossibility of meeting the expected performances in H_2 storage materials (above 6 wt%, including the container weight), the interest in H_2 as an energy vector, particularly in cars, is declining recently. Similarly, several of the applications under study in the field of biofuels do not meet the above criteria, as will be discussed later.

Chemical research, being the element for innovation in the sector of sustainable energy, requires a blue-sky horizon, and thus not to be too limitated in the areas under investigation. However, at the same time, the complexity of the energy problem requires avoidance of the dispersion of R&D investments. For this reason, it is necessary to contextualize the R&D in a sustainable energy scenario, which should be continuously updated with close interaction between science and politics.

8.5
Nanomaterials for Sustainable Energy

The development of materials is one of the main areas of chemistry, and a significant push has been given to this sector by the development of nanotechnologies. Nanomaterials are defined as those whose properties stem from their nanoscale dimensions. This dimension refers to the characteristic dimension of the functional units in the nanometer range. Figure 8.7 illustrates this concept.

In general, the chemical approach is a bottom-up approach, which is based on understanding, at a molecular level, the characteristics and reactivity of molecules, to develop supramolecular systems and nanomaterials. For the latter sector, together with the classical areas of chemical synthesis (organic and inorganic), there is an increasing importance of the area of materials chemistry and, in

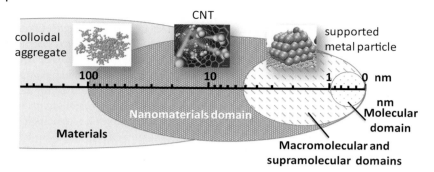

Figure 8.7 Nanomaterials: the length dimension; adapted from [41]. (Please find a color version of this figure in the color plates.)

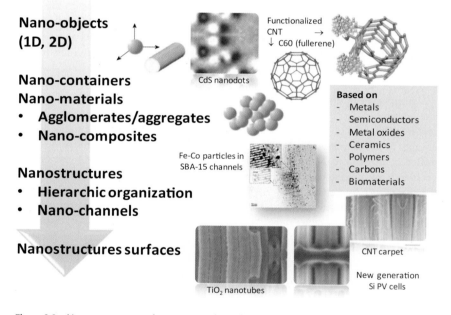

Figure 8.8 Nanostructures and nanomaterials, with some examples; adapted from [42]. (Please find a color version of this figure in the color plates.)

particular, of how to design and manipulate solid materials, and to link these aspects to their reactivity and functional properties which often depend on the nano-architecture and hierarchical organization of the material itself. Figure 8.8 reports examples of the different types of nanostructures and of nanomaterials, which can be obtained. As discussed later, these aspects are critical to the development of the next-generation sustainable energy technologies [4].

The recent developments in hydrodesulfurization (HDS) catalysts [43] are a good example of how the development in understanding and controlling the nanostructure of materials leads to to a more sustainable use of energy. HDS is a key catalytic process used in refineries to remove sulfur (S) from petroleum fractions. During the combustion of these fractions, the sulfur is oxidized to SO_2, which in the troposphere is further oxidized to SO_3 and reacts with water forming sulfuric acid (H_2SO_4). In addition, sulfur oxides deactivate the catalytic converter used in cars to reduce the emissions of NOx and HC/CO. Therefore, the content of sulfur in gasoline, and more recently in diesel, has been progressively decreased, reaching in commercial products concentrations of about 10 ppm (parts per million). This has required the development of exceptionally active catalysts able to convert selectively the sulfur compounds, that is, able to specifically cleave by hydrogenation the C–S bonds in a few molecules present between millions of other potentially reactive ones.

In refineries, thousand of tons of oil per day are processed using solid catalysts. The quality of the fuels depends on the capacity of the solid catalysts to operate continuously in severe reaction conditions and to be highly selective in the presence of a very complex feedstock. The removal of up to a few parts per million of sulfur is possible now with the new nanocatalysts developed in recent years. This is an example of controlled materials synthesis at the nanoscale level leading to large-scale applications. As an example, BRIM™ commercial (new generation) HDS catalysts [44] are used in 170 hydrotreating units with a capacity of about 6 million barrels of oil per day. These new catalysts are based on the molecular understanding of the reaction mechanism of the HDS reaction on model nanoscale catalysts [45], some aspects of which are visualized in Figure 8.9b. Figure 8.9a evidences also the concept that many properties of the catalyst depend on the nanodimension, but a catalyst for practical applications should be optimized on a multi-scale space and time domain, from the molecular level of the active site to the macro-level of the full catalyst pellet or monolith. It is thus necessary to integrate molecular design with nanoarchitecture and hierarchic organized nano/meso/micro-structure.

A further relevant example of how the chemistry of nanomaterials contributes to the development of improved materials enabling sustainable energy is related to the nano-architecture of electrodes [24, 25, 46–50]. There are several key technological areas of the energy sector, which require better design of the electrode nanostructure to overcome current limits and/or move to new levels of performance: [4, 25]

- photo-electrochemical solar cells
- water photoelectrolysis
- photoelectrocatalytic devices for the conversion of CO_2 to fuels
- advanced Li-batteries
- supercapacitors
- fuel cells

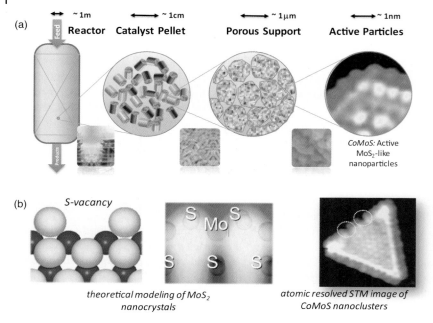

Figure 8.9 Development of hydrodesulfuriza-tion (HDS) catalysts: (a) the structure of a heterogeneous catalyst illustrating the length scales and complexity involved in a heteroge-neous catalyst and (b) atom-resolved STM image of an MoS₂ nanocluster and theoretical modeling of the active sites for HDS reaction; adapted from http://www.inano.au.dk, accessed 7 January 2011. (Please find a color version of this figure in the color plates.)

- third generation photovoltaic cells
- nanostructured thermoelectric devices

In all these electrodes, the common problem is how to control/optimize the mass and charge transport (i.e., electronic and ionic mobility), the electron-transfer kinetics in multi-phase boundaries, and the modifications of these processes, which occur upon application of a potential between the electrodes. Therefore, optimization of the performance requires the ability to control a complex reaction environment, where many kinetic aspects simultaneously concur in determining the performance.

Many of these controlling aspects are size dependent. Recently, it was recognized that not only the nano-dimension would be relevant, but also the nano-architecture of these materials is critical. In general, there is a design problem related to finding the optimal compromise between nanosize, nanoarchitecture, robustness and stability, and performances per unit weight or volume, together with a method of producing the electrode, which is cost-effective and scalable. The concept of *hierarchical organized materials* is a useful approach to reaching the above objectives with the development of a multilevel 3D organization based on a host macrostructure, which allows the right tridimensional organization necessary for a fast mass-transport, for example. On this host macrostructure, a secondary guest micro- and/or nano-scale sub-structure is built in order to take advantage of the properties of nanometer-sized building blocks and micron-sized assemblies.

Lithium-ion batteries (LIB) are rechargeable devices in which lithium ions move from the negative electrode to the positive electrode during discharge, and move back when charging [51–53]. Unlike Li primary batteries (which are disposable), Li-ion cells use an intercalated Li compound as the electrode material instead of metallic Li. LIBs are common in consumer electronics, because they have the best energy-to-weight ratios, no memory effect, and a slow loss of charge when not in use. Their relevance is increasing for military, electric vehicle, and aerospace applications, but the energy density storage should be further improved.

Traditional electrode materials for LIBs are based on micrometer-sized materials which have both mixed electron and ion transport (for Li^+), such as (i) layered metal oxides that have high redox potentials, and act as positive electrodes, and (ii) graphitic carbons capable of reversible uptake of Li at low potentials which act as negative electrodes. The use of nanostructured solid-state materials, due to the intimate nanoscale contact, allows one not only to increase the power density, but also to enhance Li reversibility and thus life cycle. In fact, the cycle ability of Li^+ ion transfer depends on the dimensional stability of the host material during insertion and deinsertion of Li^+ ions. Mechanical stress occurs during charge/discharge cycles, causing cracks, and finally loss of performance.

The necessary breakthrough to increase specific storage capacities relies, on the one hand, on an in-depth understanding of the complex interface processes of the charge-carrying chemical species (Li) during discharge and charge of the battery, and, on the other hand, on the development of synthetic procedures to prepare the nano-designed electrodes. Developing oriented nanostructures on a nanoscale is one of the directions of research actively being explored. Figure 8.10 exemplifies this concept showing how the control of the nanodimension and the nanoarchitecture of Li-ion batteries based on V_2O_5 supported over a structured nanocarbon material led to a very large improvement in the performance of the device (note that it is a logarithmic scale) [24]. This would reflect in LIBs having higher energy density, being lighter and more stable.

There are many advanced methods to prepare tailored hierarchical organized structures for electrodes. Ordered metal nanostructures with hierarchical porosity (macropores in combination with micro- or meso-pores) can be prepared using colloidal crystals (or artificial opals), that is, an ordered array of silica or polymer microspheres, as the template on which the metal particles are deposited. By careful removal of the template material, an ordered metal nanostructure may then be obtained. The extension of the concept of template synthesis for tailored nanomaterials is the synthesis of replica mesostructures by nanocasting. Nanocasting [54], using highly ordered mesoporous silica as a template, has led to incredible possibilities in the preparation of novel mesostructured materials, and to a great number of ordered nanowire arrays with:

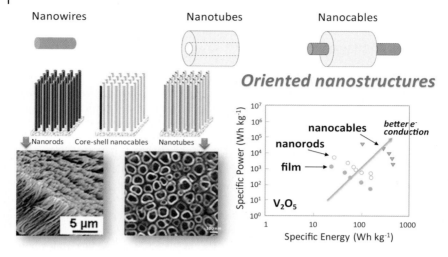

Figure 8.10 Example of the role of oriented nanostructures (vanadium oxide supported over nanocarbon materials) in improving the performances of LIBs; adapted from [24]. (Please find a color version of this figure in the color plates.)

- small diameter (<10 nm)
- large surface areas (up to 2500 m^2 g^{-1}) and uniform mesopores (1.5–10 nm) diameter
- tunable 2D or 3D mesostructures
- controlled morphology such as spheres, rods, films, and monoliths
- different components including carbons, metals, metal oxides and metal sulfides

The method was initially used in the preparation of ordered mesoporous carbon materials [55, 56]. Recent breakthroughs in the preparation of porous materials have resulted in the development of methods for the preparation of mesoporous carbon materials with extremely high surface areas and ordered mesostructures. Current syntheses can be categorized as either hard-template or soft-template methods, but the synthesis essentially involves C precursor infiltration in the template pores, its carbonization, and subsequent template removal. Consecutive surface functionalization of the carbon materials obtained, to give further optimization of their characteristics, is possible. The carbon replica material can then also be the template for other replica materials.

The nanocasting method can be used for the preparation of a variety of mesostructured and mesoporous materials, including mesostructured metal and semiconductor nanowires. Although powerful, the nanocasting method has limits in terms of cost and scale-up. The alternative approach is based on the synthesis of small-size particle units using the various available physico-chemical methods, such as colloidal, sol–gel and micelle methods, as well as other wet or gas-phase

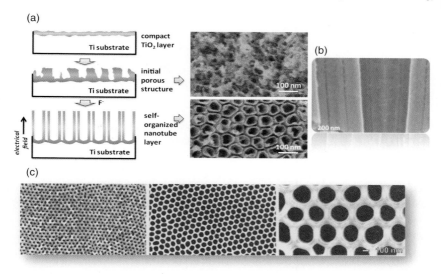

Figure 8.11 (a) Evolution of the TiO₂ morphology during AO of a Ti foil in the presence of fluoride ions, (b) example of the TiO₂ nanotube structure produced by AO in ethylene glycol electrolyte, (c) different alumina nanomembranes produced by AO; adapted from [25]. (Please find a color version of this figure in the color plates.)

procedures. These nanounits may then be organized in 2D and 3D super-structures by self- or directed-assembling, seeded or field-induced growth, epitaxial growth or other nano-structuring procedures. An interesting area with fast development is also that of assisting the self- assembling (in particular, to avoid a fractal-type growth) with an electric field. An example of these electrochemical methods is anodic oxidation. The electric field created at the interface between the electrode and electrolyte is a powerful factor for orientation of nanostructure growth, as shown for titania (TiO₂) electrodes [57].

An example is presented in Figure 8.11a, which summarizes schematically the mechanism of anodic oxidation (AO) of Ti foils to produce nanostructured titania thin films formed by an ordered array of titania nanotubes. Using organic electrolytes, well ordered and straight TiO₂ nanotubes with inner diameter of about 50 nm can be prepared (Figure 8.11b). The same methodology could be applied in the case of alumina, producing well-ordered alumina (Al₂O₃) nanomembranes with straight aligned nanochannels with diameter ranging from 30 to 300 nm, depending on the conditions of preparation (Figure 8.11c). These materials, besides their already interesting properties, are often also used for nanocasting procedures discussed before [58].

There are several other methods to prepare nanotailored materials for energy application, and the above discussion was not intended to discuss exhaustively the topic, but to give a glimpse of the possibilities. This is a very fast emerging area of chemistry S&T with high relevance to the production of new sustainable solutions for energy.

8.6
Biofuels

Producing fuels from biomass is and probably will remain more expensive than from oil, at least for an oil barrel costing below about 80–120 US$ (forecasts indicate that probably this will be the average future cost in the next decade, see http://www.eia.doe.gov, accessed 6 January 2011) and apart from special situations such as in Brazil (for ethanol). However, there are many motivations which push the development in this area, and the continuation of subsidies make the production of biofuels more economically attractive :

- biofuels reduce the carbon footprint of mobility which requires liquid fuels which can be integrated into the actual infrastructure (distribution, compatibility with actual fuels and engines, etc.); the alternative is to produce fuels from natural gas or coal (GTL and CTL processes), but these processes are also more expensive than fuels from oil

- the possible use of biomass wastes, with the double benefit of reducing their environmental impact and efficient valorization of unused resources

- the benefits for local agriculture and land preservation, with the creation of energy districts where biomass is used more efficiently through integration between the production of food, energy and raw materials for the chemical industry

- significant opportunity for creating and maintaining employment

- reduced dependence on fossil fuels and thus a positive impact on energy economy and geopolitical strategies

There are thus significant social motivations to push the production and use of biofuels. However, it should be considered that this is a short to medium term solution (for about the next 20 years) and which will realistically not go beyond about 10–15% of the market share of fuels for the transport sector (this accounts for about one third of the global energy consumption). It is always difficult to make a precise forecast in a sector like that of energy, which depends on many different factors. However, after the excitement of the last five years, many researchers and policy makers now agree that the contribution of biofuels in future energy scenarios will be limited, and that the strategies and technologies to produce biofuels should be reconsidered.

The introduction of biofuels had a significant impact on the energy market, particularly that related to fossil fuels, which was dominated by relatively few companies. One of the reasons was the need for very large investment. First generation biofuels (ethanol from sugar cane or corn, biodiesel from vegetable oils) do not require very large investment for the production plant (fixed capital investment (FCI)), particularly in the case of biodiesel. This has created the opportunity for a large number of new investors to enter the market, favoring a fast spreading of the technology. However, the raw materials for these first generation biofuels are in direct competition to their use as food or other valuable products (for example, many interesting and higher value oleochemical products could be produced from vegetable oils) [5b].

The second generation of biofuels is instead, in principle, not in competition with food, and in addition is characterized by a much larger use of the biomass [32–35]. Vegetable oils (palm oil, for example) constitute about 5% of the biomass, and it is thus necessary to be able to progressively make full use of the biomass. There are many possible routes for these second-generation biofuels (see later). Some are already close to commercialization (although often there are still many techno-economic aspects to solve), while other are at an earlier stage of development. Among the aspects often not fully taken into consideration, the FCI is an important one. The biofuels are significantly subsidized to be economically competitive and thus the interest of society over that of the companies is an important element for decision making. For a modern society, it is very important to create the conditions for true competition to avoid monopolization of the market, which will increase the costs. It is thus important for biofuels to incentivise the conditions and solutions favoring a large number of investors. In other words, solutions should be promoted (from the many possible) which can be efficient on a small to medium scale, and which do not require very large investment.

From a sustainable perspective (a main driver for biofuels is to reduce the carbon footprint), large investment in an energy market where it is difficult to make an exact forecast, will be limited and will take a long time for a decision to be reached. Thus technologies requiring smaller investment and with a broader base of possible investors will be introduced onto the market much more quickly. In other words, society will benefit more quickly from the introduction of the new technologies [8].

Thus the field of biofuels, due to the many relevant socio-economic implications, cannot be evaluated with the conventional techno-economic parameters, and the societal implications and return in terms of sustainability and competiveness have to be included. This is the perspective to use in discussing the second-generation biofuels.

There are several areas of chemistry S&T critical for the development of new processes for biofuels. Among them we can cite as an example the chemical industry and chemical engineering processes, the development of enzymes and white biotechnology, catalysis and new materials, but the list is long and involves essentially all areas of chemistry.

Biofuels currently used are those defined as first-generation biofuels, since their production is from raw materials such as sugar cane and corn (starch crops) for bioethanol, and vegetable oils for biodiesel. The energy use of these raw materials is in direct competition with their food use, not an acceptable sustainable perspective. There has been much public discussion on the negative effect on their costs as food, particularly for poor countries, and the consequent social tensions. We should remark that speculation was the main cause of the peaks in the cost of starch crops, but there are clearly several interconnected aspects to consider. For example, increased agricultural prices will give incentives for farmers to stray away from producing other less profitable grains, causing a shift in the crop production structure, leading to a decrease in agricultural diversity, subsequently diverting food away from the human food chain. In general, it is reasonable that shifts in crop production and the changes in the world price of agricultural commodities due to the expansion of the biofuel market will have global impacts on consumers. However, most of the studies consider full substitution of fossil fuels with biofuels, while current forecasts indicate a much lower impact (10–20% at the maximum for the transport sector, that is, 3–6% of the total energy consumption). The effective impact of biofuels, together with the shift from first to second-generation biofuels, will thus be much less than that initially supposed. At the same time, particularly for several developing countries, it will be a great opportunity to use natural resources and act as a stimulus for the quality of life in rural areas. For this reason, international organizations such as UNIDO (United Nations Industrial Development Organization) are promoting the use of biofuels in developing countries [59]. Biofuels play an important role in realizing UNIDO's objectives and addressing different *Millennium Development Goals*:

- biofuels for electricity, heat and transport are key options to ensure access to energy, especially for rural areas;
- biofuels provide an opportunity to promote the establishment of local small businesses, and thereby create economic revenue, and employment;
- biofuels contribute to diversifying energy supply, and hedging risks of impacts from global fossil fuel price increases;
- biofuels provide a means for reducing greenhouse gas emissions.

Therefore, biofuels are an important factor to promote the economy and quality of life in developing countries by enhancing access to energy, diversifying fuel supply sources, creating income opportunities in rural areas, reducing greenhouse gas emissions, and promoting sustainable industrial conversion processes and productive uses of biofuels.

As for any industrial change, it is important to anticipate the possible impact on the environment and sustainability of biofuels, and the twin relationship with the socio-economic context [33]. The market growth of these raw materials has led in the tropics to a change in land use from forest to crop (for example, reduction of tropical forests in Malaysia to increase palm plantations to produce palm oil for biodiesel use). This has caused a negative effect rather than a positive one on CO_2 emissions, and on local ecosystems. When marginal land is available, or when biomass residues can be used (the actual amount can cover about one third of the forecast biofuel production), the carbon footprint is positive (e.g., there is a net saving on energy, considering the full life cycle assessment (LCA)), but the case is different when forests are substituted by plantations, as in tropical countries.

Life cycle assessment and carbon footprint

Life Cycle Assessment(LCA) is the investigation "cradle-to-grave" and evaluation of the environmental impacts of a given product or service caused by or deriving from its existence. The goal of LCA is to compare the full range of environmental and social damage related to products and services, to be able to choose the least burdensome one.

A carbon footprint is the total set of greenhouse gas (GHG) emissions caused by an organization, event or product. It is often expressed in terms of the amount of carbon dioxide, or its equivalent of other GHGs, emitted.

The carbon footprint is a subset of the ecological footprint and of the more comprehensive LCA.

The second generation of biofuels will be based on lignocellulosic raw materials or algae- and micro-algae. The former may result from different sources, such as dedicated energy cultures (plants and herbs), agro-food production waste, and municipal residuals. The current processes for bioethanol can be adapted to use certain types of lignocellulosic biomass (especially, high growth rate dedicated energy crops) by introducing a pre-processing stage. This is an expensive process step, which accounts for over one third of the entire production cost. There are alternatives to lignocellulosic bioethanol, and an active R&D area concerns the growing of algae and microalgae, a source of lipids, which can be converted to biodiesel after harvesting and appropriate treatments.

There is intense research in various other production routes to obtain hydrocarbons from biomass because the energy density of hydrocarbons is greater than that of alcohols such as ethanol [60]. C-efficiency and energy density are two important elements to consider in biofuels, as well as the integration with the current energy infrastructure. The fermentation of sugars to ethanol leads to the formation of two molecules of CO_2 for every molecule of glucose converted (glucose is the unit building block of the cellulosic component of lignocellulosic biomass). Alternative fuels, maintaining as much as possible the integrity of the molecule, allow improvement in the C- and energy-efficiency. While in glucose fermentation to ethanol (EtOH) only four of the sixth carbon atoms of glucose are present in the fuel, the production of 2,5-dimethylfuran (DMF) from glucose maintains all the six initial carbon atoms in the fuel molecule. Consequently, DMF has an intrinsic energy content about 40% higher than EtOH, which is reflected in lower CO_2 emissions per km when used in the car. The analysis is more complex, because these two fuels are produced in different ways and H_2 is needed in the DMF route (production of H_2 also implies CO_2 emissions, even though it may be produced from the biomass itself). However, the example introduces the concept that hydrocarbon biofuels (such as DMF) are preferable over the alcohol fermentation route in terms of C-efficiency and other aspects (compatibility with current engines and infrastructure, etc.). There are thus alternatives to bioethanol (first and second generation).

Catalysis and a bio-based economy

A bio-based economy defines the new production scheme based on the use of biomass as raw material instead of fossil fuels as currently. It defines not only the production of biofuels, that is, fuels derived from biomass, but also the production of materials and chemicals starting from biomass. Most of the processes to enable this bio-based economy are catalytic, similar to over 90% of the current refinery and petrochemistry processes, that is, those based on fossil fuels. Biorefineries are the equivalent of current oil refineries, but a main difference is that for the higher complexity of converting biomass it is necessary to integrate the production of fuels, chemicals and materials.

There are three main primary approaches to biomass conversion, namely: (i) thermochemical, (ii) biochemical and (iii) chemical catalytic. The thermochemical approach consists in the pyrolytic treatment of biomass to produce solid, liquid or gaseous products that can be subsequently upgraded to fuels (synthetic biofuels). These primary treatment methods produce intermediates that should be first purified, often in multiple steps, and then further upgraded to fuels or chemicals through catalytic treatments, for example, hydroprocessing, cracking, steam reforming, methanation, FT synthesis, and so on. In principle, all types of biomass can be used for the thermochemical treatment.

The biochemical approach is that currently most used. The first-generation processes start from biomass in competion with food, such as sugarcane and corn grain, while second-generation processes use ligno-cellulose materials. The process is based on the enzymatic hydrolysis of cellulose and hemicellulose (after pretreatment to increase the rate of reaction), while lignin is not converted by enzymes and can be combusted. Its conversion using new advanced biochemical methods is gaining attention. Following the hydrolysis of cellulose and hemicellulose to hexoses and pentoses, further enzymatic transformation (fermentation) produces mainly alcohols. For example, bioethanol is produced via fermentation of C5–C6 sugars. Using different enzymes, other products (butanol, for example) can also be synthesized.

In the chemical or chemical catalytic approach, the cellulosic biomass undergoes catalytic hydrolysis with acids. The acid is used either in aqueous solution (e.g., diluted sulfuric acid, commercial method) or as a heterogeneous phase (e.g., solid catalysts and ionic liquids) to lower the impact on the environment. Another example of primary chemical catalytic treatment is the transesterification of triglycerides, that is, the technology used in the production of first-generation biodiesel from vegetable oils. In the case of transesterification of waste oils or algae oils the biodiesel obtained is referred to as a next generation biofuel. Chemical catalytic routes are then necessary also for conversion of the intermediates (platform molecules) formed in the primary treatment steps to synthetic fuels and biohydrogen.

Other important aspects to consider are the energy needed for the process or the stages of pretreatment, for separation, and effluent treatment (the various options differ significantly from this point of view). In general, biofuels have about one order of magnitude higher impact on water than current fuels. This is an issue scarcely addressed, but critical for eco-sustainability. However, the different routes do not have the same impact on water. In general, fermentation routes produce the largest volumes of effluents (wastewater) to be treated.

The size necessary for efficient biofuel processes is another key element for evaluation. In fact, it will determine the distance from which it is necessary to transport the biomass to allow the constant feed of the plant (e.g., the penalty in CO_2 emissions due to transport of the feedstock to the plant), and other relevant aspects for sustainability such as the investment costs, the possibility of location and so on. Flexibility in feed, for example, the possibility to use multiple types of raw materials, is another key parameter. The possibility to use different lignocellulosic sources (agro-food and wood production residues, sorted municipal solid wastes, herbaceous energy crops) makes it possible to:

1) Limit the distance from which the raw materials are transported,

2) guarantee more constant use throughout the year, avoiding the storage of large volumes of biomass (and related problems of fermentation, emission of odors and wastewater, etc.),

3) integrate better and more efficiently in agro-food districts.

There are thus several parameters determining the optimal choice and these are significantly dependent on local conditions. An optimum technology for one country may not necessarily be valid in a different country. Due to the presence of multiple aspects, advanced analytical techniques such as LCA, though necessary, are not sufficient. Each country should develop its own preferred mix of technologies, or at least invest significantly in research and development to enable the optimal choices that best suit the country needs.

Figure 8.12 reports a general outline of various alternatives in the conversion of biomass to biofuels [61]. It is possible to identify four different types of approach:

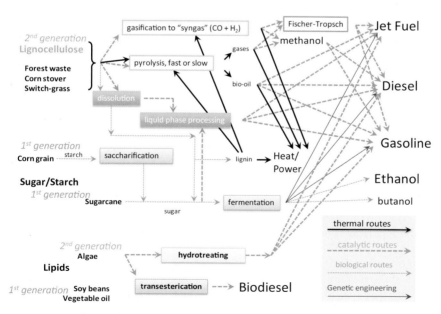

Figure 8.12 Biofuel production alternatives; adapted from [61]. (Please find a color version of this figure in the color plates.)

1) thermal processes
2) catalytic processes
3) enzymatic processes
4) new processes based on genetic engineering.

The stages of development are different, as are their efficiency, adaptability to various types of biomass, production complexity, influence on the environment, and so on. An in-depth discussion of the various alternatives is beyond the scope of this chapter. Some further insight can be found elsewhere [22, 23, 33, 34, 60–63], but it is necessary to remark that sometimes the comparison has been made from a too limited perspective, as commented above. On the other hand, it is also necessary to remark that the energy problem is so complex that it is necessary to have a broad range of possible options, each of which is best suited to specific cases.

8.7
Towards Solar Fuels

The need for solar energy to meet a renewable energy scenario can be clarified from the following data. If the irradiance on only 1% of the Earth's surface could be converted into electric energy with 10% efficiency (e.g., the average efficiency of current photovoltaic (PV) cells), it would provide a resource base of 105 TW, for example, ten times the estimated world energy increase by the year 2050 with respect to the current value. By contrast, the estimated amount of energy extractable by wind is about 2-4 TW, by tides about 2–3 TW, by biomass 5–7 TW and by geothermal energy 3–6 TW [64]. Therefore, the future (long-term) energy scenario should be based on a significant contribution from solar energy, but the question is: in which form?

The world energy consumption was about 500 EJ in 2009 [6]. Only about half of this energy was used in services and the remainder was lost as low temperature heat, or other unusable forms. For example, only about 15–20% of the input energy (as fuel) in cars is effectively used for motion and the remainder is rejected energy, such as low temperature heat, friction, air resistance, and so on, due to the low efficiency of chemical to mechanical conversion (engine). There are thus ample margins to save energy (for example, new low friction tires, new nanostructured paints, new thermoelectric devices based on the Peltier effect to transform heat directly to electrical energy, and so on), but the significant step could occur only by changing completely the technology of engines. This is a very slow process due to the very large investment necessary. For the next 30 years, therefore, the technology in cars will still be based on the present type of engines, with some improvements (hybrid cars, etc.). The transport sector is actually using about one third of the fuel production, while the remaining two thirds are related to household, tertiary (heating, electrical energy) and industry (industrial production, etc.) uses. Electrical energy accounts for about one third of the energy consumption.

Very large progress has been made in PV cells, for example, those able to convert solar energy directly into electrical energy. PV cells are already close to grid parity [65], while only a decade ago the cost was one order of magnitude higher. In other words, today it is already economically valuable to produce electrical energy using PV cells, when economic incentives (subsidies) exist. The conventional PV technology now makes it possible to obtain very high conversion yields, up to 40% for prototype multi-junction devices, although commercially available solar cells ensure 12–18% yields [66]. However, the photovoltaic technology currently available, based on p–n junctions of inorganic semiconductors, still has high production costs due to the cost of materials (the same as used for the production of microprocessors), the high purity of these and the high energy content of the materials (which are produced at high temperatures and in ultrahigh vacuum conditions). These characteristics result in a still high cost per watt of peak power produced by PV technology, although there is a rapid decrease in costs. There are also some other constraints in silicon solar panels related to their architectural integration. They are dark in color, rigid and should be positioned at 45° to the incidence of sunlight for optimum efficiency.

The maximum theoretical efficiency of a single solar cell is limited to 33% [67], but this limit could be overcome using multiple cells, with varying band gaps, in a serial (tandem) arrangement. The first generation of PV cells was based upon silicon wafers, in which material costs accounted for over 70% of the total cost, and overall efficiency was around 20%. Thin-film techniques were used in the second generation of PV cells. The primary benefit is a reduction in production costs at the expense of an efficiency limited to between 5 and 10%. So-called third-generation PV cells seek to achieve efficiencies greater than 50% [68, 69], with the hope of approaching the thermodynamic limit of 93%, without radically changing the manufacturing process of second-generation cells. If manufacturing costs can be kept comparable to those of thin film cells, and efficiency greatly improved, then the overall cost per kWh can be dramatically decreased.

The operating principle of a PV cell is simple. When a photon with energy higher than the band gap of the absorber material hits a p-n junction, it creates an electron–hole pair and, due to the inbuilt electric field, photo-produced carriers are transported to the electrodes. Thus for making a solar cell, the basic requirements are an absorber material and a junction with an electric field that assists the transport of the photo-generated carriers to the electrodes. There is also a probability that photo-generated carriers can recombine before reaching the electrodes. Therefore, for crystalline and polycrystalline solar cells, the charge separation time of photogenerated carriers must be smaller than the minority carrier lifetime.

The energy associated with photons greater than the energy band gap of silicon is converted into heat. This is one of the reasons for the limit in the efficiency. In tandem solar cells [70–72], already used for manufacturing amorphous silicon solar cells and III–V-based high efficiency concentrator solar cells, materials with different band-gaps are used to improve this efficiency. In multiple-junction solar cells, several different materials are used to reach efficiencies up to 40–42%, but with exponentially increasing costs.

A different approach is based on the use of nanostructures to build new, highly efficient, solar cells, based on the following concept:

- Hot carriers generated by absorbing photons of energy more than the energy band gap can be employed to generate additional current and or voltage,
- energy states in the energy band gap can be used to trap carriers generated from photons of energy lower than the energy band gap.

Various approaches based on the one-, two-, and three-dimensional nanostructures (quantum dots and semiconductor nanocrystals) are currently being investigated [73]. For example, well-ordered arrays of nanopillars in Si, to be used as the inorganic semiconductor-phase in bulk-heterojunction PV cells, could be prepared by a modification of the nanocasting method described previously [74]. The method could be extended to develop nanocrystal-based tandem solar cells, based also on other semiconductors. It was also shown recently that it is possible to fabricate highly ordered nanopillar-based solar-cell modules on flexible substrates using techniques compatible with roll-to-roll processing, which potentially brings down costs [75]. It is thus a rapidly expanding area, which is driven by the increasing knowledge in nanotechnology and the chemistry of nanomaterials.

A different approach in third-generation solar cells is that of the dye-sensitized solar cells (DSCs, often called Grätzel-type cells) [76–78] and of organic and conductive polymers, in particular to develop ultra-low-cost solar cells. DSCs utilize three main components to perform photo-electro conversion: a metal oxide semiconductor (TiO_2), a dye and an electrolyte. A dye is used to create electron–hole pairs and the other two materials transport these photogenerated charge carriers to the electrodes. The best efficiency reported is about 11%, and a number of technical problems still have to be solved, in terms of electrolyte, stability, sealing, and so on. However, there are a number of aspects that make these materials attractive, besides the potential to decrease the cost of production: lightness, transparency and flexibility, the possibility to deposit on any surface (including fabrics), the possibility to produce in different colors by varying the chemical nature of the dye sensitizer. These PV devices thus have high potential in terms of adaptability and integration with the environment and architecture, for example, the creation of PV windows or roofs in historic centers. The long-term stability, however, has to be improved.

Producing renewable electrical energy using PV cells is thus attractive and rapidly expanding, but is only part of the energy problem. There is a peak in energy consumption during the day, when PV cells have the maximum power efficiency. PV cells could thus efficiently integrate the extra energy required during these peaks in consumption, when also the cost of production of energy is higher. Nevertheless, there is a mismatch between power consumption and solar irradiation during the year and in terms of geographical distribution. Current methods to store electrical energy are not very efficient, and, in general, a still critical issue is the storage of electrical energy, notwithstanding the R&D progress in nanomaterials for these applications. Therefore, the conversion of solar to chemical energy is a necessary integration to current solar options for storing and transporting solar energy.

PEM fuel cells and electrolysers

Proton exchange membrane (PEM) fuel cells are types of fuel cell developed especially for transport applications and portable devices. Their distinguishing features include lower temperature/ pressure ranges (50 to 100°C) and a special polymer electrolyte membrane. The water electrolyzer is the reverse of a PEM fuel cell. It produces gaseous hydrogen and oxygen from water. Electrolyzer technology may be implemented on a variety of scales. There are two main types of electrolyzer (alkaline and proton-exchange-membrane) and both are well proven and long-lived. Modern electrolyzers usually have high conversion efficiency (up to 90%) and yield very high purity oxygen and hydrogen. The principal drawback of electrolyzers is the high unit cost.

PV cells may be integrated with water electrolyzers to produce H_2 (and O_2) which can be then used in PEM fuel cells to produce electrical energy again, when it is needed [79]. The efficiency of the system is low (2–6%), but optimizing the system (by matching the voltage and maximum power output of the PV device to the operating voltage of PEM electrolyzers) it is possible to increase the efficiency to 10–12% [80]. This solar powered PV-PEM electrolyzer could supply enough hydrogen to operate a fuel cell vehicle. However, the possible further improvement of the efficiency and decrease in the costs are limited. In addition, the problem of efficient H_2 storage is still not solved. As hydrogen is a gas, the density per volume is low. It is thus necessary either to liquefy H_2 (however, this requires very low temperatures, about −250°C at 1 atm) or to use extremely high pressures. Intense research has been carried out on H_2-storage materials (metal hydrides, amine borane complexes, MOF and other nanostructured materials, etc.) [30, 31, 51], but the prospect of reaching the necessary densities now appears remote. Therefore, the use of alternative energy vectors to H_2 is now of increasing interest [10].

The energy density (per unit volume or weight) of liquid fuels (hydrocarbons and oxygenated compounds such as ethanol) is by far larger larger than that possible for H_2 and for electrical energy. The use of H_2 as a clean energy vector has thus the main drawbacks in transport/storage with respect to liquid fuels. In addition, realistic energy scenarios should consider, besides sustainability, the continuity in usage of existing technologies (and infrastructure) wherever possible. Switching to H_2 as energy vector (the so-called "hydrogen economy") [81–83] would require such large investment in infrastructure as to make unlikely this solution as a primary future energy vector, although more niche applications will be possible. Other energy vector alternatives, such as ammonia, have limits in terms of safety and toxicity, even if there are some interesting characteristics [2]. However, a renewable source of H_2 could be used to convert CO_2 back to liquid fuels, which can then be easily stored/transported and will integrate into the existing energy infrastructure. Liquid products such as methanol (the so-called "methanol economy") [84] or better longer-chain alcohols or hydrocarbons [85], are certainly preferable options as energy vectors in terms of energy density, low toxicity, easy and safe storage/transport, and, especially, capability of integration into the existing energy infrastructure for both mobile and stationary applications. These *C-based energy vectors* could be proposed as the preferred and more sustainable options towards solar fuels [10]. This solution also provides a viable solution to greenhouse gas emissions and climate change [86].

Syngas

Syngas (from synthesis gas) indicates a gas mixture containing varying amounts of carbon monoxide (CO) and hydrogen (H_2). It is typically produced by steam or authothermal reforming (or catalytic partial oxidation) of natural gas or liquid hydrocarbons, but could be produced also by gasification of coal, biomass and in some types of waste-to-energy gasification facilities. Syngas can be converted to methanol (CH_3OH) or higher alcohols, dimethyl ether (DME) or synthetic hydrocarbons via the Fischer–Tropsch process. Methanol or DME can also be converted to hydrocarbons via the methanol to gasoline (MTG) process. Methanol is also an intermediate for various other petrochemistry products, and in the energy sector is used for producing octane-booster compounds (MTBE) and in vegetable oil transesterification to biodiesel.

There are different possibilities in converting CO_2 back to fuels [39]. The most investigated area is the hydrogenation of CO_2 to form oxygenates and/or hydrocarbons. Methanol synthesis from CO_2 and H_2 has been investigated up to pilot-plant stage with promising results. The alternative possibility is the production of DME, a clean-burning fuel that is a potential diesel substitute. Ethanol formation, either directly or via methanol homologation, or the conversion of CO_2 to formic acid are also potentially interesting routes. Methanol, ethanol, and formic acid may be used as feedstock in fuel cells, providing a route to store energy from CO_2 and then produce electricity. Alcohols are, in principle, preferable to hydrocarbons because their synthesis requires less hydrogen per unit of product. In fact, the key problem in this route is the availability of H_2. If the latter is produced from hydrocarbons (the main current route is by steam reforming of methane) there are no real advantages in converting CO_2. H_2 must thus be derived from renewable sources. The possible options are the following:

- *Water electrolysis* coupled with a renewable source of electrical energy (photovoltaic cells, wind or waves, etc.) (see also previous discussion). This technology is already available, but the need for multiple steps, the overpotential in the electrolyzer and other issues limit the overall efficiency. The technology is mature with a limited degree of further possible improvements. Although, in principle, this system could be adapted also to the electrochemical conversion of CO_2, there are many technical problems (low productivity, deactivation, type of products and recovery from the solution) making this route not very effective.

- Thermal water splitting using *concentrated solar energy* [87–90]. Concentrated solar radiation (in solar high-temperature furnaces) could be used to produce H_2 (and O_2) from water or CO (and O_2) from CO_2. Thermochemical cycles are necessary to lower the temperatures necessary. An example is the use of a metal oxide, which spontaneously reduces at high temperature. The reduced metal oxide is then reoxidized by interaction with H_2O to form H_2 or with CO_2 to produce CO. Temperatures above 1200–1400 °C are necessary in the spontaneous reduction step, while they are much lower (in the 700–800 °C range) for the reoxidation step. This creates a number of issues in terms of materials and stability, cost-effectiveness and productivity. The syn gas (CO/H_2) should then be catalytically upgraded to fuels (methanol, Fischer–Tropsch hydrocarbons). The approach is essentially suitable for solar plants, while it may be difficult to adapt in a delocalized production of solar fuels. In addition, scale-up problems are relevant.

- A variation of the concept is to use the solar concentrators and reactors to drive the CO_2 reforming with methane (solar dry reforming of methane) [91] to produce syngas (mixtures of CO/H_2), which can then be converted to methanol and FT products. The advantage is the possibility of continuous operation, instead of cyclic, while the disadvantage is the need for a methane feed. The solar illumination in this case provides the heat necessary for the endothermic dry reforming of methane with CO_2. The main issues are related to the difficulty of having a homogeneous temperature in the monolith, problems of materials stability, and of carbon formation. Scale-up of the solar system to larger production of solar fuels remains an issue.

- Therefore, although the potential gain in energy efficiency of solar reformers over the conventional catalytic process to generate syngas or H_2 is attractive, the low productivity and limited scale economy are the main critical issues, together with the problem of materials. The feasibility of concentrated solar power (CSP) for producing solar fuels was proven, but not the stability of operation or its economics. The expansion of the market for CSP (mainly to produce electrical energy) will probably be an incentive in future, as also their use in producing solar fuels.

- *Biomass conversion*, preferably using waste materials and in conditions which require low energy consumption [92]. An example is the catalytic production of H_2 directly in the liquid phase from aqueous solutions (for example, ethanol waste streams). This option could be a way for valorization of side waste streams in a biorefinery, but it is not an efficient way if considered alone. In fact, if we consider the whole life cycle from growing the plant, harvesting, fermentation, and so on, and finally H_2 production (from bioethanol, for example), the overall energy consumption (and thus amount of CO_2 produced) is higher than the advantage in the hydrogenation of CO_2 back to fuels.

- *Production of H_2 via biogas* produced by anaerobic biomass fermentation. In this case, it could be a valuable option using waste biomass, but it is a quite complex process considering the whole production chain. There are also problems of purification of biogas.

- Production of H_2 using *cyanobacteria or green algae* [93]. This is an interesting option, but still with low productivity, and it is currently under development. Using algae, the efficiency in using solar light is high (around 10%) and CO_2 (from power plants) could even be fed directly to the photobioreactors or open ponds. However, the process of producing biofuels from the (micro)algae is quite complex as well as there being critical problems with the controlled growing of the algae and, in general, the cost of the process. Biofuels from microalgae are considered the third-generation biofuels, after the second-generation processes based on biofuels produced from lignocellulosic materials. However, from a conceptual point of view, direct routes of solar energy use for conversion of CO_2 to fuels which do not pass through the creation of complex molecules (cellulose, hemicellulose, lignine, starch, lipids, oils, etc.), which then need to be further deconstructed, are preferable.

- Direct H_2 production by *water photoelectrolysis* [94–96], which suffers still from low productivity and, in some cases, the need for further separation/recovery of hydrogen. In this low-temperature approach, solar energy is used by a suitable semiconductor to generate, by charge separation, electrons and holes, which further react with water to generate H_2 and O_2. Many studies have been dedicated recently to the splitting of water on semiconductor catalysts under solar irradiation. Remarkable progress has been made since the pioneering work by Fujishima and Honda, but the development of photocatalysts with improved efficiencies still faces major challenges. Recent efforts have focused on new materials and synthesis methods for efficient photocatalysts. Good quantum efficiencies (>50%) have been obtained with ultraviolet light, but the use of visible light still poses major problems.

In water photoelectrolysis, H_2 and O_2 are produced in the same place and this creates two main issues:

1) H_2 and O_2 must be then separated for use, and this is an energy-intensive process, moreover, the H_2 and O_2 mixture is explosive in a large range of conditions.

2) These gases may react back together to water and quenching the photoactivity.

In order to solve these issues, a photoelectrochemical (PEC) approach is necessary [97]. The PEC path to water splitting involves the separation of the oxidation and reduction processes into half-cell reactions. Different approaches are possible. One of them is shown schematically in Figure 8.13, which could be considered an evolution of a *solar fuel cell* [98]. The design and characteristics are similar to those of PEM fuel cells, with the anode and cathode separated by a proton-conducting membrane (for example, Nafion®, but other membranes could be used). The electrodes, in the form of a thin film, are deposited over a porous conductive

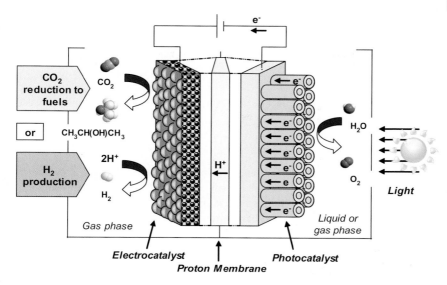

Figure 8.13 Scheme of the PEC device for the CO_2 reduction to fuels using solar energy and the H_2 production; adapted from [98]. (Please find a color version of this figure in the color plates.)

substrate, which allows the efficient collection/transport of the electrons over the entire film as well the diffusion of protons to/from the membrane. This solar fuel cell may be used either for water splitting with physically separated formation of H_2 and O_2, or for solar conversion of CO_2 to fuels. The advantage is that, in the latter case, the electrocatalytic reduction of CO_2 occurs in the gas phase and thus the recovery of the products of reaction occurs by simple condensation from the gas phase in mild operational conditions. However, up to now, only the proof-of-the-concept has been demonstrated, and thus still large effort would be necessary before implementation.

Producing solar fuels is a topic of current high scientific and industrial interest. We have discussed here some of the possible options to produce solar fuels from water and CO_2, but we need to remark that most of them are still far from possible application. Nevertheless, they are part of the R&D activities necessary in going to a more sustainable energy scenario based on renewable energy. From this perspective, it is necessary to evidence the need to produce fuels easy to transport and store, which can be integrated into the existing energy infrastructure. Liquid fuels produced from carbon dioxide and water using solar energy are preferable, notwithstanding the difficulties in the reaction.

While the stage of development of solar fuels is definitely less advanced compared with solar to electrical energy conversion, the latter is only part of the future sustainable energy scenario. There is thus an urgent need to invest in R&D on solar fuels. Direct bio-routes and concentrated solar thermal are certainly valuable and interesting options, but we suggest that low temperature conversions using

PEC solar cells, due to the lower investment cost and easier delocalization on the territory would be a preferable option. However, it must be remarked that the problem of sustainable energy is complex and, therefore, there is not only one single preferable solution, but different technologies are necessary for the various cases.

8.8
Conclusions

Chemistry, catalysis and chemical engineering have played, and will continue to play, a pivotal role in enabling societal needs of energy [99, 100], in terms not only of amount and quality of the products, but also of protection of the environment from their use. We have discussed here only a few examples due to the complexity of the topic. They give a glimpse of the holistic role of chemistry. Mass mobility was enabled by the availability of quality fuels meeting the increasing demands for performance of the cars. Chemistry was also able to find solutions to mitigate the negative effects of this mass mobility in terms of its impact on the environment. We have briefly discussed the case of elimination of sulfur from fuels with also a brief consideration of catalytic converters. With these devices, it is possible to eliminate over 95% of pollutants from gasoline cars. This means that without these systems, we would have to cut by a factor of at least 20 the number of circulating vehicles. This gives a feeling of how chemistry (and catalysis in this specific case) has contributed to make possible a societal need such as mass mobility. Chemistry has also allowed reduction in the weight of the vehicles (thus limiting CO_2 emissions and fuel consumption), made cars more confortable, reduced the noise, and so on.

The new challenge is to reduce the carbon footprint related to mobility. We have discussed this aspect in relation to some of the possibilities for improving the efficiency in energy use by cars, and the effort towards producing renewable fuels. Biofuels are an ongoing transition, although the preferred solution from the many possible routes (which were outlined schematically) is still under debate. We have evidenced, however, the relevance, which should be given in this analysis to the social return from the different options.

The next challenge is the transition from fossil to solar energy, which requires not only new materials to increase the effectiveness of using renewable energy (we discussed briefly the cases of wind turbines and PV cells, but more examples are possible). We noted the need to find viable solutions to storing the primary energy in chemical carriers, due to the need to match seasonal and geographic disparities with the demand structure for energy. This requirement is independent of the existence of an energy distribution grid but may require different solutions for grid-based and stand-alone energy supply structures.

Chemistry is thus a strategic science and technology for solving the energy challenge, providing tailored structural and functional materials, processes and technologies for energy saving, for a better and more efficient use of natural resources by diversification of those actually in use, and for the use of renewable energy. This will also require improving the educational effort in training chemists for the interdisciplinary approach needed to provide creative and scalable solutions to the energy issues.

References

1 Reddy, A.K.N. (2000) Energy and social issues, in *World Energy Assessment: Energy and the Challenge of Sustainability* (ed. J Goldemberg), United Nations, New York, Ch. 2, pp. 40–60.

2 Orecchini, F. (2006) The era of energy vectors. *Int. J. Hydrogen Energy*, **31** (14), 1951–1954.

3 Schlögl, R. (2010) The role of chemistry in the energy challenge. *ChemSusChem*, **3** (2), 209–222.

4 Garcia-Martinez, J. (ed.) (2010) *Nanotechnology for the Energy Challenge*, Wiley-VCH Verlag GmbH, Weinheim, Germany.

5 (a) Barbaro, P. and Bianchini, C. (eds) (2009) *Catalysis for Sustainable Energy Production*, Wiley-VCH Verlag GmbH, Weinheim, Germany; (b) Centi, G., and van Santen, R.A. (2007) *Catalysis for Renewables*, Wiley-VCH Verlag GmbH, Weinheim, Germany.

6 IEA (2009) *World Energy Outlooks 2009*, International Energy Agency, Paris, France.

7 Ozren, O. (2005) *Oil Refineries in the 21st Century*, Wiley-VCH Verlag GmbH, Weinheim, Germany.

8 Cavani, F., Centi, G., Perathoner, S., and Trifiró, F. (2009) *Sustainable Industrial Chemistry*, Wiley-VCH Verlag GmbH, Weinheim, Germany.

9 Robinson, P.R. (2006) Petroleum processing overview, in *Practical Advances in Petroleum Processing* (eds C.S. Hsu and P.R. Robinson), Springer, Heidelberg, Germany, pp. 1–31.

10 Centi, G. and Perathoner, S. (2010) Towards solar fuels from water and CO_2. *ChemSusChem*, **3** (2), 195–208.

11 (a) Centi, G. and Perathoner, S. (2009) Catalysis: role and challenges for a sustainable energy. *Top. Catal.*, **52** (8), 948–961; (b) Centi, G. and Perathoner,

S. (2008) Catalysis, a driver for sustainability and societal challenges. *Catal. Today*, **138** (1–2), 69–76.

12 European Technology Platform for Sustainable Chemistry (6 January 2011) http://www.suschem.org.

13 Smart Energy Home (SEH) Initiative (6 January 2011) http://www.smartenergyhome.eu.

14 Rao, C.N.R., Mller, A., and Cheetham, A.K. (eds) (2007) *Nanomaterials Chemistry: Recent Developments and New Directions*, Wiley-VCH Verlag GmbH, Weinheim, Germany.

15 Vielstich, W. (ed.) (2009) *Handbook of Fuel Cells*, Wiley-VCH Verlag GmbH, Weinheim, Germany.

16 Höök, M., and Aleklett, K. (2009) A review on coal-to-liquid fuels and its coal consumption. *Int. J. Energy Res.*, **34**, 848–864.

17 Hanawa, M., Onozaki, M., and Mochida, I. (2009) Technology vision and projects of synthetic liquid fuel production. *J. Jpn. Inst. Energy*, **88** (6), 473–479.

18 Goodenough, J.B., and Kim, Y. (2010) Challenges for rechargeable Li batteries. *Chem. Mater.*, **22** (3), 587–560.

19 Bruce, P.G., Scrosati, B., and Tarascon, J.-M. (2008) Nanomaterials for rechargeable lithium batteries. *Angew. Chem. Int. Ed.*, **47** (16), 2930–2946.

20 (a) Cocks, F.H. (2009) *Energy Demand and Climate Change: Issues and Resolutions*, Wiley-VCH Verlag GmbH, Weinheim, Germany; (b) Soga, T. (2006) *Nanostructured Materials for Solar Energy Conversion*, Elsevier Science, Amsterdam, The Netherlands.

21 (a) Nocera, D.G. (2006) The global energy future: the challenge for science in the 21st century. *Daedalus*, **135**, 112–115; (b) Eisenberg, R. and Nocera, D.G. (2005) Preface: overview of the

forum on solar and renewable energy. *Inorg. Chem.*, **44**, 6799–6801.

22 Stöcker, M. (2008) Biofuels and biomass-to-liquid fuels in the biorefinery: catalytic conversion of lignocellulosic biomass using porous materials. *Angew. Chem. Int. Ed.*, **47** (48), 9200–9211.

23 Langeveld, H., Sanders, J., and Meeusen, M. (2009) *The Biobased Economy: Biofuels, Materials and Chemicals in the Post-Oil Era*, Earthscan Ltd, London, UK.

24 Liu, J., Cao, G., Yang, Z., Wang, D., Dubois, D., Zhou, X., Graff, G.L., Pederson, L.R., and Zhang, J.G. (2008) Oriented nanostructures for energy conversion and storage. *ChemSusChem*, **1** (8–9), 676–697.

25 Centi, G. and Perathoner, S. (2009) The role of nanostructure in improving the performance of electrodes for energy storage and conversion. *Eur. J. Inorg. Chem.*, **26**, 3851–3878.

26 Krüger, P. (2009) Nanotechnologies: turning small ideas into something big. Presented at Workshop New Technologies: The Future, Dec 8 2009, Brussels, Belgium.

27 Chu, H., Wei, L., Cui, R., Wang, J., and Li, Y. (2010) Carbon nanotubes combined with inorganic nanomaterials: preparations and applications. *Coord. Chem. Rev.*, **254** (9–10), 1117–1134.

28 Centi, G. and Perathoner, S. (2010) Problems and perspectives in nanostructured carbon-based electrodes for clean and sustainable energy. *Catal. Today*, **150** (1–2), 151–162.

29 Su, D.S. and Schlögl, R. (2010) Nanostructured carbon and carbon nanocomposites for electrochemical energy storage applications. *ChemSusChem*, **3** (2), 136–168.

30 Züttel, A., Borgschulte, A., and Schlapbach, L. (2008) *Hydrogen as a Future Energy Carrier*, Wiley-VCH Verlag GmbH, Weinheim, Germany.

31 Serrano, E., Rus, G., and García-Martínez, J. (2009) Nanotechnology for sustainable energy. *Renew. Sust. Energy Rev.*, **13** (9), 2373–2384.

32 Naik, S.N., Goud, V.V., Rout, P.K., and Dalai, A.K. (2010) Production of first and second generation biofuels: a comprehensive review. *Renew. Sust. Energy Rev.*, **14** (2), 578–597.

33 Zinoviev, S., Müller-Länger, F., Das, P., Bertero, N., Fornasiero, P., Kaltschmitt, M., Centi, G., and Miertus, S. (2010) Next generation biofuels: survey of emerging technologies and sustainability issues. *ChemSusChem.*, **3**, 1106–1133.

34 Kamm, B., Gruber, P.R., and Kamm, M. (2008) *Biorefineries-Industrial Processes and Products*, Wiley-VCH Verlag GmbH, Weinheim, Germany.

35 Hermann, B.G., Blok, K., and Patel, M.K. (2007) Producing bio-based bulk chemicals using industrial biotechnology saves energy and combats climate change. *Environ. Sci. Technol.*, **41** (22), 7915–7921.

36 Omer, A.M. (2008) Focus on low carbon technologies: the positive solution. *Renew. Sust. Energy Rev.*, **12** (9), 2331–2357.

37 Nielsen, S.K. and Karlsson, K. (2007) Energy scenarios: a review of methods, uses and suggestions for improvement. *Int. J. Glob. Energy Issues*, **27** (3), 302–322.

38 Jiang, Z., Xiao, T., Kuznetsov, V.L., and Edwards, P.P. (2010) Turning carbon dioxide into fuel. *Phil. Trans. R. Soc. LOndon, Ser. A*, **368** (1923), 3343–3364.

39 Centi, G. and Perathoner, S. (2009) Opportunities and prospects in the chemical recycling of carbon dioxide to fuels. *Catal. Today*, **148**, 191–205.

40 Rand, D.A.J., and Dell, R.M. (2007) *Hydrogen Energy Challenges and Prospects*, RSC Energy Series, Royal Chemical Society, Cambridge, UK.

41 Zecchina, A. (2007) Nanoscience and catalysis. Presented at VI INSTM Conference, Perugia (Italy), 12–15 June 2007.

42 Krüger, P. (2009) Nanotechnology along the value chain – benefits and challenges. Presented at NanoConvention, Zurich, 06 July 2009.

43 Fujikawa, T. (2010) Hydrodesulfurization catalyst and

process of vacuum gas oil. *J. Jpn. Inst. Energy*, **89** (3), 212–217.

44 Skyum, L. and Topsøe, H. (2009) Catching up on catalysts. *Hydrocarbon Eng.*, **14** (6), 48–50.

45 (a) Temel, B., Tuxen, A.K., Kibsgaard, J., Topsøe, N.-Y., Hinnemann, B., Knudsen, K.G., Topsøe, H., Lauritsen, J.V., and Besenbacher, F. (2010) Atomic-scale insight into the origin of pyridine inhibition of MoS2-based hydrotreating catalysts. *J. Catal.*, **271** (2), 280–289; (b) Besenbacher, F., Brorson, M., Clausen, B.S., Helveg, S., Hinnemann, B., Kibsgaard, J., Lauritsen, J.V., Moses, P.G., Nørskov, J.K., and Topsøe, H. (2008) Recent STM, DFT and HAADF-STEM studies of sulfide-based hydrotreating catalysts: Insight into mechanistic, structural and particle size effects. *Catal. Today*, **130** (1), 86–96.

46 Eftekhari, A. (2008) *Nanostructured Materials in Electrochemistry*, Wiley-VCH Verlag GmbH, Weinheim, Germany.

47 Rolison, D.R., Long, J.W., Lytle, J.C., Fischer, A.E., Rhodes, C.P., McEvoy, T.M., Bourg, M.E., and Lubers, A.M. (2009) Multifunctional 3D nanoarchitectures for energy storage and conversion. *Chem. Soc. Rev.*, **38**, 226–252.

48 Yi, C., Liu, D., and Yang, M. (2009) Building nanoscale architectures by directed synthesis and self-assembly. *Curr. Nanosci.*, **5** (1), 75–87.

49 Wallace, G.G., Chen, J., Mozer, A.J., Forsyth, M., MacFarlane, D.R., and Wang, C. (2009) Nanoelectrodes: energy conversion and storage. *Mater. Today*, **12** (6), 20–27.

50 Aricò, A.S., Bruce, P., Scrosati, B., Tarascon, J.-M., and van Schalkwijk, W. (2005) Nanostructured materials for advanced energy conversion and storage devices. *Nat. Mater.*, **4**, 366–377.

51 Liu, C., Li, F., Lai-Peng, M., and Cheng, H.-M. (2010) Advanced materials for energy storage. *Adv. Mater.*, **22** (8), E28–E62.

52 Fergus, J.W. (2010) Recent developments in cathode materials for lithium ion batteries. *J. Power Sources*, **195** (4), 939–954.

53 Ozawa, K. (ed.) (2009) *Lithium Ion Rechargeable Batteries: Materials, Technology, and New Applications*, Wiley-VCH Verlag GmbH, Weinheim, Germany.

54 Lu, A.-H. and Schüth, F. (2006) Nanocasting: a versatile strategy for creating nanostructured porous materials. *Adv. Mater.*, **18** (14), 1793–1805.

55 Liang, C., Li, Z., and Dai, S. (2008) Mesoporous carbon materials: synthesis and modification. *Angew. Chem. Int. Ed.*, **47** (20), 3696–3371.

56 Ryoo, R., Joo, S.H., Kruk, M., and Jaroniec, M. (2001) Ordered mesoporous carbons. *Adv. Mater.*, **13** (9), 677–681.

57 (a) Grimes, C.A. (2009) *Mor GK, TiO₂ Nanotube Arrays: Synthesis, Properties, and Applications*, Springer, Heidelberg, Germany; (b) Shankar, K., Basham, I., Allam, N.K., Varghese, O.K., Mor, G.K., Feng, X., Paulose, M., Seabold, A., Choi, K.-S., and Grimes, C.A. (2009) Recent advances in the use of TiO_2 nanotube and nanowire arrays for oxidative photoelectrochemistry. *J. Phys. Chem. C*, **113** (16), 6327–6359.

58 Piao, Y., Lim, H., Chang, J.Y., Lee, W.-Y., and Kim, H. (2005) Nanostructured materials prepared by use of ordered porous alumina membranes. *Electrochim. Acta*, **50** (15), 2997–3013.

59 United Nations Industrial Development Organization (2009) Bioenergy Strategy: Sustainable Industrial Conversion and Productive Use of Bioenergy, http://www.unido.org/fileadmin/media/documents/pdf/Energy_Environment/rre_bioenergyStrategy_latest.pdf (accessed 7 January 2011).

60 Huber, G.W. (ed.) (2008) *Breaking the Chemical and Engineering Barriers to Lignocellulosic Biofuels: Next Generation Hydrocarbon Biorefineries*, National Science Foundation, Washington, DC.

61 (a) Regalbuto, J.R. (2010) Next generation hydrocarbon biofuels. Presented at COST Strategic Initiative Workshop Sustainable Fuels and Chemicals, Oostende (Belgiu, 27–28 April 2010); (b) Regalbuto, J.R. (2010) An NSF perspective on next generation hydrocarbon biorefineries. *Comput. Chem. Eng.*, doi:10.1016/j. compchemeng.2010.02.025.

62 Gallezot, P. (2008) Catalytic conversion of biomass: challenges and issues. *ChemSusChem*, **1** (8–9), 734–737.

63 Huber, G.W., Iborra, S., and Corma, A. (2006) Synthesis of transportation fuels from biomass: chemistry, catalysts, and engineering. *Chem. Rev.*, **106** (9), 4044–4098.

64 Lewis, N.S., Crabtree, G., Nozik, A., Wasielewski, M., and Alivisatos, P. (2005) *Basic Research Needs for Solar Energy Utilization*, US Department of Energy, Washington, DC.

65 Bhandari, R. and Stadler, I. (2009) Grid parity analysis of solar photovoltaic systems in Germany using experience curves. *Solar Energy*, **83** (9), 1634–1644.

66 Ginley, D., Green, M.A., and Collins, R. (2008) Solar energy conversion toward 1 terawatt. *MRS Bull.*, **33** (4), 355–364.

67 Shockley, W. and Queisser, H.J. (1961) Detailed balance limit of efficiency of p-n junction solar cells. *J. Appl. Phys.*, **32**, 510–519.

68 Konagai, M. and Kurokawa, Y. (2008) High-efficiency novel solar cells – present status of third generation photovoltaics. *J. Jpn. Inst. Energy*, **87** (3), 193–198.

69 Conibeer, G. (2007) Third-generation photovoltaics. *Mater. Today*, **10** (11), 42–50.

70 Gaudian, R. (2010) Third-generation photovoltaic technology – the potential for low-cost solar energy conversion. *J. Phys. Chem. Lett.*, **1** (7), 1288–1289.

71 Tanabe, K. (2009) A review of ultrahigh efficiency III-V semiconductor compound solar cells: multijunction tandem, lower dimensional, photonic up/down conversion and plasmonic nanometallic structures. *Energies*, **2** (3), 504–530.

72 Yamaguchi, M., Takamoto, T., and Araki, K. (2005) III-V compound multi-junction and concentrator solar cells. *Adv. Solar Energy*, **16**, 97–12.

73 Ameri, T., Dennler, G., Lungenschmied, C., and Brabec, C.J. (2009) Organic tandem solar cells: a

review. *Energy Environ. Sci.*, **2** (4), 347–363.

74 Conibeer, G., Green, M., Corkish, R., Cho, Y., Cho, E.-C., Jiang, C.-W., Fangsuwannarak, T., Pink, E., Huang, Y., and Puzzer, T. (2006) Silicon nanostructures for third generation photovoltaic solar cells. *Thin Solid Films*, **511–512**, 654–662.

75 Fan, Z., Ruebusch, D.J., Rathore, A.A., Kapadia, R., Ergen, O., Leu, P.W., and Javey, A. (2009) Challenges and prospects of nanopillar-based solar cells. *Nano Res.*, **2**, 829–843.

76 Fan, Z., Razavi, H., Do, J.-W., Moriwaki, A., Ergen, O., Chueh, Y.-L., Leu, P.W., Ho, J.C., Takahashi, T., Reichertz, L.A., Neale, S., Yu, K., Wu, M., Ager, J.W., and Javey, A. (2009) Three-dimensional nanopillar-array photovoltaics on low-cost and flexible substrates. *Nat. Mater.*, **8**, 648–653.

77 Grätzel, M. (2007) Photovoltaic and photoelectrochemical conversion of solar energy. *Phil. Trans. R. Soc. Ser. A*, **365** (1853), 993–1005.

78 Lin, H., Li, X., Liu, Y., and Li, J. (2009) Progresses in dye-sensitized solar cells. *Mater. Sci. Eng. B Solid-State Mater. Adv. Technol.*, **161** (1–3), 2–7.

79 Li, B., Wang, L., Kang, B., Wang, P., and Qiu, Y. (2006) Review of recent progress in solid-state dye-sensitized solar cells. *Solar Energy Mater. Solar Cells*, **90** (5), 549–573.

80 Tributsch, H. (2008) Photovoltaic hydrogen generation. *Int. J. Hydrogen Energy*, **33** (21), 5911–5930.

81 Gibson, T.L., and Kelly, N.A. (2008) Optimization of solar powered hydrogen production using photovoltaic electrolysis devices. *Int. J. Hydrogen Energy*, **33** (21), 5931–5940.

82 Zhang, Y.-H.P. (2009) A sweet out-of-the-box solution to the hydrogen economy: is the sugar-powered car science fiction? *Energy Environ. Sci.*, **2** (3), 272–282.

83 Farrauto, R.J. (2009) Building the hydrogen economy. *Hydrocarbon Eng.*, **14** (2), 25–26.

84 Neef, H.-J. (2008) International overview of hydrogen and fuel cell research. *Energy (Oxford, UK)*, **34** (3), 327–333.

85 Olah, G.A., Goeppert, A., and Surya Prakash, G.K. (2009) *Beyond Oil and Gas: The Methanol Economy*, 2nd edn, Wiley-VCH Verlag GmbH, Weinheim, Germany.

86 Centi, G., Perathoner, S., Winè, G., and Gangeri, M. (2007) Electrocatalytic conversion of co_2 to long carbon-chain hydrocarbons. *Green Chem.*, **9** (6), 671–678.

87 Mikkelsen, M., Jørgensen, M., and Krebs, F.C. (2010) The teraton challenge. A review of fixation and transformation of carbon dioxide. *Energy Environ. Sci.*, **3** (1), 43–81.

88 Tamaura, Y. (2009) Solar hydrogen production from concentrated solar thermal energy using reactive ceramics. *J. Jpn. Inst. Energy*, **88** (5), 391–395.

89 Kodama, T. and Gokon, N. (2007) Thermochemical cycles for high-temperature solar hydrogen production. *Chem. Rev.*, **107** (10), 4048–4077.

90 Abanades, S., Charvin, P., Flamant, G., and Neveu, P. (2006) Screening of water-splitting thermochemical cycles potentially attractive for hydrogen production by concentrated solar energy. *Energy (Oxford, UK)*, **31** (14), 2805–2822.

91 Steinfeld, A. (2005) Solar thermochemical production of hydrogen – a review. *Solar Energy*, **78**, 603–615.

92 Klein, H.H., Karni, J., and Rubin, R. (2009) Dry methane reforming without a metal catalyst in a directly irradiated solar particle reactor. *J. Solar Energy Eng.*, **131**, 021001.1–02100114.

93 Tanksale, A., Beltramini, J.N., and Lu, G.M.A. (2010) Review of catalytic hydrogen production processes from biomass. *Renew. Sust. Energy Rev.*, **14** (1), 166–182.

94 Lee, H.-S., Vermaas, W.F.J., and Rittmann, B.E. (2010) Biological hydrogen production: prospects and challenges. *Trends Biotechnol.*, **28** (5), 262–271.

95 Navarro, R.M., Sánchez-Sánchez, M.C., Alvarez-Galvan, M.C., Valle, F.D., and Fierro, J.L.G. (2009) Hydrogen production from renewable sources: biomass and photocatalytic opportunities. *Energy Environ. Sci.*, **2** (1), 35–54.

96 Woodhouse, M. and Parkinson, B.A. (2009) Combinatorial approaches for the identification and optimization of oxide semiconductors for efficient solar photoelectrolysis. *Chem. Soc. Rev.*, **38** (1), 197–210.

97 Serpone, N., Emeline, A.V., and Horikoshi, S. (2009) Photocatalysis and solar energy conversion (chemical aspects). *Photochemistry*, **37**, 300–361.

98 Arakawa, H. (2009) Solar hydrogen production by photoelectrochemical cell composed of semiconductor photoelectrode. *J. Jpn. Inst. Energy*, **88** (5), 405–412.

99 Ampelli, C., Centi, G., Passalacqua, R., and Perathoner, R. (2010) Synthesis of solar fuels by a novel photoelectrocatalytic approach. *Energy Environ. Sci.*, **3** (3), 292–301.

100 Vlachos, D.G., and Caratzoulas, S. (2009) The roles of catalysis and reaction engineering in overcoming the energy and the environment crisis. *Chem. Eng. Sci.*, **65** (1), 18–29.

9
Ozone Depletion and Climate Change
Glenn Carver

9.1
Introduction

One of the most notable events in atmospheric chemistry in recent decades has been the discovery of the Antarctic Ozone Hole: a significant reduction in the stratospheric ozone layer over Antarctica during the Southern Hemisphere spring. It caused concern for life on the planet since stratospheric ozone protects biological life from the harmful portion of sunlight. Not only did the discovery of the Antarctic Ozone Hole lead to rapid developments in our understanding of atmospheric chemistry, it also captured the public interest in the impact of humans on the environment, fostered international collaboration between scientists and determined global policy on the manufacture of the chemicals responsible in a way not seen before. The story of its discovery is a fascinating one, not just in scientific terms but in human terms as well. It was not discovered by the latest technology of the time, but instead by the earliest invented instrument for measuring ozone in the atmosphere. And over the subsequent years it formed careers for many scientists, some who went on to receive Nobel Prizes, many international awards, and, for Joe Farman, the scientist who led the team that discovered it, the Order of the British Empire.

This chapter looks at the role of ozone in the atmosphere and presents the story of the Ozone Hole and ozone depletion in general. The role of ozone and the Ozone Hole in climate change is also very relevant and fortuitous in some ways as the reader will discover. Yet ozone at the Earth's surface is a pollutant and can be harmful to life, for example causing respiratory problems and damaging plants. It is perhaps ironic that ozone can harm us but at the same time we could not live without it. Although climate change and ozone depletion are largely separate problems they do influence each other. It is impossible to predict what will happen to ozone in the atmosphere without considering the effect of global warming.

The year this chapter was written coincides with the 25th anniversary of the discovery of the ozone hole.[1] It also coincides with the next publication of the

---

1) See: http://www.antarctica.ac.uk/press/press_releases/press_release.php?id=1192

The Chemical Element: Chemistry's Contribution to Our Global Future, First Edition.
Edited by Javier Garcia-Martinez, Elena Serrano-Torregrosa
© 2011 Wiley-VCH Verlag GmbH & Co. KGaA. Published 2011 by Wiley-VCH Verlag GmbH & Co. KGaA.

definitive quadrennial scientific assessment on ozone depletion by the World Meteorological Organisation (WMO) and the United Nations Environment Programme (UNEP); a volume which summarizes the collective knowledge of ozone and its relation to climate change (see resources at the end of this chapter). Notably, it is also the year that many of the ozone depleting substances (CFCs and their replacements) are to be phased out completely.

Ozone (O_3) is a triatomic form or "allotrope" of atomic oxygen. It is a strongly oxidizing gas and slightly blueish in color. It is a "bent" molecule with a permanent dipole moment. Its dipole property allows it to absorb and emit electromagnetic radiation [1]. Ozone is considered a greenhouse gas equivalent to 25% the warming effect of carbon dioxide in the troposphere. Ozone (Image from Wikipedia)

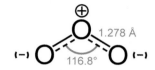

9.2
Ozone in the Atmosphere

Ozone, O_3, is a gas that occurs naturally in the atmosphere. Before describing ozone in more detail, it is important to understand something about the structure of the atmosphere [2] as the role of ozone in the atmosphere changes with height.

Figure 9.1 illustrates the various layers of the Earth's atmosphere. The first distinct layer from the surface up is called the "troposphere". It is this layer that

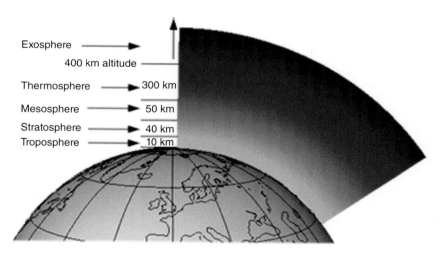

Figure 9.1   The basic structure of the Earth's atmosphere. Provided by the Center for Atmospheric Science, University of Cambridge, UK.

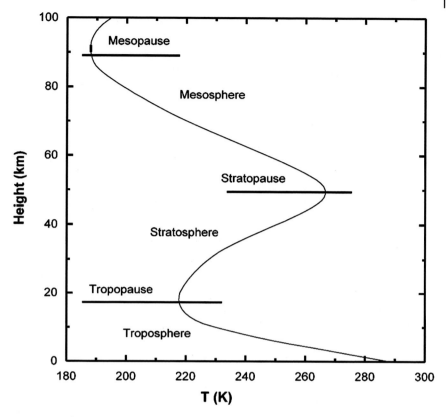

Figure 9.2 Change in temperature with altitude above the Earth's surface up to 100 km. Adapted from the US standard atmosphere.

we live in and the one that gives us our weather. At the equator, the height of the troposphere is approximately 10 km but this reduces towards the poles. For comparison, the height of Mt. Everest is about 8.8 km. Over 90% of the mass of the atmosphere lies between the surface and 10 km. Above the troposphere lies the stratosphere, a layer approximately 40 km thick between approximately 15–50 km altitude. The atmosphere behaves quite differently in this layer to that of the troposphere. In particular, the temperature in the stratosphere increases with height rather than decreases as in the troposphere (see Figure 9.2). This increase in temperature is largely due to heating by absorption of solar ultraviolet radiation by the "ozone layer". The stratosphere is heated from above, unlike the troposphere which is heated from below (by the Earth's surface).

Most of the ozone in the atmosphere lies in a layer between 10 and 50 km above the Earth's surface. The peak in ozone concentration can be found at about 30 km with the largest amounts of ozone found over the polar regions and with less over the tropics. The total amount of ozone in the atmosphere is, however, very small.

If all the ozone in a column above the Earth's surface was compressed to the pressure of 1 atm uniformly, it would form a layer just 3 mm thick. Nitrogen and oxygen compressed the same way would give layers 6 km and 2 km thick, respectively.

Although this chapter will discuss ozone in the stratosphere it is worth briefly describing the effects of ozone in the troposphere; the part of the atmosphere in which we live. Ozone near the surface is a pollutant [3] and forms when reactions involving hydrocarbons and nitrogen oxides (produced by car exhausts for example) are initiated with sunlight. Ozone levels are monitored in urban areas as exposure to high levels can cause respiratory problems and harm the lining of the lungs, leading to increased risks for asthma sufferers. Elevated ozone levels are also harmful to plant life as ozone interferes with photosynthesis [1]. As ozone has a lifetime of approximately several weeks in the troposphere (it is not the same in different layers of the atmosphere), it can be produced in urban areas but carried by air currents to non-urban areas.

It is well known that an important role of ozone is filtering harmful solar ultraviolet (UV) radiation. Living cells are damaged by solar radiation at wavelengths shorter than about 300 nm and little solar radiation of this wavelength and shorter is incident at the Earth's surface. The human eye is sensitive to light wavelengths of about 400–700 nm. Oxygen (O_2), is effective at filtering solar UV wavelengths of 230 nm or less [1] but ozone is the only gas that filters solar UV in the range 230–300 nm. Without the formation of ozone in our atmosphere, life on earth in the form we know it would not be able to exist.

9.2.1
Chapman Reactions

Ozone is formed in the stratosphere by a process in which the bond in molecular oxygen is broken by absorption of solar UV at wavelengths below 240 nm; a process known as photolysis or photodissociation. The atomic oxygen produced quickly combines with another molecule of oxygen to form ozone:

$$O_2 + h\nu \rightarrow O + O \tag{9.1}$$

$$O + O_2 + M \rightarrow O_3 + M \tag{9.2}$$

In Eq. (9.1), $1/\nu$ represents a wavelength <240 nm and h is the Planck constant. $h\nu$ then represents one photon at a wavelength of sufficient energy to break the molecular bond. M in Eq. (9.2) represents any other molecule and is unchanged by the reaction. M is needed for two reasons: to conserve angular momentum in the reaction and to quench any excited molecules produced. Otherwise the ozone formed would rapidly dissociate back to atomic and molecular oxygen. It should be noted that the formation of ozone in this way depends on the presence of oxygen, which itself depends on biological activity. This synergy between ozone and life is one of the unique characteristics of the Earth's atmosphere [4]. Life started in the Earth's oceans because the top layers of water filtered the harmful

UV but still allowed enough sunlight to penetrate. As oxygen was released into the atmosphere, atmospheric ozone formed and began filtering UV, allowing life to progress and develop elsewhere.

Ozone is itself destroyed by photolysis [5]:

$$O_3 + hv \rightarrow O + O_2 \tag{9.3}$$

but can reform through Eq. (9.2). Ozone is also destroyed by reaction with atomic oxygen:

$$O + O_3 \rightarrow O_2 + O_2 \tag{9.4}$$

These reactions collectively are known as the *Chapman reactions* after Chapman in 1930 [6] offered them as the first explanation of the ozone layer based purely on oxygen photochemistry. Equations (9.2) and (9.3) are rapid and interconvert ozone and atomic oxygen. Both are dependent on pressure; Eq. (9.2) becomes slower with increasing altitude as the pressure reduces, while Eq. (9.3) becomes faster with height as the amount of solar UV at the right wavelengths increases. The concentration of ozone in the stratosphere is, therefore, largely a balance between these reactions, and the layer of ozone so formed in the stratosphere is sometimes called the Chapman layer [1].

Comparison of observations of ozone and the ozone layer predicted by the Chapman reactions show there are notable differences, even when taking into account the variation in solar intensity with latitude, temperature and so on. Photolysis rates will be greatest at the equator yet the maximum in ozone is towards polar regions. There is also a marked asymmetry in the distribution of stratospheric ozone between the two hemispheres. This suggests that the motion of the air plays an important role in the distribution of ozone, something first noted by a scientist named Dobson [7]. Another factor was that the Chapman reactions do not fully describe the chemical reactions controlling ozone in the atmosphere.

9.2.2
Catalytic Cycles

By the early 1960s, improvements in atmospheric observations and laboratory measurements of reaction rates, led to the realization that Eq. (9.4) was too slow to be solely responsible for the destruction of ozone. Other chemical mechanisms were required: we now know that Eq. (9.4) only accounts for about 25% of the chemical destruction of ozone in the stratosphere. Catalytic cycles were proposed in which ozone could be destroyed through a series of chemical reactions involving another chemical species but without a significant change or loss of that species. A simple catalytic cycle, enabled by reaction with a species "X" would be:

$$X + O_3 \rightarrow XO + O_2 \tag{9.5}$$

$$XO + O \rightarrow X + O_2 \tag{9.6}$$

$$\text{Net:} \quad O + O_3 \rightarrow O_2 + O_2 \tag{9.7}$$

The net of these reactions Eq. (9.7) is the same as Eq. (9.4). In order for this to be efficient both Eqs. (9.5) and (9.6) must be exothermic, which places constraints on the molecules that "X" could be. Suitable species in the stratosphere are oxides of nitrogen and hydrogen (OH, HO_2, NO and NO_2), chlorine (Cl), and bromine (Br). Catalytic cycles involving some of these species will be discussed later when the theory behind the ozone hole is discussed. The interested reader can find more detailed descriptions of catalytic cycles and their influence on stratospheric ozone in the texts given as references at the end of this chapter.

9.2.3
How Ozone is Measured

Ozone is measured in the atmosphere using a variety of techniques. Remote methods from the ground or onboard satellites use optical methods. Measurements *in situ* (in the air immediately around the instrument) can be made using chemical and optical means on aircraft, balloons or even rockets.

G.M.B. Dobson was the scientist at the University of Oxford, UK, who in the 1920s developed the first instrument, the "Dobson spectrometer" to measure ozone in the atmosphere remotely: this instrument is still used today. It works by measuring the intensity of solar UV at four wavelengths; two of which are absorbed by ozone, two that are not. In honor of the pioneering work by Dobson, the amount of ozone in a column above the Earth is often quoted in "Dobson Units" (DU). As previously stated, if the ozone in a column of air was compressed to 1 atm pressure it would give a layer 3 mm thick, or 300 DU where a DU is 0.01 mm.

An *in situ* method for determining ozone concentrations at altitudes up to about 30 km is with small and lightweight ozonesondes carried aloft on balloons (Figure 9.3). Electrochemical sondes were first developed in the 1960s but with the advent

Figure 9.3 Preparations for a balloon launch. The large balloon is carrying an ozonesonde and other instruments to measure several chemical species as it ascends through the atmosphere. The smaller balloon is used to lift the cable, between the large balloon and payload clear of the ground so the payload does not get dragged as the large balloon rises. The bundles of large black cylinders hold the helium gas used to fill both balloons. (Please find a color version of this figure in the color plates.)

of the ozone hole regular ozone profiles have been taken from an extensive global network [8].

Satellites were proposed in the early 1960s as a means of mapping global ozone column amounts and test instruments were launched soon after. The total ozone mapping spectrometer (TOMS) is perhaps the most well known as it was one of the first to provide ozone column amounts over the entire sunlit part of the globe each day. The first TOMS instrument was launched in 1978 on the Nimbus 7 satellite [9]. A number of other TOMS instruments and others for measuring ozone and other species of interest have been launched since. A more detailed description of how these instruments work is outside the scope of this chapter but the interested reader will find a great deal of information online; a number of starting points on the internet are given in the list of resources at the end.

9.3
The Antarctic Ozone Hole

Chloroflurocarbons (CFCs) were invented in the 1930s by General Motors, chiefly as a replacement for refrigerants. They had the useful properties of being inert and insoluble and became used extensively in industry, for example in refrigeration, air conditioners, aerosols, solvents and packaging. They are essentially carbon compounds with hydrogen atoms replaced by chlorine or fluorine and the two most common were CFC-11 ($CFCl_3$) and CFC-12 (CF_2Cl_2). Inert in the troposphere they can be transported into the stratosphere to be photolyzed by shorter solar wavelengths:

$$CF_2Cl_2 + h\nu \rightarrow CF_2Cl + Cl \tag{9.8}$$

thus providing a stratospheric source of chlorine. Scientists had been concerned for a while about the possibility of catalytic cycles involving nitrogen [10] and chlorine [11] impacting on stratospheric ozone, but focus was on nitrogen oxides from high altitude aircraft and chlorine from volcanoes and the Space Shuttle. A landmark scientific paper by Molina and Rowland [12] in 1974 was the first to make the link between chlorine from CFCs and destruction of stratospheric ozone.[2] Using gas-phase chemistry they found only the upper stratospheric ozone would be depleted at a slow but increasing rate. With the increasing emissions of CFCs this was potentially a serious problem for the future and the United Nations Environment Programme (UNEP) hosted talks that led to the Vienna Convention for the Protection of the Ozone Layer. Even before the discovery of the ozone hole, the threat posed to stratospheric ozone from CFCs was starting to develop, though the Vienna convention did not require limits on production of ozone-depleting chemicals.

---

2) It was this work that earned the pair, together with Professor Paul Crutzen for his related work, the 1995 Nobel Prize in Chemistry.

What's in a name?

Manmade emitted compounds that contain carbon and chlorine are known as chlorocarbons. Compounds that also include fluorine are known as chlorofluorocarbons (or CFCs). Where a hydrogen atom is bonded to a carbon atom these compounds are known as hydrochlorocarbons (HCFCs). Hydrofluorocarbons (HFCs) are another replacement for CFCs. The short form of these compounds is constructed by taking the general abbreviation: CFC-*vwxyza*, where:

v = number of double bonds (omitted if zero).

w = number of carbons atoms minus 1.
x = number of hydrogen atoms plus 1.
y = number of fluorine atoms.
z = number of chlorine atoms replaced by bromine.
a = letter (a, b or c) identifying isomers.

For example,
CFC-12, CCl_2F_2 – dichlorodifluoromethane.
HCFC-22, $CHClF_2$ – chlorodifluromethane.

A team of scientists, Joe Farman, Brian Gardiner and Jonathan Shanklin, from the British Antarctic Survey, Cambridge, UK, were based at Halley station in Antarctica (Figures 9.4 and 9.5) taking measurements of ozone using a Dobson spectrometer. As would become apparent later, they were in a perfect geographical location in Antarctica to observe the rapid decline in ozone associated with the ozone hole as it developed above them. Ironically, Farman and colleagues were doing ground truth comparisons for the recently launched satellites, yet had observed significant decreases in ozone during the Antarctic spring (October) that

Figure 9.4 Aerial view of Antarctica at low sun. Although a region of superb natural beauty it was and still is also a region of significant scientific interest. Photograph courtesy of P. Bucktrout, British Antarctic Survey, Cambridge, UK. (Please find a color version of this figure in the color plates.)

Figure 9.5 Halley station in Antarctica where the Dobson spectrometer is housed. Photograph courtesy of C. Gilbert, British Antarctic Survey, Cambridge, UK. (Please find a color version of this figure in the color plates.)

had not been reported by anyone else, particularly the satellite data scientists. They were concerned their measurements were real but had the advantage over satellite data that their records of ozone at Halley went back to the 1950s (Figure 9.6). Nevertheless there was, at first, a reluctance to publish their results. The instrument might have been at fault, but confirmation came from a second instrument based on what was then called the Argentine Islands station (later to be known as Faraday and now Vernadsky) [13] that also saw decreasing ozone during southern hemisphere spring. The BAS team also wrote to the NASA satellite team to ask if they had seen the decrease. There was, however, no reply; the decreases in ozone had gone unnoticed. This was in part due to the large amounts of data being received from the monitoring satellites (TOMS and SBUV) that the technology of the time was dealing with; it took nearly two years to process the data before scientists could study it. The early processing software simply flagged the unexpectedly low values as bad data.

Some two years after the decrease in ozone was first observed the scientific paper by Farman *et al.* [14] was finally published. It was to become a landmark paper that changed much. It took scientists in new directions, made careers for some and brought a much increased awareness of human impact on the environment to policy-makers and the world in general. This was clearly a global problem. Confirmation of the decrease came quickly from a long term ozone column record at a Japanese station (not so well sited compared to Halley so the data was less clear) and NASA confirmed the TOMS satellite had seen low ozone in 1986 [15].

Figure 9.7 shows the dramatic change in ozone from August to October; August shows normal values yet several months later virtually all of the ozone is removed in a layer some 3–4 km thick over an area the size of Antarctica.

As a scientist it was an exciting time. This was completely unexpected, existing gas-phase theory did not explain such a large ozone decrease; something unknown

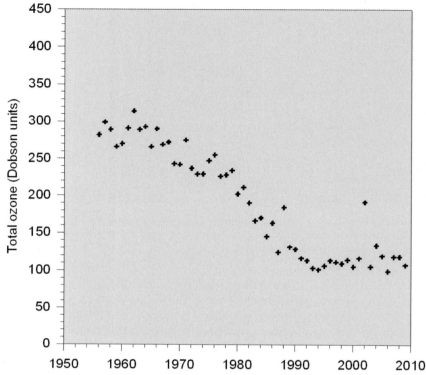

Figure 9.6 Minimum values of column ozone at Halley station, British Antarctic Survey, from 1950s to present day in Dobson units. The onset of the ozone hole in the late 1960s / early 1970s is apparent. This coincided with a measured increase in CFCs in the stratosphere. Courtesy of J. Shanklin, British Antarctic Survey, UK.

was responsible. Possible mechanisms were put forward in the scientific literature, both in terms of chemical processes and changes in the transport of air. Field campaigns took place to measure more chemical species and, crucially, measurements of small particles called aerosols and stratospheric cloud particles were made. Solomon and colleagues [16] were among the first to point out that existing gas-phase theories of ozone depletion could not explain the ozone hole and showed that "heterogeneous" reactions (i.e., those between chemical species that take place on a surface of some kind) on "polar stratospheric clouds" (PSCs) were likely to be responsible. Not a lot was known then about PSCs and their composition. It is now known that there are various forms and the exact processes of reaction on their surfaces is still not completely understood. If temperatures fall low enough, below −85 °C, then ice particles can form. Smaller particles can form at −80 °C but these are mainly composed of nitric acid trihydrate ($HNO_3 \cdot 3H_2O$). As these form at warmer temperatures this extends the duration of chemical processing on their surfaces and enlarges the areal extent over which conversion can take place.

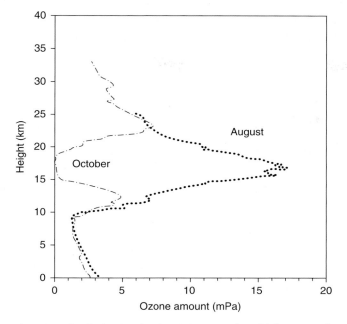

Figure 9.7 Graph showing the change in ozone above Halley station from August to October for present day. Note how the ozone is almost completely removed in a layer from 15 to 18 km. Figure courtesy of J. Shanklin, British Antarctic Survey, UK.

Various theories were proposed for the formation of the ozone hole but it was 1987 before the science could provide an explanation that accounted for the large decrease in ozone. Molina and Molina [17] showed how chlorine released from CFCs in the upper stratosphere in the form of chlorine monoxide (ClO) could react with itself. Fundamental laboratory experiments were carried out to measure these new reaction rates.

9.3.1
The Steps to the Ozone Hole

We now understand the steps in the formation of the ozone hole (see Figure 9.8). The CFCs are transported into the stratosphere, some 10 years or more after they are emitted at the surface and are then broken down by solar UV. The shorter UV wavelengths required to break down the CFCs are not present in the troposphere due to the filtering effect of ozone. The main long-lived species that form from the breakdown products are hydrochloric acid (HCl) and chlorine nitrate (ClONO$_2$). In order to produce more reactive forms of chlorine (and bromine) the temperature in the stratosphere above Antarctica must fall low enough for polar stratospheric clouds (PSCs) to form. This happens during the southern hemisphere winter when the south polar region is in polar night. During this time strong

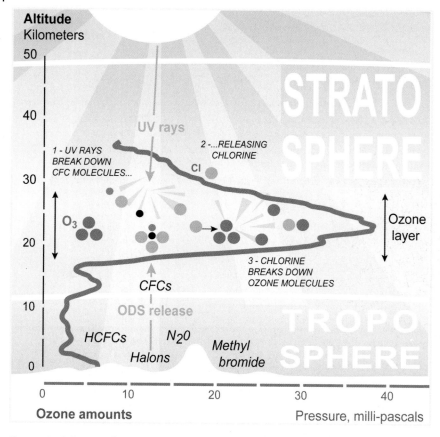

Figure 9.8 Schematic illustration of the steps that led to the Antarctic ozone hole. Source: UNEP/GRID-Arendal Maps and Graphics library (http://maps.grida.no/go/graphic/, accessed 18/7/2010). (Please find a color version of this figure in the color plates.)

westerly winds form a circular vortex of air over Antarctica in the stratosphere, within which temperatures fall to about −95 °C, low enough for the PSCs to form. This vortex also keeps the air inside it chemically isolated. More reactive forms of chlorine can then be produced by reactions on the surface of the PSCs.

A number of reactions are thought responsible but the main one is:

$$HCl_{(s)} + ClONO_{2(g)} \rightarrow HNO_{3(s)} + Cl_{2(g)}$$

where s denotes solid phase, g denotes gas phase. It is important to appreciate that all of these reactions are rapid and take place on the surface of aerosol particles in the cold polar vortex. The nitric acid (HNO_3) formed in these reactions was found to stay within the PSC particles, reducing the gaseous concentrations of nitrogen oxides. As the ice PSCs are sufficiently large, they fall out from the stratosphere effectively removing nitrogen from the air. This "denitrification" process is key to keeping the active chlorine around longer in the vortex. This prevents the active chlorine forming the long-lived, inactive chlorine nitrate through the termolecular reaction:

$$ClO + NO_2 + M \rightarrow ClONO_2 + M$$

and ensures levels of chlorine are kept high. With the lower stratospheric air kept isolated the air now has high concentrations of reactive chlorine compounds. By keeping it isolated high levels of active chlorine species do not mix readily with air at lower latitudes that would otherwise reduce active chlorine. However, ozone destruction not only requires chlorine atoms but the return of sunlight to the polar stratosphere.

Molecular chlorine from the above reactions is readily photolyzed to give:

$$Cl_2 + hv \rightarrow Cl + Cl$$

and then forms chlorine monoxide (ClO). The catalytic cycle given by Eqs. (9.5)–(9.7), with Cl as X, would not be sufficient to destroy ozone as there is little atomic oxygen in the lower stratosphere. Other catalytic cycles are required. The main cycles responsible for depleting the ozone are:

(A) $ClO + ClO + M \rightarrow Cl_2O_2 + M$

$\quad Cl_2O_2 + hv \rightarrow Cl + ClO_2$

$\quad ClO_2 + M \rightarrow Cl + O_2 + M$

then $2 \times (Cl + O_3 \rightarrow ClO + O_2)$

Net: $2 \times O_3 \rightarrow 3 \times O_2$

and similarly for bromine

(B) $ClO + BrO \rightarrow Br + Cl + O_2$

$\quad Cl + O_3 \rightarrow ClO + O_2$

$\quad Br + O_3 \rightarrow BrO + O_2$

Net: $2 \times O_3 \rightarrow 3 \times O_2$

The chlorine dimer (Cl_2O_2) is thermally unstable (the dimer will disassociate at warmer temperatures) and so cycle A is effective at the low temperatures found in the polar stratospheric vortex and is responsible for about 70% of the ozone loss. Cycle B likely to be more effective in the Arctic during northern hemisphere spring where the temperatures are warmer. The low temperatures not only produce the polar stratospheric clouds which allow heterogeneous reactions to release more active forms of chlorine, they also increase the efficiency of the reactions that destroy the ozone. Also note that cycle A involving the chlorine dimer results in ozone loss that is proportional to the square of the ClO concentration. Stratospheric chlorine was about 1 ppb in 1970 and increased to about 3.5 ppb in 1990. This more than three-fold increase in chlorine would produce a nine-fold increase in the rate of ozone loss; largely explaining the rapid ozone loss in Antarctica. The unique conditions of the Antarctic polar stratosphere produce a chemical environment that is primed to rapidly destroy ozone once sunlight returns to the region in the spring.

Though its concentration is lower in the stratosphere, bromine is more effective at destroying ozone than chlorine. This is partly because bromine-containing halocarbons are photolyzed at lower altitudes in the stratosphere, but also because there are catalytic cycles involving both bromine and chlorine together (B). The inactive forms of bromine such as HBr and $BrONO_2$ are photolyzed more readily than their chlorine counterparts. It is for these reasons that increases in bromine in the stratosphere are of concern, though concentrations of bromine are much lower than chlorine at present. The sources of bromine in the stratosphere come from the naturally occurring methyl bromide CH_3Br and brominated CFCs or halons.

9.4
Arctic Ozone

The Arctic polar stratosphere is similar to the Antarctic. It also experiences a polar night, with the onset of a circumpolar vortex and cold temperatures. It seemed likely then that ozone could be depleted in the same way, though temperatures in the Arctic polar vortex do not become as low as in the Antarctic polar vortex [18] (Figure 9.9). This is because the vortex is more disturbed, the winds are more variable and it changes morphology more readily. The northern hemisphere has more mountainous land mass than the southern hemisphere and this can create waves in the atmosphere [19] that disturb the air in the polar lower stratosphere and do not allow such a strong isolated vortex to form.

This not only keeps the temperatures in the Arctic polar vortex warmer it also allows more mixing between the air circulating inside the vortex over the pole and the air at lower latitudes. Increasing the mixing of air from inside to outside the vortex acts to reduce any activated chlorine (ClO).

Figure 9.9 shows the average temperature at the north and south poles in the lower stratosphere. Recall that, in order for ozone depletion to occur to produce an "ozone hole", the temperatures must first fall low enough to allow the formation of PSCs. On average, temperatures are only low enough in the Arctic from late December into January to form the PSCs and convert HCl and $ClONO_2$ to active chlorine.

In the 1980s and 1990s a series of measurement campaigns were carried out to look at the Arctic and the same chemical signatures were detected as seen in Antarctica [20, 21]: elevated levels of ClO were found and PSCs were detected. Figure 9.10 shows more recent measurements taken in the stratospheric air above Sweden. Elevated levels of chlorine monoxide are clearly present. Although PSCs are not present for as long nor over as wide an area as in the Antarctic, the conver-

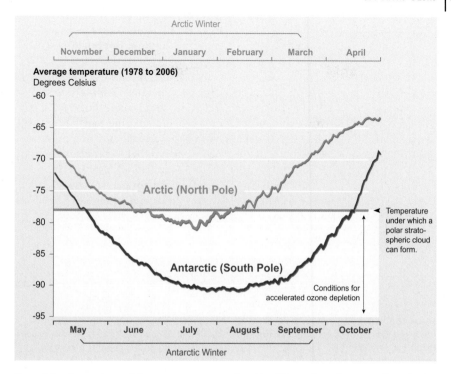

Figure 9.9 Comparison of the temperatures found at the Antarctic pole compared to the temperatures at the Arctic pole during winter in both hemispheres. Source: Emmanuelle Bournay, UNEP/GRID-Arendal Maps and Graphics Library (http://maps.grida.no/go/graphic/, accessed 18/7/2010). (Please find a color version of this figure in the color plates.)

sion or processing of HCl and ClONO$_2$ to more active chlorine is very rapid; it does not require PSCs to be present over the complete vortex to fill it with active chlorine. Low concentrations of NO$_2$ were also observed so the "denitrification process" seen in the Antarctic ozone hole also occurs. However, HNO$_3$ is less affected as the warmer temperatures mean the PSCs do not persist and fall out, they evaporate before falling very far. Also the Cl$_2$O$_2$ dimer decomposes back to ClO at warmer temperatures so the ozone loss cycles become less efficient. However, quantifying the amount of ozone loss by ozone hole processes is complicated as ozone is replenished by movement of air down through the vortex from the upper stratosphere. Nevertheless, at altitudes of 15–20 km, ozone depletion of about 15–20% does occur [22, 23].

One method used to quantify the amount of ozone loss is to use highly detailed computer models of the atmosphere and atmospheric chemical processes [24]. These models not only include an accurate treatment of the way in which the

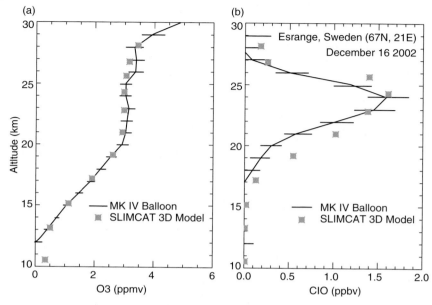

Figure 9.10 Observed and modeled profiles of O₃ and ClO at the Arctic station Esrange (67N, 21E) on December 16, 2002. The observations of the balloon-borne Mk IV instrument (courtesy of Dr. G. Toon, NASA JPL) are compared with output from the SLIMCAT 3D model. Panel (a) shows ozone and panel (b) shows chlorine monoxide (ClO). Figure courtesy of Wuhu Feng and Martyn Chipperfield, University of Leeds.

air moves in the troposphere and stratosphere with a detailed set of chemical reactions for gas-phase chemistry, but also include complex representations of heterogeneous chemistry and the processes that control formation of PSCs. It takes many man-years of effort to develop these models and test them and con- siderable computer resources are required to run them. However, these models are now capable of capturing the behavior of the chemistry in the Antarctic and Arctic polar stratosphere very well. Figure 9.10 shows how well one particular model, SLIMCAT [23] reproduces the observations of ozone and chlo- rine monoxide. Figure 9.11 demonstrates how these models can be used to deter- mine the amount of ozone lost to "processing" on PSCs. By including a pseudo-ozone in the model (so-called "passive ozone" in the figure), which behaves as ozone except it does not experience reactions on PSCs, it is possible to quantify the rapid ozone loss by comparing the two ozone representations in the model. The accuracy with which models like this can capture the behavior of complex chemistry and physical processes is a testament to the skill and dedication of the many scientists who not only developed the models, and carried out the measure- ments in remote places of the planet, but also those who investigated the funda- mental reactions in the laboratory without which these computer models could not be created.

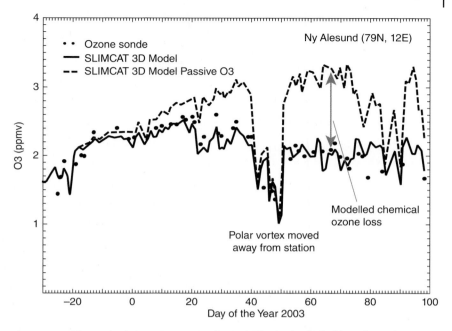

Figure 9.11 Observed volume mixing ratio of ozone (parts per million) at around 18 km altitude in the winter 2002/2003 at the Arctic station Ny Alesund (79N, 12E). Also shown are results from the SLIMCAT 3D model (solid line). The dashed line shows the model simulation without the effects of ozone loss from PSC-related chemistry. Figure courtesy of Wuhu Feng and Martyn Chipperfield, University of Leeds.

9.5
Montreal Protocol and Beyond

As noted earlier in this chapter, even before the ozone hole was discovered concern was raised by the possibility of the reduction in the ozone layer by CFCs. The discovery of the ozone hole led to a new impetus in controlling ozone-depleting substances with the signing of the Montreal Protocol in 1987 [25], though there were considerable disagreements between developed countries on the actions to take. As the scientific understanding of the processes responsible for the depletion of ozone grew, it became clear that further action was needed, not just to slow production of CFCs, but to stop production of them completely. A series of amendments to the Montreal Protocol followed (see Figure 9.12), that continued the process of ensuring a complete removal of these compounds from the atmosphere. The year 2010 is when developing countries are scheduled to phase out CFCs, halons and carbon tetrachloride (another ozone-depleting gas). To date, the Montreal Protocol is the only UN protocol all nations have signed up to [26, 27].

The HCFCs were seen as interim substitutes that do not contribute significantly to the depletion of the ozone layer. These shorter-lived compounds react with OH

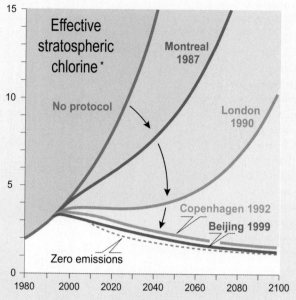

Predicted abundance
Thousand parts per trillion

Effective stratospheric chlorine*

No protocol

Montreal 1987

London 1990

Copenhagen 1992

Beijing 1999

Zero emissions

* Chlorine and bromine are the molecules responsible for ozone depletion. "Effective chlorine" is a way to measure the destructive potential of all ODS gases emitted in the stratosphere.

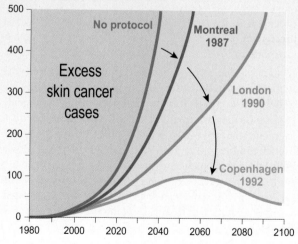

Cases per million people per year

No protocol

Montreal 1987

Excess skin cancer cases

London 1990

Copenhagen 1992

Figure 9.12 The effects of the Montreal Protocol amendments and their phase-out schedules. Source: Emmanuelle Bourney, UNEP/GRID-Arendal Maps and Graphics Library (http://maps.grida.no/go/graphic/, accessed 18/7/2010). (Please find a color version of this figure in the color plates.)

in the troposphere and generally do not reach the stratosphere, mainly releasing their chlorine in the troposphere instead. However, HCFCs and HFCs create another problem as they are effective greenhouse gases. HCFCs are included in the Montreal Protocol and are due to be phased out, but over a longer timescale than CFCs. The HFCs are also being used as replacements. Although their use only accounts for 2% (in 2010) of the total reported greenhouse gas emissions, demand is expected to grow rapidly with the continuing phasing out of CFCs and HCFCs. HCFCs are expected to be phased out by 2030 while regulatory controls and reduction plans are in development for HFCs. The global warming implications of the continued use of these super-greenhouse gases, ozone-depleting substances, are well known and there is concern that unregulated use of replacement substances will undo the reprieve in global warming the Montreal Protocol's phase-out of CFCs has provided. It is a complicated problem; there are considerable stocks of CFCs and HCFCs that need to be safely destroyed in such a way that none is leaked to the atmosphere. Safe removal and collection of CFCs from insulation in buildings for example is a concern and a difficult problem for the authorities. Also, replacement of these substances by less efficient gases in uses such as refrigeration and air conditioning has to be weighed against increases in energy use (and increased CO_2 emission as a result).

The ozone hole is expected to recover back to pre-1980 levels sometime in the latter half of this century given the reduction in chlorine in the stratosphere from the phase-out of the CFC and related compounds [28]. However, the ozone layer will then be under threat from increases in nitrous oxide (N_2O) [29]. This gas is emitted naturally from the marine environment, decomposition of deforestation and also from human activities, particularly farming. The atmospheric concentration of N_2O is increasing by 1–2% every four years and is now the main future threat to the ozone layer. N_2O was not included in the Montreal Protocol, but because it is also a greenhouse gas it was included in the Kyoto Protocol. It does not destroy ozone directly but, like the CFCs, is stable in the troposphere and as a long-lived gas is transported into the stratosphere where it is photolyzed:

$$N_2O + hv \rightarrow N_2 + O$$

with about 1% converted into nitrogen oxide (NO) by:

$$N_2O + O(^1D) \rightarrow NO + NO$$

that can react further to produce NO_2 (NO and NO_2 are collectively known as NOx). These species can then take place in the catalytic cycle shown earlier to destroy ozone:

$$NO + O_3 \rightarrow NO_2 + O_3$$

$$NO_2 + O \rightarrow NO + O_2$$

$$\text{Net:} \quad O_3 + O \rightarrow O_2 + O_2$$

This cycle is responsible for approximately 50% of the removal of odd oxygen $(O + O_3)$ in the stratosphere. Imposing controls on N_2O will be challenging; its uses are diverse though agricultural use accounts for roughly 80% of N_2O emissions by humans. Improved farming methods, such as more efficient use of fertilizer, will therefore be important.

9.6
Ozone and Climate Change

Earlier in this chapter, it was noted that ozone is a radiatively important gas which exerts a control on the behavior of the stratosphere. It also acts as a greenhouse gas itself. Figure 9.13 illustrates the complex processes controlling ozone changes and global warming and how they can link together. This is a complex scientific problem with many uncertainities on the magnitude of some of the changes. However, some processes are understood. Large changes in ozone will impact on the circulation and behavior of the stratospheric winds and temperatures; the Antarctic ozone hole has been shown to have resulted in a change to the stratospheric circulation [28]. As carbon dioxide (CO_2) increases in the atmosphere, the stratosphere will cool as the troposphere warms. This cooling may lead to an increase in the frequency and area of PSCs. The Antarctic is unlikely to be sensitive to even lower temperatures but the Arctic lower stratospheric temperatures are often close to the PSCs formation threshold, so ozone depletion in the Arctic stratosphere could increase. Lower temperatures in the stratosphere could also result in a change in the circulation of air there that will change the distribution of ozone and then, potentially, impact on stratospheric ozone depletion. Decreases in ozone could possibly stabilize the polar vortex more, leading to sustained ozone loss [30]. Lower temperatures also give enhanced solubility of HCl in stratospheric aerosol, increasing the efficiency of production of active Cl, and again leading to more efficient ozone loss. It is possible that the recovery of the ozone hole could be delayed due to lower stratospheric temperatures leading to further ozone loss and so on; a feedback.

Stratospheric ozone in mid-latitudes has also been observed to be steadily decreasing by about 1% per year as a result of gas-phase catalytic cycles, consistent with our current understanding of chemistry in the atmosphere. Global warming will cool the stratosphere and alter the balance of reactions controlling ozone. This could lead to a "super-recovery" of lower latitude ozone because lower temperatures favor formation of reservoir species (less chemically active and longer lived), such as nitric acid (HNO_3), over the formation of more active forms capable of destroying ozone. However, this is not all good news. The rate at which the air is moved from the stratosphere to the troposphere is predicted to increase with climate change. If lower stratospheric ozone increases, and more is moved into the troposphere, this will add to concerns about ozone levels near the surface affecting human health.

Ozone in the troposphere, and particularly at the surface, is a great concern for a future climate, and understanding the processes controlling surface ozone levels is important [31]. A number of different effects can impact on the ozone near the

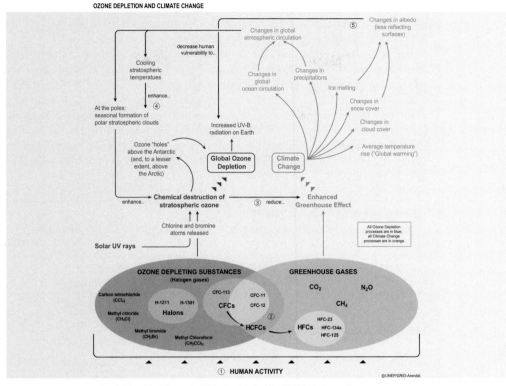

Figure 9.13 The relation between ozone and climate change. Source: Emmanuelle Bourney, UNEP/GRID-Arendal Maps and Graphics Library (http://maps.grida.no/go/graphic/, accessed 18/7/2010). (Please find a color version of this figure in the color plates.)

surface. For example, as the troposphere warms the amount of water vapor in the air will increase. More water vapor will mean more ozone though the impact is mitigated if cloud cover increases with the extra water vapor. Another effect is the response of plants and trees. Certain species emit significant amounts of a class of chemicals called volatile organic compounds (VOCs). Like nitrogen oxides, these react with other chemical species to produce ozone. Warmer temperatures can lead to increased production of VOCs and increases in ozone. The conditions under which increased ozone occurs are complicated and depend on the amount of nitrogen oxides present. Highly detailed and complex computer models of the atmosphere are a valuable tool in understanding the feedbacks and impacts of all these different effects. Equally valuable are monitoring and observations in the field that help to develop the processes described in these models.

One of the most important aspects of the ozone hole and climate change relates to the removal of the CFCs (and their replacement HCFC and HFCs). These are very potent greenhouse gases. They are much more effective per molecule than carbon dioxide and they absorb in a less opaque region of the spectrum. Removing them from the atmosphere has averted even more significant greenhouse warming than is currently underway. In terms of greenhouse warming it has been calculated that the Montreal Protocol is equivalent to removing a total of 135 billion tonnes of carbon dioxide since 1990 or delaying an increase in global temperatures by 7 to 12 years [28].

Global Warming Potential (GWP)

The GWP is the ratio of the warming caused by a substance to the warming caused by a similar mass of carbon dioxide. CO_2 has a GWP of 1. The CFCs and replacements are very effective greenhouse gases: CFC-11 has a GWP of 5000, CFC-12 is 8500. The HCFC and HFC replacement compounds have GWPs ranging from 90 to 12 000.
(source: http://www.epa.org/)

9.6.1
The World Avoided

It is interesting to consider what would have happened had the Montreal Protocol not been effected and CFCs levels had increased without any controls [32]. Several scientific studies have used detailed computer models to simulate the atmosphere assuming increases in ozone-depleting substances. In one such study [33] a concentration of 9 ppbv of equivalent stratospheric chlorine (combining the expected chlorine and bromine concentrations) was input into the stratosphere of a three-dimensional chemistry-climate model. Stratospheric chlorine reached levels of 3.5 ppbv in the late 1990s and 9 ppbv could have been present in the stratosphere between 2010 and 2020, given growth rates typical before the implementation of the Montreal Protocol. This increased level of chlorine had a significant impact on the modeled ozone (Figure 9.14).

Modeled losses of ozone up to 40% occur in the upper stratosphere in both hemispheres. Column ozone decreases everywhere, with significant loss of tropical ozone. In another similar study [34] levels of stratospheric chlorine were allowed to increase by 3% annually until the year 2065, to give a total chlorine loading of 45 ppbv. By this time two-thirds of the total ozone would have been destroyed and changes in the atmospheric circulation would lead to heterogeneous processes in the tropics (not just at the poles) almost completely destroying the ozone in the tropical lower stratosphere; while low ozone in Antarctica is less of a threat to biological life, the same cannot be said of the tropics. UV levels at the surface would almost double compared to what they are today. As the CFCs are powerful greenhouse gases there would have been significant changes to the temperatures in the troposphere; not modeled in these studies. The Montreal

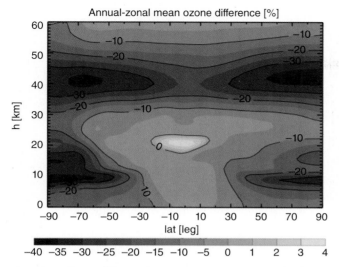

Figure 9.14 Ozone difference in percent between model calculations assuming 9 ppbv and 3.5 ppbv of stratospheric chlorine. The figure is an average of 30 years and represents the change in height of the ozone averaged around a latitude circle. From O. Morgenstern, NIWA, New Zealand.

Protocol, therefore, not only saved the planet from catastrophic ozone loss, it also reduced the warming due to greenhouse gases. The world owes the British Antarctic Survey team of Farman, Gardiner and Shanklin a significant debt.

9.7
Perspectives

The formation of the ozone hole was without question man-made. It was a surprise to the science of the time in many ways, yet shows how responsive scientists and policy makers can be, and how well they *can* work together to create a solution. The way in which the ozone hole problem was tackled might be considered as a model for solving the issues of man-made climate change, however, the issues are quite different. Although CFCs released at ground level take time to reach the stratosphere, the timescale involved is shorter than those pertinent to climate change. The effects of the CFCs were clearly visible year on year and easy to measure: their impact was obvious and largely understood by the public. Manufacturers could readily find alternatives to CFCs with little extra cost and there was little impact on everyday life. Sadly, the same is not true for climate change; the science is more complex, the impacts take longer to appear, the solution will involve significant lifestyle changes and public perception is nowhere near as clear [35]. As Jonathan Shanklin, one of the discoverers of the ozone hole [36], points out, the most startling lesson from the ozone hole is how quickly the planet can change and respond to the changes we all make to it. We should not become complacent and continue to monitor the planet closely.

9.8
Resources

There is a vast amount of information online regarding ozone in our atmosphere and ozone depletion, ranging from material aimed at schools to detailed scientific assessments. The highest quality material is generally from international organizations, research institutes and universities. This list, by no means complete, should provide the reader with a good starting point for further discovery.

World Meteorological Organisation (http://www.wmo.int)

The WMO provide a large amount of information regarding ozone and the ozone hole. For their ozone-related science products see http://www.wmo.int/pages/themes/wmoprod/science_en.html. A useful resource for younger readers can be found at http://www.wmo.int/youth/ozone_en.html with links to videos about ozone.

United Nations Environment Programme (http://www.unep.fr/ozonaction)

UNEP OzonAction is another large collection of resources and information. This site covers the science, and monitoring, and provides links to resources on social networking websites.

United Nations Environment Programme – Ozone Secretariat (http://www.unep.ch/ozone/)

The UNEP Ozone Secretariat is responsible for the Vienna Convention for the Protection of the Ozone Layer and for the Montreal Protocol. The definitive Scientific Assessment of Ozone Depletion publications from the UNEP/WMO (from which much of the material for this chapter came) can be found here as well as other official documents.

British Antarctic Survey (http://www.antarctica.ac.uk)

The scientists who discovered the ozone hole worked for the British Antarctic Survey in Cambridge, UK. This site has some useful information about the ozone hole as well as a lot more information and images of Antarctica.

US Environmental Protection Agency (http://www.epa.gov)

The US EPA have numerous resources on ozone in the atmosphere which can be found by searching the site. Their resources on ozone layer depletion can be found at http://www.epa.gov/ozone/strathome.html.

Environment Canada (exp-studies.tor.ec.gc.ca/e/index.htm)

The Experimental Studies Unit of the Science and Technology branch of Environment Canada have provided maps and data as well as documents on links between ozone and climate change, Arctic ozone and the ozone layer in general.

TOMS satellite data (toms.gsfc.nasa.gov)

The Total Ozone Mapping Spectrometer website is one of the best places to go for data and maps of the ozone layer. The Ozone Hole Watch website can be found at http://ozonewatch.gsfc.nasa.gov/.

University of Cambridge Ozone Hole Tour (http://www.atm.ch.cam.ac.uk/tour/)

Developed as an educational resource for students and teachers this guide to the ozone hole is one of the most popular resources on the web.

Scientific Visualisation Studio (svs.gsfc.nasa.gov/search/Keyword/Ozone.html)

The Scientific Visualization Studio at the NASA Goddard Space Flight Center is well worth a visit to see the range and quality of the images and animations they have produced related to ozone in the atmosphere.

Acknowledgments

The author would like to express his thanks to his colleagues Drs Nicola Warwick and Peter Braesicke for proof-reading this chapter.

References

1 Wayne, R.P. (2000) *Chemistry of Atmospheres*, 3rd edn, Oxford University Press, Oxford. ISBN 019850375X.

2 Andrews, D.G. (2010) *An Introduction to Atmospheric Physics*, 2nd edn, Cambridge University Press. ISBN 0521693187.

3 World Health Organization (2004) Europe, Health Aspects of Air Pollution, http://www.euro.who.int/__data/assets/pdf_file/0003/74730/E83080.pdf (accessed 10/6/2010).

4 Gaidos, E.J. and Yung, Y.L. (2003) Evolution of the earth's atmosphere, in *Encyclopedia of Atmospheric Sciences*, vol. 3 (eds J.R. Holton, J.A. Curry, and J.A. Pyle), Academic Press, pp. 762–767. ISBN 0122270908.

5 Ravishankara, A.R. (2003) Photochemistry of ozone, in *Encyclopedia of Atmospheric Sciences*, vol. 4 (eds J.R. Holton, J.A. Curry, and J.A. Pyle), Academic Press, pp. 1642–1649. ISBN 0122270908.

6 Chapman, S. (1930) A theory of upper-atmosphere ozone. *Mem. Roy. Meteorol. Soc.*, **3**, 103.

7 Dobson, G.M.B. (1931) Ozone in the upper atmosphere and its relation to meteorology. *Nature*, **137** (3209), 668–672.

8 Smit, H.G.J. (2003) Ozonesondes, in *Encyclopedia of Atmospheric Sciences*, vol. 4 (eds J.R. Holton, J.A. Curry, and J.A. Pyle), Academic Press, pp. 1469–1476. ISBN 0122270908.

9 Stolarki, R.S., Bloomfield, P., McPeters, R.D., and Herman, J.R. (1991) Total ozone trends deduced from Nimbus 7 TOMS data. *Geophys. Res. Lett.*, **18** (6), 1015–1018.

10 Crutzen, P.J. (1970) Influence of nitrogen oxides on atmospheric ozone content. *Q. J. R. Meteorol. Soc.*, **96**, 320–325.

11 Stolarski, R.S. (2003) A hole in the earth's shield, in *A Century of Nature* (eds L. Garwin and T. Lincoln), University of Chicago Press, pp. 281–298. ISBN 022628413.

12 Molina, M.J. and Rowland, F.S. (1974) Stratospheric sink for chlorofluormethanes: chlorine atom-catalysed destruction of ozone. *Nature*, **249**, 810–812.

13 Riffenburgh, B. (2007) *Encyclopedia of the Antarctic*, Routledge, Taylor-Francis Group, ISBN 100415970245.

14 Farman, J.C., Gardiner, B.G., and Shanklin, J.D. (1985) Large losses of total ozone in Antarctica reveal seasonal ClOx/NOx interaction. *Nature*, **315**, 207–210.

15 Stolarski, R.S. *et al.* (1986) Nimbus 7 satellite measurements of the springtime Antarctic ozone decrease. *Nature*, **322**, 808–811.

16 Solomon, S., Garcia, R.R., Rowland, F.S., and Wuebbles, D.J. (1986) On the depletion of Antarctic ozone. *Nature*, **321**, 755–758.

17 Molina, L.T. and Molina, M.J. (1987) Production of Cl_2O_2 from the self-reaction of the ClO radical. *J. Phys. Chem.*, **91**, 433–436.

18 Brune, W.H. *et al.* (1991) The potential for ozone depletion in the Arctic polar stratosphere. *Science*, **252**, 1260–1266.

19 Carslaw, K.S. *et al.* (1998) Increased stratospheric ozone depletion due to mountain induced atmospheric waves. *Nature*, **391**, 675–678.

20 Amanatidis, G.T. and Ott, H. (1995) European commission research on stratospheric ozone depletion. *Phys. Chem. Earth*, **20** (1), 13–19.

21 Newman, P.A. *et al.* (2002) An overview of the SOLVE/THESEO 2000 campaign. *J. Geophys. Res.*, **107**, 8259.

22 Waibel, A.E. *et al.* (1999) Arctic ozone loss due to denitrification. *Science*, **283**, 2064–2069.

23 Sinnhuber, B.-M. *et al.* (2000) Large loss of total ozone during the Arctic winter of 1999/2000. *Geophys. Res. Lett.*, **27**, 3473–3476.

24 Lefevre, F., Figarol, F., Carslaw, K.S., and Peter, T. (1998) The 1997 Arctic Ozone depletion quantified from three-dimensional model simulations. *Geophys. Res. Lett.*, **25**, 2425–2428.

25 UNEP (2007) Ozone secretariat, brief primer on the Montreal Protocol: the treaty, chemicals controlled, achievements to date and continuing challenges. http://www.unep.ch/ozone/Publications/MP_Brief_Primer_on_MP-E.pdf (accessed June 2010).

26 Roan, S.L. (1989) *Ozone Crisis. The 15yr Evolution of Sudden Global Emergency*, Wiley Science Editors, New York.

27 UNEP/DTIE (2010) OzonAction Newsletter and Publications. http://www.uneptie.org/ozonaction/ and also see http://www.unep.fr/ozonaction/information/mmc/main.asp (accessed June 2010).

28 United Nations Environment Programme / World Meteorological Organisation (2006) The Scientific Assessment of Ozone Depletion: 2006, Global Ozone Research and Monitoring Project–Report No. 50. http://www.unep.ch/ozone/Assessment_Panels. (accessed June 2010) Please note that a new version of this report is due out in 2010.

29 Codispoti, L.A. (2010) Interesting times for marine N_2O. *Science*, **327**, 1339–1340.

30 Chubachi, S. (1997) Annual variation of total ozone at Syowa station, Antarctica. *J. Geophys. Res.*, **102**, 1349–1354.

31 Zeng, G. and Pyle, J.A. (2003) Changes in tropospheric ozone between 2000 and 2100 modeled in a chemistry-climate model. *Geophys. Res. Lett.*, **30**, 1392.

32 Prather, M., Midgley, P., Rowland, F.S., and Stolarski, R. (1996) The ozone layer: the road not taken. *Nature*, **381**, 551–554.

33 Morgenstern, O. *et al.* (2008) The world avoided by the Montreal Protocol. *Geophys. Res. Lett.*, **35**, L16811, 5pp, doi: 10.1029/2008GL034590.

34 Newman, P.A. *et al.* (2009) What would have happened to the ozone layer if chlorofluorocarbons (CFCs) had not been regulated? *Atmos. Chem. Phys.*, **9**, 2113–2128.

35 Ahuja, D.R. and Srinivasan, J. (2009) Why controlling climate change is more difficult than stopping stratospheric ozone depletion. *Curr. Sci.*, **97** (11), 1531–1534.

36 Shanklin, J. (2010) Reflections on the ozone hole. *Nature*, **465** (7294), 34–35.

Epilogue

To read these chapters is to share in humanity's greatest triumph: the mobilization of scientific knowledge for human betterment. Chemistry, in particular, has been foundational, providing the tools, models, insights, and techniques for every major area of endeavor: health, agriculture, energy, transport, water, and more. The stories in this marvelous book are exhilarating, powerful, and forward looking. They draw on chemistry's past challenges and triumphs to shine a light on the future needs and potential achievements.

The challenge of translating science into human advance is more urgent than ever before. In many ways, we are of course victims of our own success. Chemistry's past accomplishments – in creating nitrogen-based fertilizers, harnessing large-scale energy sources, controlling infectious diseases, cleaning the urban water supplies – have led to a burgeoning of the world's population as well as an unprecedented scale of production per person on the planet. We are now 7 billion human beings tightly packed into the Earth's fragile ecosystems. Average output per person per year has reached $10,000 (in a common purchasing-power adjusted measurement). The consequence is a human impact on the environment of unprecedented scale and scope. The dangers for future wellbeing abound, whether in human-induced climate change, natural resource depletion, unchecked pollution, or other adverse consequences of the global economy.

Our tasks are highly complex. Not only is humanity imposing unprecedented burdens on the Earth's ecosystems, but is doing so in the context of unprecedented inequalities of conditions and risks on the planet. Around 1 billion people, one sixth of humanity, continue to live in extreme poverty, fighting daily for their survival, basic health, and dignity. Another billion at the top of the income distribution consume roughly half of the planet's annual economic output, and often elbow the poor aside to ensure their own continued disproportionate access to the world's goods and services.

Our challenges are therefore multiple and complex, and at all scales from the local to the global. The world, and notably the poor world, yearns for economic improvement, which in the short term often entails even greater demands on the Earth's ecosystem services and depleting resource base. Yet the planet as a whole needs to live within global ecological and biophysical boundary conditions to sustain life and wellbeing. And in the midst of these unprecedented challenges,

The Chemical Element: Chemistry's Contribution to Our Global Future, First Edition.
Edited by Javier Garcia-Martinez, Elena Serrano-Torregrosa
© 2011 Wiley-VCH Verlag GmbH & Co. KGaA. Published 2011 by Wiley-VCH Verlag GmbH & Co. KGaA.

the specific, urgent, life-and-death problems of extreme poverty must be addressed and solved. Our first rendezvous with global solutions is 2015, the target date for fulfillment of the Millennium Development Goals.

Every scintillating chapter in this book makes clear that chemistry must play a central role if we are to succeed in facing these complexities. Sustainable development – combining poverty reduction, global economic improvement, and ecosystem health – will require new ways of accomplishing our most basic economic tasks: growing food, preserving public health, mobilizing safe and plentiful energy, and converting materials safely for human comfort and safety.

This volume is unique in offering a comprehensive and cutting-edge perspective on the future of sustainable development through the vision of some of the world's leading chemists. As a policy strategist, I was riveted page by page, as the technological possibilities for the future were authoritatively conveyed. Each chapter offers remarkable clarity, breadth, technical precision, and a deep sense of humanity. If there is a theme that runs in common, it is that the highest flights of science are bound up intimately with the highest human aspirations. Chemistry is not a dry subject of equations and reactions. It is a science of human purpose as well, with the drama and illustrious history of breakthroughs and contributions of monumental importance.

Professor Peter Mahaffy is compelling in advocating a new way to teach science, one that grips the students through the drama of the human condition. I was convinced and entranced. He recounts the dramatic "call to action" made by British chemist William Crookes in 1898 to find a new chemical pathway to mobilize nitrogen for food production, lest the world succumb to hunger in a global nitrogen shortfall. Crookes' call to action was answered a decade later by the world changing Haber–Bosch process for the industrial manufacture of nitrogen-based fertilizer. This book is our generation's call to action, to find new chemical pathways for supporting food production and nutrition, the mitigation of human-induced climate change, the supply of safe water, the development of new medicinal compounds, and the greening of chemical processes so that the vast benefits of industry are not undone by tragic and often unforeseen side effects.

During my quarter century of work on the challenges of sustainable development, I have seen repeatedly the essential, and indeed overpowering, role of technical knowledge in sound policy making. When science is brought to bear on our great challenges, solutions are found. When politics succumbs to special interest groups, public prejudices, and even outright ignorance, we are dangerously led astray. The remarkable chapter by Dr. Glen Carver on ozone depletion reminds us indeed that some of humanity's greatest challenges and risks will be uncovered first by cutting-edge science. Bringing top science to bear on public policy will be increasingly vital for our very survival.

Fittingly, the United Nations has designated 2011 as the International Year of Chemistry. If there is any question as to why this choice was made, this book answers it fully. Chemistry is key to human wellbeing. The appreciation of chemistry's contributions is vital to emerging the next generation of scientists, policy makers, and informed citizens. This book makes a unique and important contribution to that task. It will be widely read around the world, and provide a path and inspiration for sustainable development in the years ahead. I personally would like to express my profound appreciation to the editors and authors of this book for this important contribution.

Prof. Jeffrey Sachs
Director of The Earth Institute
Quetelet Professor of Sustainable Development, and Professor of Health
Policy and Management at Columbia University

Color Plates

(a)

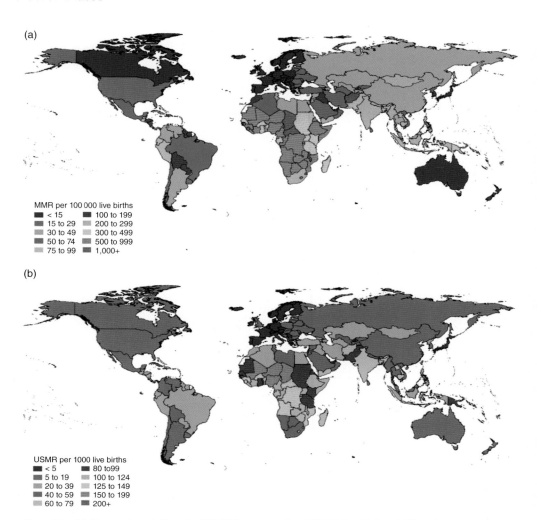

MMR per 100 000 live births
- < 15
- 15 to 29
- 30 to 49
- 50 to 74
- 75 to 99
- 100 to 199
- 200 to 299
- 300 to 499
- 500 to 999
- 1,000+

(b)

USMR per 1000 live births
- < 5
- 5 to 19
- 20 to 39
- 40 to 59
- 60 to 79
- 80 to 99
- 100 to 124
- 125 to 149
- 150 to 199
- 200+

Figure 1.3 (a) Maternal mortality ratio (MMR) by country (per 100 000 live births), 2008 and (b) under-5 mortality rate (U5MR) by country (per 1000 live births), 2008 (from [56]).

The Chemical Element: Chemistry's Contribution to Our Global Future, First Edition.
Edited by Javier Garcia-Martinez, Elena Serrano-Torregrosa
© 2011 Wiley-VCH Verlag GmbH & Co. KGaA. Published 2011 by Wiley-VCH Verlag GmbH & Co. KGaA.

Figure 1.4 IOCD scientists meeting at Berkeley, California in 1986. From left to right: Carlos Rius, IOCD's first secretary; Pierre Crabbé, founder; Elkan Blout, first treasurer and one of three founding vice presidents; Carl Djerassi, one of the inspirations behind IOCD; Sune Bergström, a founding vice president; Sydney Archer, leader of the Tropical Diseases Working Group; (unknown); Glenn Seaborg, IOCD's first president and associate director of the Lawrence Berkeley National Laboratory; C.N.R. Rao, a founding IOCD vice president; and Joseph Fried, leader of the Male Fertility Regulation Working Group.

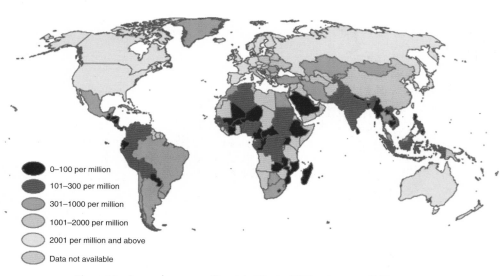

Figure 1.7 Researchers per million inhabitants, 2007 or latest available year.

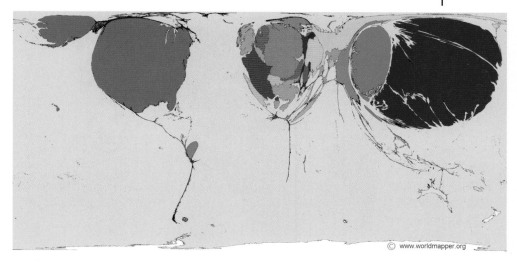

Figure 1.8 Scientific publications by countries, 2001. Territory size shows the proportion of all scientific papers published in 2001 written by authors living there. Scientific papers cover physics, biology, chemistry, mathematics, clinical medicine, biomedical research, engineering, technology, and earth and space sciences [111].

Figure 1.9 Patents granted by countries, 2002. Territory size shows the proportion of all patents worldwide that were granted there [111].

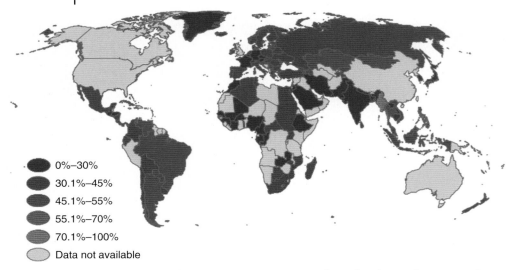

0%–30%
30.1%–45%
45.1%–55%
55.1%–70%
70.1%–100%
Data not available

Figure 1.12 The gender gap in science. Women as a share of total researchers, 2007 or latest available year [55].

(a)

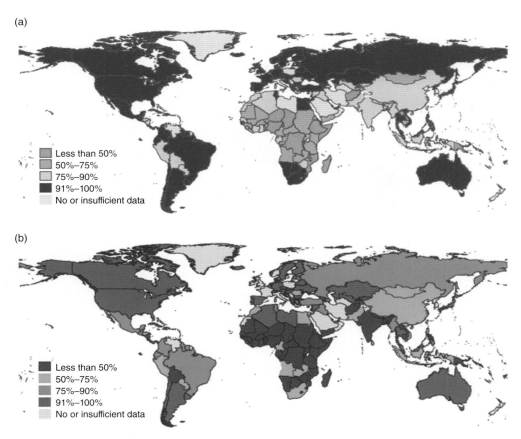

Less than 50%
50%–75%
75%–90%
91%–100%
No or insufficient data

(b)

Less than 50%
50%–75%
75%–90%
91%–100%
No or insufficient data

Figure 1.14 Global water and sanitation coverage [236]. (a) Improved drinking-water coverage, 2006. (b) Improved sanitation coverage, 2006.

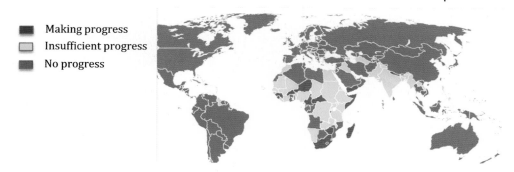

Figure 2.1 Country progress in meeting the MDG1 indicator for prevalence of children underweight (source [9c]).

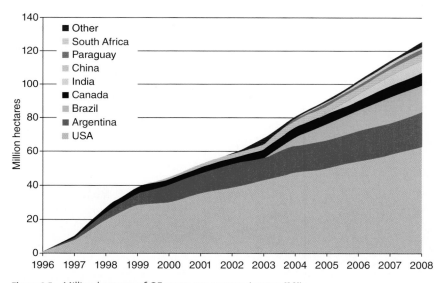

Figure 2.5 Million hectares of GE crops per country (source [39]).

Figure 2.6 The Larger Grain Borer (source [45]).

Figure 2.7 Maize destroyed by the Large Grain Borer to flour (source [45]).

(a)   (b)

Figure 2.8 Traditional granary (a) and improved granary/crib (b) in the Millennium Village Project Mwandama, Malawi (source: MDG Centre East and Southern Africa).

(a)
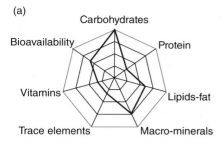

(b)
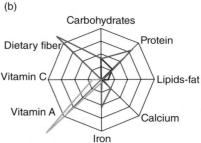

Figure 2.9 Ecological spider web presenting diversity requirements in a human diet.
(a) Nutrient composition of an ideal diet that meets all nutritional needs is shown in pink. An example of nutrient composition of a diet that meets carbohydrate demand but lacks protein and micronutrients or trace elements is shown in blue. (b) Nutrient composition data of three food crops are shown as % of daily requirement (100%). The blue line represents one cup of white corn (166 g), the green line one cup of black beans (194 g), and the orange line one cup of pumpkin (116 g) (nutrition facts from http://www.nutritiondata.com). The spiderdiagram shows the complementarity between the three food crops for carbohydrates, proteins, dietary fiber, and vitamin A (source [49]).

Figure 2.11 Cassava in Africa (source: Nestle).

Figure 4.1 Vanity Fair portrait of chemist Sir William Crookes, 1903. Courtesy of the Chemical Heritage Foundation.

Figure 4.3 The learning environments, curriculum, pedagogy, and physical spaces for effective science learning need to be appropriate to local cultures and education systems. Reproduced with permission © UNICEF/KENA2010-00321/Noorani.

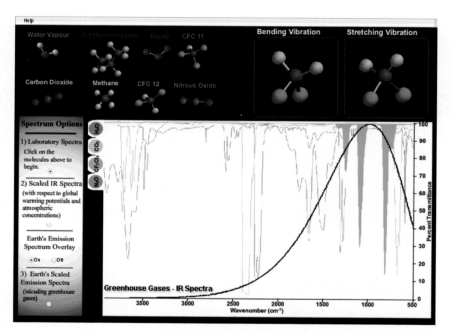

Figure 4.6 Interactive visualization showing the connection between the infrared spectra of gases and their global warming potential.

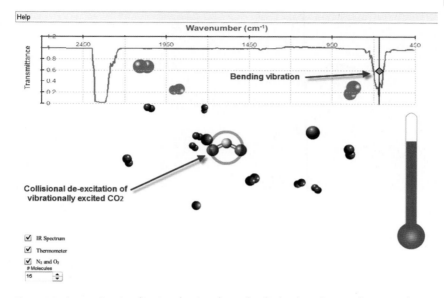

Figure 4.7 Interactive visualization showing the molecular-level mechanism for tropospheric warming by greenhouse gases.

Figure 5.1 Medicine man in American Indian healing [1].

Figure 5.2 A moment in the life of an Egyptian physician of the 18th dynasty (1500–1400 BC) [2].

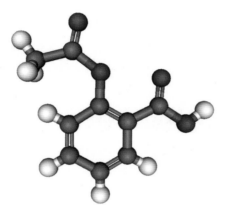

Figure 5.3 Structure of Aspirin™ shown as ball and stick representation.

Figure 5.5 The hypertension drug Capoten™ as a ball and stick 3D model showing chemical space available for binding interactions.

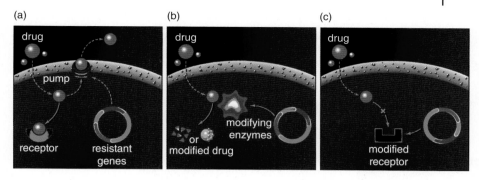

Figure 5.9 Mechanisms of bacterial drug resistance: (a) active efflux; (b) enzymatic modification of the drug; (c) modification of the target receptor or enzyme. Taken from [31].

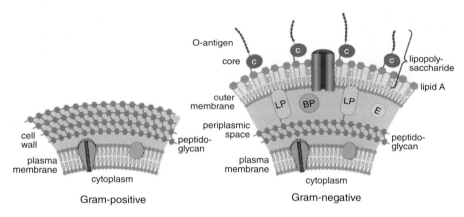

Figure 5.10 The cell envelope of Gram-positive and Gram-negative bacteria. Gram-positive bacteria have a thicker peptidoglycan layer than Gram-negative bacteria, but lack the outer-membrane at the cell surface. Taken from [31].

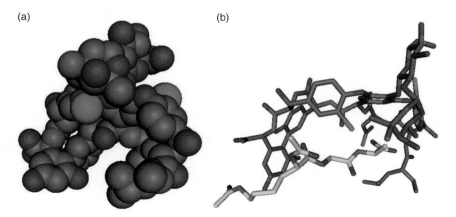

Figure 5.12 CPK model of (a)Vancomycin and (b)Vancomycin bound to the *N*-acetyl-D-Ala-D-Ala residue (PDB No. 1FVM).

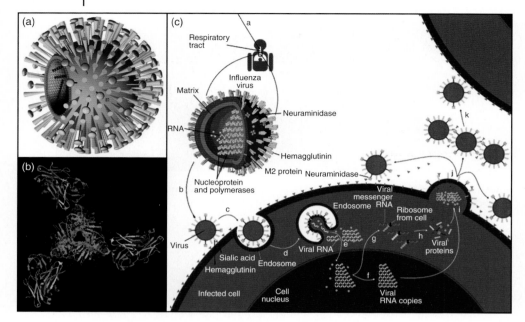

Figure 5.13 Typical representation of a flu virus virion 100 nm wide (a) and covered with hundreds of copies of two surface proteins: a neuraminidase enzyme responsible for infected cell detachment and a hemagglutinin trimers (b) responsible for host receptor adhesion. (c) Influenza flu virus life cycle [48].

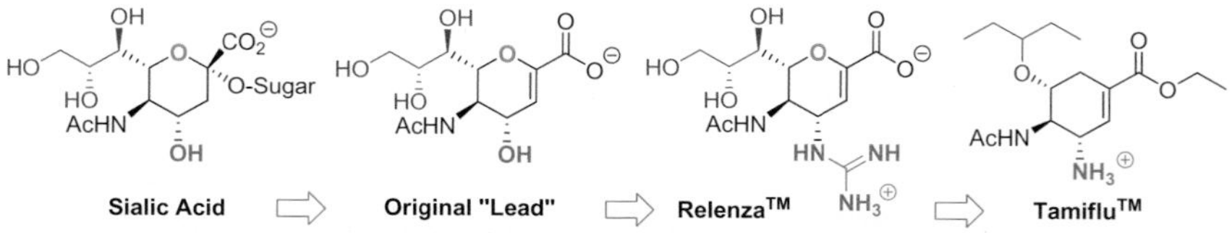

Figure 5.14 A good example of a lead optimization that led to the first anti-flu drug Tamiflu™.

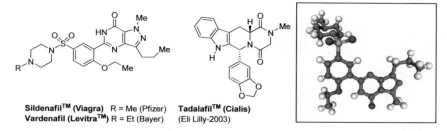

Sildenafil™ (Viagra) R = Me (Pfizer)
Vardenafil (Levitra™) R = Et (Bayer)

Tadalafil™ (Cialis)
(Eli Lilly-2003)

Figure 5.15 Chemical structures of Viagra and related drugs together with its 3D depiction.

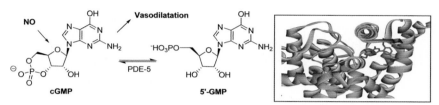

Figure 5.16 Viagra works by blocking the action of the PDE-5 enzyme leading to restoration of nitric oxide (NO) in vasodilatation. Viagra in the active site of PDE-5, the zinc and magnesium ions are indicated by blue and green spheres, respectively (PDB-3JWQ).

(a)　　　　　　　　　　　(b)

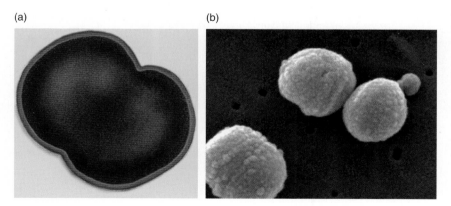

Figure 5.18 (a) The bacteria *Neisseria meningitidis* is surrounded by a capsular polysaccharide against which antibodies can kill the bacteria. (b) *Streptococcus pneumonia* is presently the number one killing bacteria worldwide.

(a)

(b)

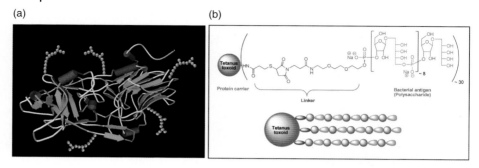

Figure 5.19 (a) Schematic representation of a bacterial polysaccharide antigen (yellow balls) linked in several copies to an immunogenic protein carrier (here Tetanus toxoid). (b) Chemical structure and linkage of the synthetic CPS repeating unit of *Haemophilus influenza* type b bound to T. Toxoid.

(a)

(b)

(c)

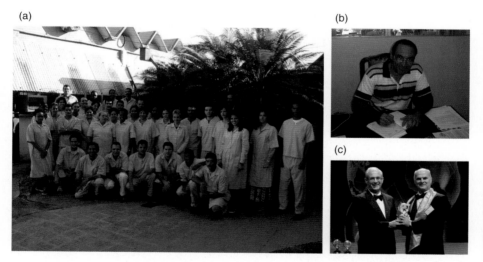

Figure 5.20 (a) Members of the Cuban team that have contributed to the first worldwide semi-synthetic antibacterial vaccines. (b) Dr. Vicente Verez Bencomo heading the Cuban team in his office at the University of Havana, Cuba. (c) Professor René Roy receiving a Tech Museum Award on behalf of the team in 2005.

Figure 6.1 Picture representing chemistry in the service of the environment (picture taken from the cover of "Introduzione alla Chimica Verde (Green Chemistry)", eds P. Tundo, S. Paganelli–Lara Clemenza. In Italian).

Figure 6.2 Tenth edition of the summer school on green chemistry organized by INCA (Cà Dolfin, Cà Foscari University, Venezia, 2008).

Figure 6.4 1st, 2nd and 3rd International IUPAC conferences on Green Chemistry [6].

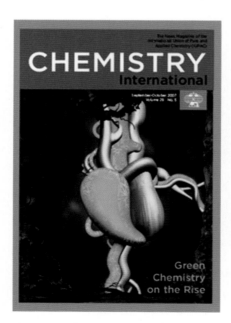

Figure 6.5 A heart representing the concept of Circular Economy: a heart that recycles the waste into new useful green products. This image taken from the cover of Chemistry International [10] the newsmagazine of IUPAC.

Figure 6.6 Biomass is biological material derived from living, or recently living organisms, such as wood and animal waste.

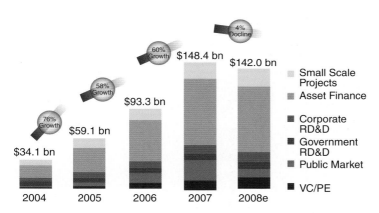

Figure 6.9 Total global new investment in clean energy, 2004–2008, US$ billions.

Substainability of Chemical Industry:

Where we are...   ...where we want to go

Figure 6.12 From the current industrial scenario to green industrial chemistry [13].

National/International Organizations

IUPAC–Subcommittee on Green
Chemistry Organic and Biomolecular
Chemistry Division (III)

Interuniversity National Consortium
"Chemistry for the Environment"
(Italy)

Green Chemistry Network (UK)

Green & Sustainable
Chemistry Network (Japan)

Environment Protection Agency
(USA)
The US EPA's Green Chemistry
Program

Green Chemistry Institute (USA)

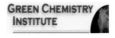

Canadian Green Chemistry Network

European Association for Chemical
and Molecular Science (EuCheMS)
WP on Green and Sustainable
Chemistry

 European Association for Chemical and Molecular Sciences

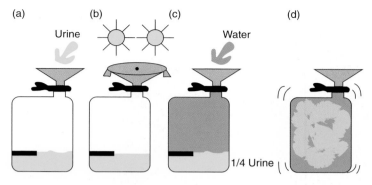

Figure 7.8 Four-step process to collect urine to use as agricultural fertilizer. (a) Fill 20-L can 25% full with human urine (in this case 5 L of urine); (b) let urine stand for 2 days to ensure destruction of the pathogen, *Schistosoma Haematobium*; (c) fill remainder of container with water (in this case 15 L); (d) mix and apply directly to crops (reprinted with permission from [74]).

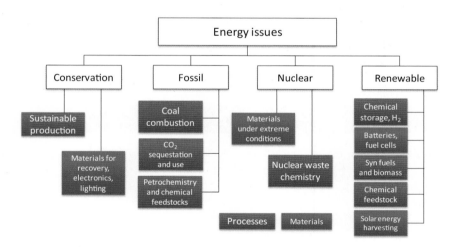

Figure 8.2 Multiple contributions of chemistry to the energy challenge; in blue contribution from material and in red from process developments; adapted from [3].

(a)

(b)

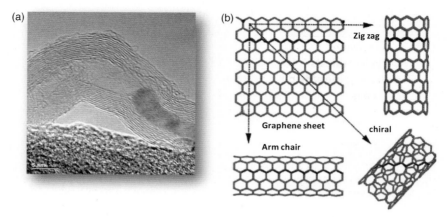

Figure 8.3 (a) Transmission electron microscopy (TEM) image of a multiwall carbon nanotube (MWCNT) which grows by chemical vapor deposition (CVD) from a catalytic nanoparticle. (b) Different modalities in which a graphene sheet (e.g., the basic structural element of CNTs) can be rolled to form the carbon nanotubes.

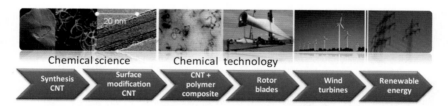

Figure 8.4 The value chain in sustainable energy: the use of CNT in wind turbines; adapted from [26].

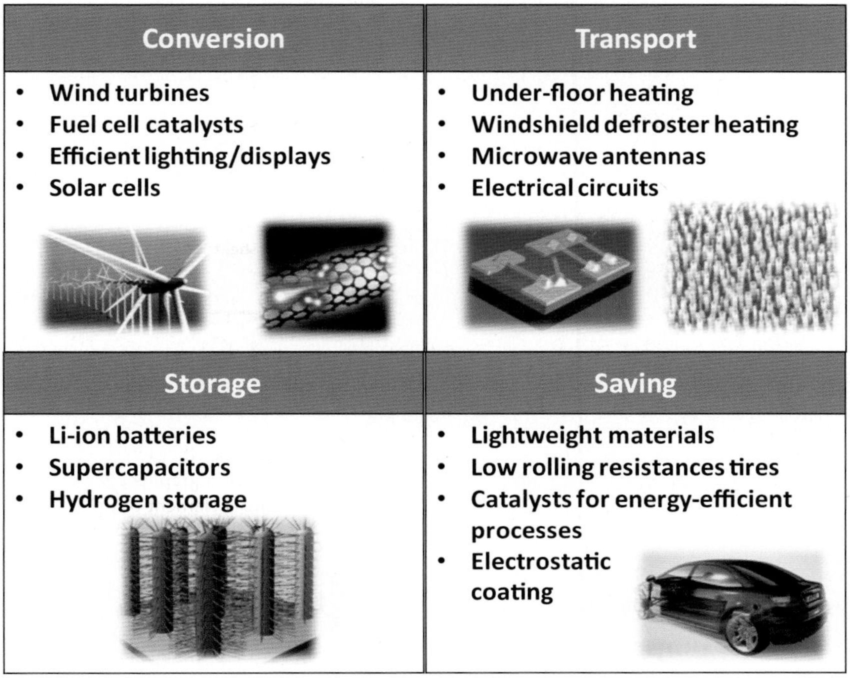

Conversion
Wind turbines
Fuel cell catalysts
Efficient lighting/displays
Solar cells
Transport
Under-floor heating
Windshield defroster heating
Microwave antennas
Electrical circuits
Storage
Li-ion batteries
Supercapacitors
Hydrogen storage
Saving
Lightweight materials
Low rolling resistances tires
Catalysts for energy-efficient processes
Electrostatic coating

Figure 8.5 Use of carbon nanotubes in sustainable energy applications; adapted from [26].

CNT
colloidal aggregate
supported metal particle
100
10
1
0 nm
nm
Nanomaterials domain
Materials
Molecular domain
Macromolecular and supramolecular domains

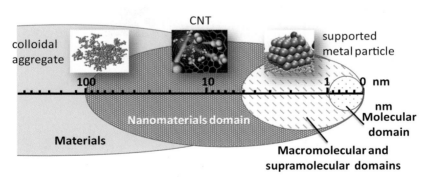

Figure 8.7 Nanomaterials: the length dimension; adapted from [41].

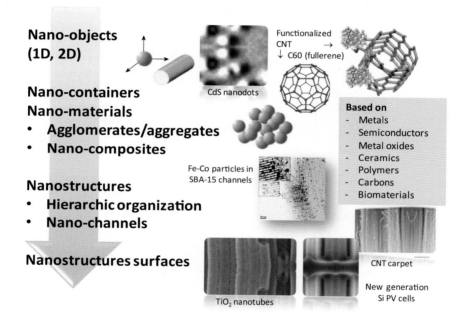

Figure 8.8 Nanostructures and nanomaterials, with some examples; adapted from [42].

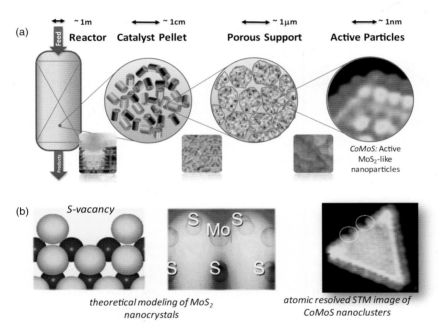

theoretical modeling of MoS$_2$ nanocrystals

atomic resolved STM image of CoMoS nanoclusters

Figure 8.9 Development of hydrodesulfurization (HDS) catalysts: (a) the structure of a heterogeneous catalyst illustrating the length scales and complexity involved in a heterogeneous catalyst and (b) atom-resolved STM image of an MoS$_2$ nanocluster and theoretical modeling of the active sites for HDS reaction; adapted from http://www.inano.au.dk, accessed 7 January 2011.

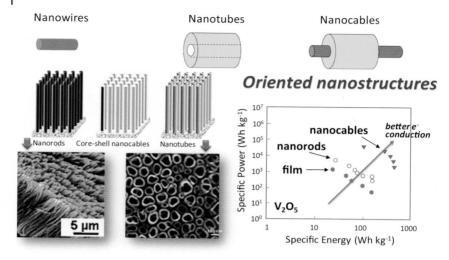

Figure 8.10 Example of the role of oriented nanostructures (vanadium oxide supported over nanocarbon materials) in improving the performances of LIBs; adapted from [24].

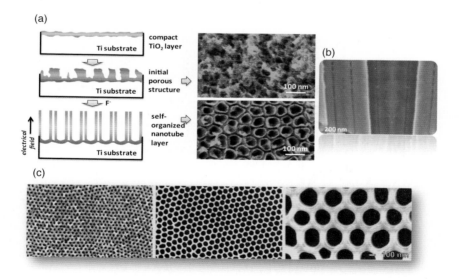

Figure 8.11 (a) Evolution of the TiO_2 morphology during AO of a Ti foil in the presence of fluoride ions, (b) example of the TiO_2 nanotube structure produced by AO in ethylene glycol electrolyte, (c) different alumina nanomembranes produced by AO; adapted from [25].

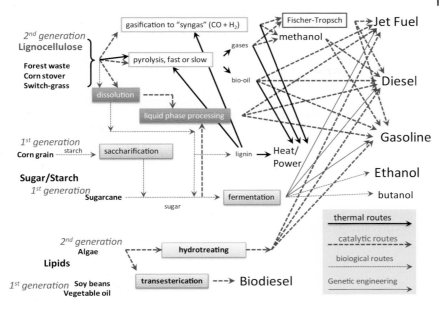

Figure 8.12 Biofuel production alternatives; adapted from [61].

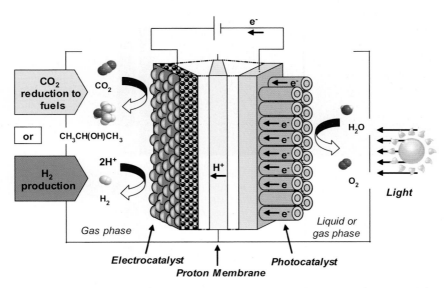

Figure 8.13 Scheme of the PEC device for the CO_2 reduction to fuels using solar energy and the H_2 production; adapted from [98].

Figure 9.3 Preparations for a balloon launch. The large balloon is carrying an ozonesonde and other instruments to measure several chemical species as it ascends through the atmosphere. The smaller balloon is used to lift the cable, between the large balloon and payload clear of the ground so the payload does not get dragged as the large balloon rises. The bundles of large black cylinders hold the helium gas used to fill both balloons.

Figure 9.4 Aerial view of Antarctica at low sun. Although a region of superb natural beauty it was and still is also a region of significant scientific interest. Photograph courtesy of P. Bucktrout, British Antarctic Survey, Cambridge, UK.

Figure 9.5 Halley station in Antarctica where the Dobson spectrometer is housed. Photograph courtesy of C. Gilbert, British Antarctic Survey, Cambridge, UK.

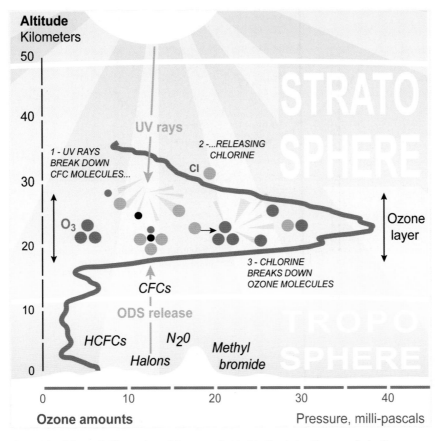

Figure 9.8 Schematic illustration of the steps that led to the Antarctic ozone hole. Source: UNEP/GRID-Arendal Maps and Graphics library (http://maps.grida.no/go/graphic/, accessed 18/7/2010).

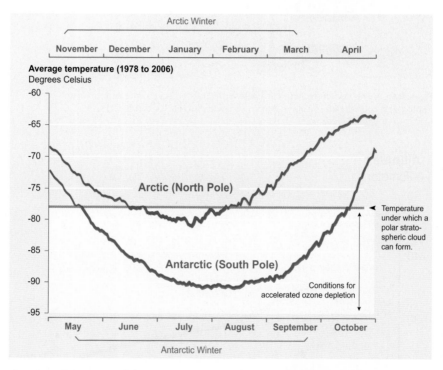

Figure 9.9 Comparison of the temperatures found at the Antarctic pole compared to the temperatures at the Arctic pole during winter in both hemispheres. Source: Emmanuelle Bournay, UNEP/GRID-Arendal Maps and Graphics Library (http://maps.grida.no/go/graphic/, accessed 18/7/2010).

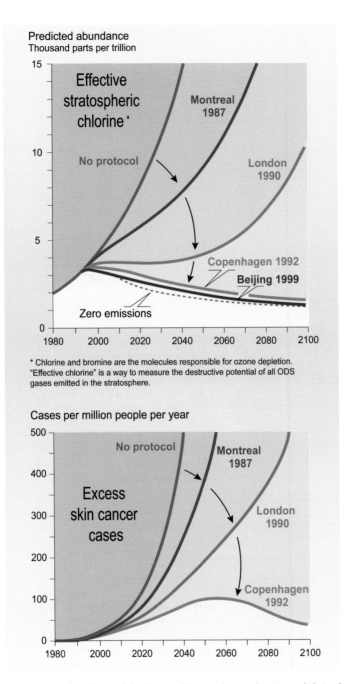

Predicted abundance
Thousand parts per trillion

Effective stratospheric chlorine*

No protocol

Montreal 1987

London 1990

Copenhagen 1992

Beijing 1999

Zero emissions

* Chlorine and bromine are the molecules responsible for ozone depletion. "Effective chlorine" is a way to measure the destructive potential of all ODS gases emitted in the stratosphere.

Cases per million people per year

No protocol

Montreal 1987

Excess skin cancer cases

London 1990

Copenhagen 1992

Figure 9.12 The effects of the Montreal Protocol amendments and their phase-out schedules. Source: Emmanuelle Bourney, UNEP/GRID-Arendal Maps and Graphics Library (http://maps.grida.no/go/graphic/, accessed 18/7/2010).

OZONE DEPLETION AND CLIMATE CHANGE

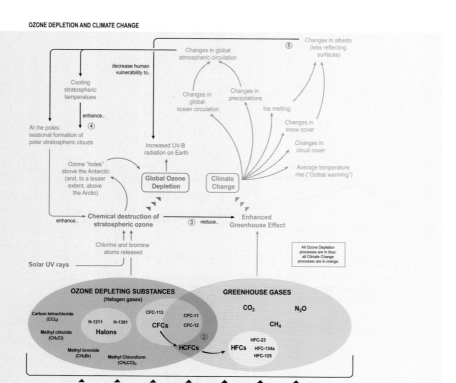

Figure 9.13 The relation between ozone and climate change. Source: Emmanuelle Bourney, UNEP/GRID-Arendal Maps and Graphics Library (http://maps.grida.no/go/graphic/, accessed 18/7/2010).

Index

The Chemical Element: Chemistry's Contribution to Our Global Future, First Edition.
Edited by Javier Garcia-Martinez, Elena Serrano-Torregrosa.
© 2011 Wiley-VCH Verlag GmbH & Co. KGaA. Published 2011 by Wiley-VCH Verlag GmbH & Co. KGaA.